The Social Insects

The Social Insects

Their Origin and Evolution

By

WILLIAM MORTON WHEELER

GYAN

Published by

GYAN

Gyan Publishing House
5, Ansari Road
Daryaganj, New Delhi-110002
Phone: 011-47034999, 9811692060
E-mail: books@gyanbooks.com

Distribution Network
gyanbooks.com
India, USA, Canada, UK, Australia, France

ISBN : 978-81-212-6055-8 (PB)
First Published, 1928

2nd Impression 2022

Printed at: Gyan Press, Delhi.

THE SOCIAL INSECTS
Author: WILLIAM MORTON WHEELER

The Social Insects

Their Origin and Evolution

By

WILLIAM MORTON WHEELER

Ph.D., Sc.D., LL.D., Professor of Entomology at Harvard University, Dean of the Bussey Institution for Research in Applied Biology

1928

TO

CHARLES THOMAS BRUES

Le destin des fourmis, des abeilles, des termites, si petit dans l'espace, mais presque sans bornes dans le temps, c'est un beau raccourci, c'est, en somme, tout le nôtre que nous tenons un instant, ramassé par les siècles, dans le creux de la main. C'est pourquoi il est bon de le scruter. Leur sort préfigure le nôtre, et ce sort, malgré des millions d'années, malgré des vertus, un heroisme, des sacrifices qui chez nous seraient qualifiés d'admirables, s'est-il amélioré ? Il s'est quelque peu statilisé et assuré contre certains dangers, mais est-il plus heureux et le pauvre salaire paie-t-il l'immense peine ?

MAETERLINCK, " La Vie des Termites."

PREFACE

THE twelve lectures which make up the present volume were delivered during the spring of 1925 at the University of Paris while I was occupying a Harvard exchange professorship established by Mr. James Hazen Hyde, and were published by Gaston Doin & Co., during the summer of 1926 under the title " Les Sociétés d'Insectes, leur Origine, leur Evolution " in the " Encyclopédie Scientifique," edited by Professor M. Caullery. The following year the prix Dollfus was conferred on the work by the Societé Entomologique de France. In preparing this English edition I have retained the original lecture form, but several passages, which had to be omitted in order to bring the volume within the requirements of the French publisher, have been restored to the text, a number of typographical and other errors have been corrected and a small amount of new material has been added. I have also introduced several new illustrations and have omitted a few of those in the French edition. In its present form the work has been brought up to date so far as this was possible without unduly expanding the text and the bibliography.

I am greatly indebted to Professor Caullery for the care and labour he devoted to seeing the French edition through the press. For typing the English manuscript my thanks are due to Miss Julia C. Foley and Miss Frances R. Rust. Mrs. Francena Taylor has given me much aid in revising the manuscript and the references to the bibliography and illustrations.

FOREST HILLS,
Boston, Mass.

CONTENTS

LIST OF ILLUSTRATIONS

INSECT SOCIETIES

I

THE SCOPE AND MEANING OF THE SOCIAL AMONG THE INSECTS

THE evolution of the social insects was selected as the topic of this course of lectures for several reasons: First, because lectures on an important group of insects seemed to me to be eminently appropriate for a laboratory founded by the illustrious Giard for the study of the evolution of organic beings. Second, I could think of no subject more interesting to the young biologists of a country which has produced a Réaumur, a Latreille, a Lepelletier de St. Fargeau, a Dufour, a Fabre, a Pérez, a Ferton, a Paul Marchal, and a Bouvier, not to mention a host of other brilliant hymenopterists. Third, having myself devoted more than a quarter of a century to the study of a single family of the social insects, with ample opportunities for travel, and as the recipient of much aid from enthusiastic entomologists in all parts of the world, it seemed to me that I might be able to suggest or emphasize some lines of thought worthy of your consideration. Fourth, it occurred to me that you could not be expected to be familiar with all the work that has been accomplished by my countrymen in entomology, and that by briefly presenting as much of it as it pertinent to my subject, I might be furthering to a slight degree that intellectual *entente cordiale*, which we are so desirous of maintaining between France and the United States. Fifth, I believe that the study of the social insects has, at the present time, a peculiar interest to the serious student of philosophy, sociology, and animal behaviour. Since we ourselves are social animals—I had almost said social insects—the philosophically inclined cannot fail to find food for thought in the strange analogies to human society, which continually reveal themselves among the wasps, bees, ants and termites, and the behaviourist will note that they suggest a bewildering array of fascinating facts and problems. Moreover, the very elaborate social behaviour of the insects, in that it is almost exclusively determined by the reflexes, tropisms, and the so-called instincts and not by intelligence, assumes great theoretical significance, when we contemplate the present anti-intellectualistic and relativistic tendencies and currents of European and American thought. We are beginning to see that our social as well as our individual behaviour is determined by a great background of irrational, subconscious, physiological processes. Any doubts in regard to the existence of this substratum will be dispelled by a perusal of Pareto's " Treatise of General Sociology " (1917), the first volume of which is devoted to these " residues " which condition our social activities.

For much that I have to say I shall have to draw on materials accumulated while I was preparing my Lowell lectures on " Social Life Among the Insects." In that work I endeavoured to stress the fundamental rôle of nutrition in the development of the various insect societies, but in the present lectures I wish to treat the insect societies in a different and somewhat more technical manner, and to dwell on certain matters which were merely suggested in my former course.

Before descending very far into details, I feel that I ought to attempt at least partial answers to four very general questions :

(1) What are the social insects ?

(2) Can they be shown to have had an evolution ?

(3) If so, what are the peculiarities of this evolution, and to what methods must we resort for their elucidation ?

(4) To what general causes or conditions may we assign this evolution ?

What are the social insects? The words "social" or "associative" are, of course, rather vague, and tinged with anthropomorphism. Their vagueness is due to the extraordinary multiplicity and Protean character of the phenomena, even among the insects, not to mention the various classes of Vertebrates. These phenomena range all the way from a feeble gregariousness or collectivity or the mere temporary union of the sexes to permanent and highly integrated societies remotely analogous to our own. The anthropomorphism attaching to the word "social" is easily explained and may be condemned or condoned according to the general philosophical convictions of the critic. I can only record my belief that in the discussion of matters psychological—and the phenomena in question are in part psychological—we can scarcely avoid a certain amount of anthropomorphism or teleology.

Recently Deegener (1918) has made a bold and rather unsatisfactory attempt, in a tome of 420 pages, to classify and describe all the various forms of association in the animal kingdom. He distinguishes some ninety-two different categories, fifty-three of which are represented among the insects. They are all supplied with scientific names, mostly of Greek derivation, some of which, like "heterosymphagopædium", "amphoterosynhesmium", and "heterosynepileium", have a truly Germanic ponderosity. The ninety-two different categories are divided into two groups, accidental societies or associations, in which the congregation of the organisms has no intrinsic value, i.e., serves no useful purpose for the individual, and essential societies or associations, in which the congregation has an intrinsic value or subserves in part a useful purpose. Each of these leading categories is subdivided according as the association comprises individuals of the same or of different species. The finalistic *principium divisionis* of the main groups is very dubious, to say the least, since even the most highly integrated animal associations are frequently confronted with situations in which membership in the society inevitably involves the destruction of the individual. And apart from the fact that the same animal (e.g., the sexton-beetle, Necrophorus) is cited by Deegener as belonging to several different categories, some of the latter border on the humorous as e.g., his " heterosymporium ", which comprises the insects of different species brought together by a freshet, or the various animals fleeing from a prairie fire, or his " symphotium ", which comprises the miscellaneous assortment of insects attracted by a lamp.

The classification of social forms adopted by Alverdes (1925) into associations and societies is both more logical and more useful than that of Deegener. Associations are aggregations of organisms brought about by extraneous factors whereas in societies the individuals are held together by a mutual attraction, or peculiar social appetency. In associations the single individual is oriented primarily towards stimuli emanating from its environment apart from the other individuals of its species, but in societies the stimuli are furnished by the latter and orientation towards the remaining environment is secondary. In true societies, therefore, an individual may exchange a favourable for an unfavourable environment merely in order to satisfy its craving to remain with other individuals of its kind.[1]

[1] Rabaud (1927) has rather captiously criticized Bouvier (1926) and myself on the basis of some observations on males of Halictus and on general principles. Since he adopts a strenuously physiological view of biology and seems to be unable to accord any value to the historical, or phylogenetic aspect of the science, discussion of his general position would be unprofitable. So far as his observations are concerned, it is difficult to see that he has added anything of importance to what has long been known in regard to a great many insects that form aggregations, or associations, such as the Ipid, or Scolytid and Plalypodid beetles, hibernating Coccinellids and many solitary Aculeates which make their individual nests in close proximity to one another (Bembix, Philanthus, Synagris, Andrena, etc.). At the same hour and especially in the evenings of several consecutive days he observed males of *Halictus latipes* and *nigricornis* congregated in compact clusters on a few dry grass stems in two localities, and concludes from several experiments that the insects were attracted to these particular spots as if they were nesting sites. He believes, however, that the initial attraction was due to interindividual stimuli and was not therefore merely gregarious, or associational, but social. I have myself seen very similar daily agglomerations of males of our North American *Chloralictus albipennis*

Now it seems to be obvious that the behaviour of every animal, figuratively speaking, revolves about two axes, one of which is aggressive and individualistic, the other co-operative, or social. The Darwinians took the former behaviour largely for granted and greatly stressed it, so that the latter appeared to be exceptional and in need of a special explanation. At the present time one might more properly require an explanation of the solitary mode of life, so deeply are all who study animals in their complicated living environment impressed by their social or associative proclivities. This is apparent even in the biocœnoses, since every organism, no matter how egocentric, predatory or parasitic, is social or co-operative at least to the extent of being a member of some biocœnose.

The truth of the foregoing statement was so long ago and so admirably expressed by Espinas in the introduction to his "Des Sociétés Animales" (1877) that I cannot refrain from quoting the pertinent passage: "No living being is solitary. Animals, especially, sustain multiple relations with the organisms of their environment, and without mentioning those that live in permanent intercourse with their kind, nearly all are impelled by biological necessity to contract, even if only for a brief moment, an intimate union with some other individual of their species. Even among organisms devoid of distinct and separate sexes, some traces of social life are manifested, both among the animals that remain, like plants, attached to a common stock and

on particular corymbs of *Aster corymbosa*, but have attributed it to an impregnation of the flowers with odours, left either by virgin females which had previously visited them, or by the males after they had once congregated on the spot. These possibilities are not considered by Rabaud. Furthermore, as will be shown in the fourth lecture many species of Halictus are really social bees, so that Rabaud is not dealing with a simple or incipient tendency to social life. Finally Bouvier and I were not concerned with the universal phenomena of the social in its broadest sense, but with the true *genetic* societies of insects, and Rabaud seems to overlook the fact that these societies have a definite ontogeny traceable to definite physiological relations between parent and offspring.

among the lowly beings which, before separating from the parental organism, remain for some time attached to it and incorporated with its substance. Communal life, therefore, is not an accidental fact in the animal kingdom; it does not arise here and there fortuitously and as it were capriciously; it is not, as is so often supposed, the privilege of certain isolated species in the zoological scale, such as the beavers, bees and ants, but on the contrary—and we believe we are in a position to prove this statement abundantly in the present work—a normal, constant, universal fact. From the lowest to the highest forms in the series, all animals are at some time in their lives immersed in some society; the social medium is the condition necessary to the conservation and renewal of life. This is, indeed, a biological law which it will be expedient to elucidate. Moreover, from the lowest to the highest stages in the series, we detect in the development of social habits a progression which if not uniform is at least constant, so that each zoological group carries the perfecting of these habits a little farther in one or another direction. Finally, social facts are subject to laws and these are the same everywhere where such facts appear, so that they constitute a considerable and uniform domain in nature, a homogeneous whole thoroughly integrated in all its parts."

Perhaps, if we were asked to point to a group of typically solitary animals, we should select the spiders, but when we study them in the tropics, we find that many of the species are decidedly gregarious or vaguely social. Recently, in Panama, Mr. Nathan Banks and I were impressed by the habits of several of these arachnids. One of the largest and commonest of the orb-weaving spiders, *Nephilla plumipes*, during the wet season, regularly builds its webs in such a fashion that those belonging to different females run together so that huge and elaborate structures are sometimes formed. One compound web, which we observed among the trees on the summit of Ancon Hill near the City of

Panama, was fully twenty feet broad and more than twenty-five feet high. In it were suspended at least 200 female and nearly as many male Nephillæ, besides a considerable number of *Argyrodes nephillæ*, a small parasitic Theridiid spider which lives only in the webs of Nephilla. We also encountered several species of Theridion and Uloborus which unite their webs in a similar manner, and a peculiarly social species, apparently undescribed, but allied to *Anelosimus socialis* Simon, in the webs of which dozens of females of all ages moved about freely and in all probability fed in common on at least the larger prey. Many of these spiders bore on their abdomens the larvæ of a parasitic Hymenopteron (probably Polysphincta). Even alien insects may live unmolested in the webs of some of these Panamanian spiders. On Barro Colorado Island we found numerous red nymphs and black adults of a small, ant-like Nabid bug (*Arachnocoris albomaculatus* Scott) resting or moving about in the webs of Uloborus and Theridion and apparently feeding on the prey abandoned by the spiders (see Myers (1925)). On two occasions we observed dozens of small Cecidomyid flies resting peacefully on the webs of Nephilla and Uloborus. When the webs were shaken, the tiny insects flew off only to return at once to their former station.

In 1891, Simon made similar observations on the social habits of spiders in Venezuela and cites numerous examples in other tropical countries. He gives an excellent account of *Epeira bandelieri*, *Anelosimus socialis* and *Uloborus republicanus*. On ordinary occasions *E. bandelieri* is solitary, but when the egg-laying season approaches several females congregate and together construct a large elliptical capsule in which they conceal themselves and make their egg-cocoons. *A. socialis* lives in colonies of several thousand individuals which construct a common web. "The spiders walk about on it freely, meet one another and exchange greetings with their palpi, as ants do with their antennæ, and sometimes devour the more voluminous pieces of prey in common." The association of *U. republicanus* is "far the most perfect, since it presents both a web constructed by all the associates working together and single webs constructed by individual spiders. Several hundred Ulobori live together. They spin among the trees an immense structure, consisting of a central web with rather dense meshes on which numerous individuals of both sexes, but mostly males, are stationed. This web is suspended by long threads, radiating in all directions and attached to surrounding objects. In the intervals of these stout strands, other Ulobori weave orbicular webs with radii and circles, and inhabited only by single individuals. Mating takes place on the central web, if we may judge from the number of males which we found congregated on it. . . . It is certainly the place where the eggs are laid. All the females of the colony seem to lay almost simultaneously. At this time the males have disappeared and the females, having ceased to spin regular webs, are attached to the central web, a few centimetres apart and each guarding her cocoon in complete immobility."

According to Bouvier (1918), "Semichon (1909) has observed the same fraternal sharing (of the prey as in *U. republicanus*) in a Mexican social spider, *Cœnothele gregalis*, brought to the Museum in Paris by Leon Diguet (1909a, 1909b). In order to capture insects, this species constructs on the trees great concentric webs consisting of carded strands. Thousands of individuals live in harmony in this immense alveolar sac and never leave it, except after the rainy season, when they emigrate, or rather swarm, to leave room for the young. These webs may be divided and are hung from the ceiling and used as fly-traps in certain parts of Mexico. One of them, exhibited in the galleries of the Museum, is several meters long". Diguet has the following notes on two interesting commensals which he found living in these nests of Cœnothele. "In all the internal portions of the nest very great numbers of a Latridiid of the genus Melanophthalma are found living as commensals. The rôle assumed by this minute beetle seems to be the cleaning of the nest, i.e., transporting and doing away with the detritus which may encumber or soil the galleries; its principal food seems to be the remnants of the spiders' repasts. . . . Another commensal also found living in perfect harmony with the 'mosquero' colony is a wandering spider which has become a guest in this lodging and there finds an easy and assured existence". This spider was identified by Simon (1909) as *Pœcilochroa convictrix* Simon.

In 1926, I published some notes on a couple of social spiders, *Cyrtophora citricola* and *Argyrodes argyrodes*, which I observed in the Canary Islands (Teneriffe, Palma and Gran Canaria). There is much the same relation between these two species as between the Panamanian *Nephilla plumipes* and *Argyrodes nephillæ* mentioned above. The webs constructed by the Cyrtophora are spread over trees and shrubs and are sometimes of great dimensions. At Puerto de la Luz I found a dense hedge of Opuntia cactus fully one hundred feet long and six or eight feet wide, covered by a single web estimated at more than 1,000 square feet and containing many thousands of spiders. The web consists of two parts, a very irregular structure or frame work of long, coarse, yellow and somewhat glutinous threads, running in all directions and attached to the plants, and a variable number of suborbicular, horizontal webs suspended side by side or one above the other in the frame work. These webs are three to eight inches in diameter and made of very even square meshes, of the size of those of mosquito netting, but consisting of exceedingly delicate, whitish silk. The Cyrtophoras rest on the lower, convex surfaces of these webs. Individuals of all ages live together amicably and seem to feed in common on the prey that is caught in the webs, but the adult females (15 mm. long), which are gray, with large, paired, silver spots on the dorsal surface of the abdomen, are usually few in number. The egg-cocoons are elliptical, about 15 to 20 mm. long, made of dense, coarse, gray-green silk, and are suspended vertically in or near the center of the whole structure. They vary from one to five in number and are attached to one another in a series, so that they resemble a string of minute sausages. The mother spider is usually found resting at the end of the lowermost cocoon. The Argyrodes, which are black, with pale legs and extensive silver spots on the abdomen and are very much smaller than the adult Cyrtophoras (adult female only 4.5 mm., adult male 3.5 mm.), were also present in all the webs which I examined. They seem to make no webs of their own but live in the coarse framework of the structure spun by the larger species. Like the latter, they are present in considerable numbers, of all sizes and of both sexes. They were seen to feed on small insects caught in the coarse yellow strands. When disturbed they quickly drop to the ground by letting out a thin silken filament, but the Cyrtophoras run off to the side and hide in the foliage of the plant supporting the web. The egg-cocoons of the Argyrodes resemble certain seed-capsules and are of the peculiar type seen in other species of the genus, being small sub-spherical or pear-shaped, yellowish brown, papery-walled structures. One pole of the capsule is prolonged into a stiff stem, or pedicel by which the capsule is suspended from the threads of the web and at the opposite pole there is a small circular, protruding rim.[1]

Probably most of us would agree with Petrucci (1906) who reviewed the various forms of societies among the vertebrates and pointed to their polyphyletic derivation,

[1] Interesting accounts of the habits of the social spiders of the genus Stegodyphus, represented by several species in South Africa and India are given by Simon (1892-5), Distant (1898), Marshall (1898), Pocock (1903) and Jambunathan (1908). Walsingham (1903), Marshall and Pocock also describe a peculiar Tineid moth (*Batrachedra stegadyphobius*) which lives in all its stages in the webs of these spiders. Mr. J. H. Emerton informs me that some of our New England spiders may build closely continuous webs, e.g. *Dictyna muraria* and *Amaurobius ferox*, and that he has seen more than a hundred small midges resting peacefully on the web of *Linyphia marginata*. Recently Prof. C. T. Brues has made some observations on *Uloborus republicanus* which is common near the Harvard Tropical Laboratory, at Soledad, Cuba. Sae elso Schwarz (1904) and Banks (1904).

when he says: "There is no inheritance in the social activities of animals, except a tendency to association, a tendency manifested in a predominant fashion throughout the domain of biology and realized in concrete phenomena whenever and wherever external conditions permit of its manifestation". Petrucci naturally traces his "tendance associative" to such phenomena as the cellular constitution of the Metazoa and Metaphyta and even to atomic equilibria and molecular associations in the inorganic world. But the conception, though not devoid of interesting philosophical implications, thus becomes very vague. I should prefer for the purpose of making the matter more concrete and intelligible, at least in the biological field, to regard the "tendance associative" as an "appetition" in the sense in which that term has been employed by the French thinker Fouillée and the British and American psychologists Drever (1917), Craig (1918) and Thurstone (1924). It thus takes its place with the other appetitions like hunger and sex, though it is feebler, more continuous, i.e., less spasmodic and, therefore, less obvious. It is most strikingly displayed, however, in the restless behaviour of the higher social animal when isolated from the continuous, customary stimuli of its kind.[1]

That this social appetition is clearly and very generally manifested in certain insects admits of little doubt, but it will be advantageous to narrow the conception still further by the introduction of other considerations if we are to make any use of it in the very special field which I have marked out for treatment in these lectures. There are authors, who like Fabre have hinted that even the societies of ants and bees may have arisen phylogenetically by chance associations of female insects of the same species, but this was before the actual constitution and genesis of insect societies were known. It is now

[1] I do not, of course, wish to imply agreement with the views of Becher (1917), who postulates a distinct altruistic *penchant* in certain plants and animals to serve other and quite unrelated organisms. This view has been adequately criticized by Bequaert (1924), A. Meyer (1926), and others.

unanimously admitted that all insect societies worthy of the name, and no matter how populous, are families, i.e., affiliations of the parents and, in most cases, of the mother alone, with the offspring. This view is abundantly supported by the study of the ontogeny of existing insect societies and by such fragmentary indications as we can obtain in regard to their phylogeny.

There has evidently been a very long evolution through numerous stages of constantly increasing intimacy of the mother with her progeny from the most rudimentary stage of complete or almost complete indifference to one of mutual and abiding co-operation. We may construct, e.g., such a series as the following, without stopping to enumerate concrete examples, many of which will at once occur to you.

(1) The insect mother merely scatters her eggs in the general environment in which the individuals of her species normally live (*atrophaptic* insects). In some cases the eggs are placed near the larval food (*dystrophaptic* insects).

(2) She places her eggs on some portion of the invironment (leaves, etc.) which will serve as food for the hatching larvæ (*eutrophaptic* insects).

(3) She supplies her eggs with a protective covering. This stage may be combined with (1) or (2).

(4) She remains with her eggs and young larvæ and protects them.

(5) She deposits her eggs in a safe or specially prepared situation (nest) with a supply of food easily accessible to the hatching young (mass provisioning).

(6) She remains with the eggs and young and protects and continuously feeds the latter with prepared food (progressive provisioning).

(7) The progeny are not only protected and fed by the mother, but eventually co-operate with her in rearing additional broods of young, so that parent and offspring live together in an annual or perennial society.

The insects included in categories (1) to (5) may be designated as "infrasocial"; those of (6), which are more interesting for our purposes may be called "quasi-" or "subsocial". Only the last category (7) comprises social forms *sensu stricto*. This final stage in the series is reached primarily through the development of an increased interest on the part of the mother in the later instars of her offspring and is, of course, made possible by a lengthening of her individual life-span. Had the students of human sociology been conversant with this very obvious inference we might have been spared some speculations which are constantly repeated in sociological literature. John Fiske in his "Cosmic Philosophy" (1874) maintained that the lengthening of human infancy and childhood has led to the definitive association of the parents with the offspring. He conceived this to be a new interpretation of the origin of the family, but Lovejoy (1922) has recently shown that the notion was familiar to many eighteenth century thinkers. He mentions particularly the poet Pope, and the philosophers Locke and Rousseau, and shows that Rousseau demolished Locke's argument, which was essentially that of John Fiske, in 1755, in his "Discourse on the Origin of Inequality" by pointing out that if the habit of family life had not been established by primitive man during the months preceding the birth of the first child, the human male would hardly have come to the aid of the female after the "accouchement". "Why should he aid her to rear an infant which he does not even know to be his, and the birth of which he has neither purposed nor foreseen". At the present day we should, of course, turn to a study of the Anthropoids for light on the remote and nebulous origins of the human family.

It is obvious, however, that no mere lengthening of infancy and childhood would be of value without a preceding or concomitant lengthening of the adult life of the parents. This consideration, so clearly indicated by the insects, seems to have been completely overlooked by the writers above mentioned.

Of the physiological causes for the increased adult longevity of the social insects we know nothing. It is probable, as Pearl (1924) has suggested, that the duration of an animal's life stands in reverse relation to the amount of its metabolic activity. Certainly the life-span of the three castes of ants and social bees would seem to be roughly proportional to their respective expenditures of energy. But the problem before us is concerned with the lengthening of adult life *after* the beginning of reproduction. The long lives which many insects lead before they reproduce, e.g., *Cicada septemdecim*, the larvæ of Ephemerids, Cerambycids, etc., cannot lead to the development of societies and, of course, adult longevity merely gives the parent the opportunity for association with its progeny, but does not account for its interest in the latter or the care expended on their nurture.

It seems to me that the decrease in metabolic activity which, according to Pearl, might be supposed to increase the adult life-span, especially of the fertile females, may be due to the fact that all the subsocial and social insects live in small cavities of the soil or wood, in hives or, in the more exceptional cases of social wasps and certain tropical ants, in the cavities of carton nests. The environment is, therefore, one which restricts or inhibits muscular movement and is dark, poor in oxygen, and of rather low and uniform temperature. All of these conditions would necessarily favour a lowered rate of metabolism and activity and an accumulation of fat in the insect body. The queens, or mothers of insect societies certainly impress one as having acquired their physiological and some of their morphological peculiarities as responses to just such an environment, for they are very sluggish and tend to lose the powers of flight (Meliponinæ) or even the wings (ants and termites) and to acquire an accentuated anabolism as shown in the accumulation of fat and of yolk-laden eggs. Very probably, therefore, both the lengthening of adult life and the fecundity of these insects have been gradually acquired in response to the very restricted environment in which they not only develop but also continue to live as adults. Their fecundity is to a certain extent a function of their longevity and is clearly expressed in the size of the adult colony in all the species of social insects, especially in the higher termites (Termes), in the ants of such genera as Eciton and Atta and in the honey bee. On the other hand, the small size of the adult colonies of many primitive ants (Ponerinæ) and termites (Calotermitidæ) would seem to be the effect of the rather short life of the mother queen.

I have drawn up for brief consideration the following list of the previously designated subsocial and social insects :—

LIST OF SOCIAL AND SUBSOCIAL INSECTS

Coleoptera

1. *Silvanidæ* (Coccidotrophus, Eunausibius).
2. *Scarabæidæ* (Copris, Minotaurus).
3. *Passalidæ* (Passalus, etc.).
4. *Tenebrionidæ* (Phrenapates).
5. *Ipidæ* (Xyleborus, etc.).
6. *Platypodidæ* (Platypus, etc.).

Hymenoptera (*Aculeata*)

7. *Bethylidæ* (Scleroderma).
8. *Masaridinæ* (Ceramius).
9. *Eumeninæ* (Synagris, Odynerus).
10. *Zethinæ* (Zethus).

*11. *Stenogastrinæ* (Stenogaster).
*12. *Epiponinæ* (Belonogaster, Chartergus, etc.).
*13. *Ropalidiinæ* (Ropalidia).
*14. *Polistinæ* (Polistes).

*15. *Vespinæ* (Vespa).
16. *Sphecinæ* (Ammophila).
17. *Trypoxyloninæ* (Trypoxylon).
18. *Bembicinæ* (Bembix).
*19. *Halictinæ* (Halictus).
*20. *Ceratininæ* (Allodape).
*21. *Bombinæ* (Bombus).
*22. *Meliponinæ* (Melipona, Trigona).
*23. *Apinæ* (Apis).
*24. *Formicidæ* (Formica, etc.).

OTHER ORDERS

25. *Blattoidea* (Dasypoma).
26. *Dermaptera* (Forficulidæ).
27. *Orthopera* (Gryllotalpa).
28. *Embidaria* (Embiidæ).
29. *Zoraptera* (Zorotypus).
*30. *Isoptera* (Termitidæ).

This list is, doubtless, incomplete since our knowledge of the habits of many insects, especially of the tropical species, is still very unsatisfactory. I might have included the Australian species of Perga belonging to the Tenthredinidæ and a number of Heteroptera, the females of which guard their eggs and just hatched young, but these and similar cases grade imperceptibly into the infrasocial forms. From the list as it stands, and from considerations to be developed in future lectures, it will be seen that social organization has been attempted on at least thirty different occasions, and in eight very different natural orders. At least twelve of these groups, designated by asterisks, have become definitively social; the remainder have not progressed beyond very rudimentary beginnings. Some of these subsocial insects, especially the Coleoptera, are very interesting, but their habits are so diverse that I shall have to refer you to the brief account which I have given of their activities in my "Social life among the Insects". Here a sketch of the habits of the social Silvanidæ which

I first observed during the summer of 1920 in British Guiana will have to serve as a paradigm of the whole series.

These beetles, which Messrs. Schwarz and Barber have named *Coccidotrophus socialis* and *Eunausibius wheeleri* (Fig. 1*a* and *b*) are less than a quarter of an inch in length and have long, slender, subcylindrical, red or chestnut brown bodies, with short legs and club-shaped antennæ. They occur only in the hollow leaf-petioles of a very interesting tree, *Tachigalia paniculata*, and only in young specimens, one-and-a-half to seven feet high while they are growing in the shade under the higher trees of the jungle. The older trees, which may attain a height of forty feet or more, have all their petioles inhabited by viciously stinging or biting ants, (*Pseudomyrma maligna* and *crucians*, *Azteca foveolata*). Each beetle colony is started by a male and female which bore through the wall of the petiole, clean out any pith (Fig. 1*c*) or remains of previous occupants it may contain and commence feeding on a peculiar tissue rich in proteins, which is developed in parallel, longitudinal strands in the wall of the petiole (Fig. 1*d* and *e*). As they keep gnawing out this tissue they gradually make grooves and pile their feces on the ungnawed intervening areas so that the interior of the petiole assumes a peculiar appearance. While the beetles are thus engaged, numbers of small mealy-bugs of the genus Preudococcus (*Ps. brevipes*), covered with snow-white wax, wander into the petiole through the opening made by the beetles, settle in the grooves, sink their delicate sucking mouth-parts into the nutritive tissue and imbibe its juices. The beetles soon begin to lay their small, elliptical, white eggs along the edges of the grooves, and the hatching larvæ, which are beautifully translucent, run about in the cavity, and feed on the same tissue as the parents. But incredible as it may seem, both the adult beetles and the larvæ in all stages have learned to stroke the mealy-bugs with their antennæ, just as our common ants stroke similar mealy-bugs and plant-lice and feed on the droplets of honey-dew, or saccharine excrement which they give off when their backs are properly titillated. So greedy are the Silvanids for this nectar that I have seen a beetle or a larva stroke a mealy-bug for an hour or longer, and receive and swallow a drink every few minutes. When two or more beetles or two or more larvæ or a group of beetles and larvæ happen to be engaged in stroking the same mealy-bug, they stand around it like so many pigs around a trough, and the larger or stronger individual keeps butting the others away with its head. The butted individuals, however, persistently return and renew their stroking till the knocks become too severe or the stronger individual departs and begins to stroke another mealy-bug. Thus the beetles and their progeny have discovered a rich food supply consisting in part of the proteid-containing tissues of the Tachigalia and in part of the sugar and water discharged by the mealy-bugs, which in turn imbibe the sap of the tree. The beetles lay their eggs at intervals, so that larvæ in all stages are found in the same colony. When mature each larva constructs a cocoon of minute particles bitten out of the plant tissues, creeps into it, closes the opening from the inside and pupates. When the young beetles emerge they remain with their parents and soon begin to lay eggs, so that eventually the colony consists of several dozen beetles, larvæ, pupæ and mealy-bugs in all stages and all living peacefully together, except for the little family bickerings of the beetles and larvæ over the milking of their patient, snow-white cattle. When the petiole becomes too crowded, pairs of young beetles leave it, enter other petioles of the same or other Tachigalia trees and start new colonies. As the tree grows and emerges from the undergrowth into the sunlight, the ants, which then take complete possession of it, oust the beetles from the petiolar cavities but adopt their mealy-bugs just as the invading German army appropriated the French cattle. The foregoing account applies to *Coccidotrophus socialis*, but the much rarer Eunausibius has very similar habits.

More recently Dr. W. M. Mann has discovered a second species of Coccidotrophus (*C. cordiæ* Barber), which lives in essentially the same manner, and with the same mealy-bugs in the cauline swellings of another myrmecophyte, *Cordia alliodora*, in Bolivia.

It has taken me so long to answer the first of the four questions which I propounded at the beginning of this lecture that I shall have to be very brief in my answers to the three others. Our second question was: Can the social insects be shown to have had an evolution? I take it that we all accept the theory of evolution at least in the sense of "transformism", and that we should all take an affirmative answer to the question for granted. Moreover, this whole course of lectures will, I hope, corroborate and give more definite outlines to this belief. At the same time, it must be admitted that our very meagre knowledge of the various groups to which most of the subsocial insects belong makes it impracticable at present to give any account of their evolution. They are mostly tropical and are either imperfectly known or isolated and sporadic representatives of larger groups, which have never been subjected to extensive comparative ethological study, so that we are unable to construct the probable course of their phylogenetic development. For example, the beetles above described belong to the Coleopterous family Silvanidæ, most species of which have very simple, nonsocial habits, but if we knew the tropical species most closely allied to Coccidotrophus and Eunausibius we might be able to throw some light on their peculiar behaviour. We are in the same predicament in regard to the Ipidæ (Scolytidæ), Platypodidæ, etc. Owing to this dearth of data we shall have to confine ourselves to the evolution of the truly social groups, which belong exclusively to the Aculeate Hymenoptera and the Isoptera. Fortunately they have been studied so intensively and extensively and by so many investigators that we can draw on a very considerable body of accurate observations and experiments.

Subsequent lectures should also answer the first part of the third question, i.e., as to the evolutionary peculiarities of the social insects, but the second part, concerning the methods to be employed in elucidating their evolution may be briefly discussed in this place. We are, of course, concerned with the phylogeny of insects characterized by a peculiar type of behaviour, and as there is no such thing as fossil behaviour we can only infer its presence or absence from the morphological structure and correlated behaviour of the species living at the present time. We are compelled, therefore, to proceed historically. Of late there has been considerable discussion, especially in Germany, as to the precise relations of biology to history and of history to philosophy, and what most of us older investigators have long known seems now to be acceded, namely that biology in the broad sense and including anthropology and psychology is peculiar in being both a natural science and a department of history (phylogeny).[1] But from the standpoint of the biologist *pure sangue*, who regards his science as nothing but a continuation of physics and chemistry and therefore capable of development solely by direct observation, experimentation and mathematics, the student of phylogeny or of the historical aspect of organisms is in an unfortunate position. He cannot observe or subject to experiment the functioning and behaviour of extinct forms so that all his attempts to reconstruct the living past of animals and plants must be more or less indirect, inferential and problematical. He cannot even adopt the methods of the archæologist because, except in extremely rare instances, there remain no objects constructed by extinct animals. The only examples that occur to me are the fossil nests of Eumenine wasps described by Handlirsch (1910) from the Upper Oligocene of Flörsheim and the hills of the larger Attine ants which Branner (1900) believes

[1] See especially the erudite works of H. Rickert (1921), Kroner (1913, 1919), Mehlis (1915), Dürken and Salfeld (1921), Schaxel (1922) and A. Meyer (1923, 1926).

to have persisted in some cases through geological epochs. If I may be permitted to use the language of present-day psycho-analysts, the student of phylogeny can only compensate for his inferior position as compared with that of the experimental and statistical biologist by exhausting and even overworking the meagre methodological resources at his command. These resources, which are certainly less limited in entomology than in many other fields, comprise the following :

(1) The paleontological data. These are numerous and have been of late critically reviewed and greatly increased by competent entomologists familiar with the allied living species. And although the record from certain geological periods is deplorably meagre, enough forms have been preserved to present a consistent picture of the evolution of most of the orders and of many of the families of the winged insects (Pterygogenea). These data constitute, of course, the most precious sources for the student of phylogeny.

(2) The comparative morphology, i.e., the comparative anatomy and ontogeny of existing insects.

(3) The vast accumulation of codified and classified data of taxonomy, or biotaxy, based very largely on the external morphology of existing species.

(4) The data of the geographical distribution, or biogeography of existing and fossil species.

(5) The comparative ethology, or behaviour of existing forms.[1]

[1] Schaxel (1922) has recently published a severe indictment of phylogeny. His discussion, however, reveals a singularly inadequate acquaintance with the actual technique of the subject, as A. Meyer has shown (1926).

In lectures like those I am attempting, certain of these data are easily presented, especially portions of the fossil record and comparative morphology of existing forms, but the innumerable intricate taxonomic and biogeographic data, which as evidence acquire their full force and significance very largely from their accumulated mass, cannot be satisfactorily presented even to an audience of biologists. Evaluation of these data is possible only to the entomological specialist, so that an attempt to cite a great array of generic and specific names before an audience that has never handled nor even seen the designated insects would under the most favourable circumstances induce slumber, and under the most unfavourable actual coma. I fear that I shall not be able to avoid a considerable amount of this soporific material. There are also certain unfortunate conditions attending the presentation of ethological matters. Although they are fascinating, they have a subordinate value at present, since notwithstanding our ability frequently to infer the physiological function of an organ from its structure, we tread on very insecure ground when we attempt to infer behaviour, which involves the organism as a whole, from its morphological components. Moreover, ethological facts, especially in such highly specialized forms as the social insects, require so many words for their adequate description that they are not very well suited for presentation in lecture form.

Our fourth question was concerned with the general causes or conditions of evolution in the social insects, with the ætiology of the process, as Huxley would have called it, and naturally resolves itself into a number of intricate problems. Several of these will occupy us in subsequent lectures. Obviously the evolution of the social insects cannot be treated without reference to the problems of organic evolution in general, but problems of more special interest are suggested by the unique behaviour of the insects as social organisms. They are undoubtedly descended from solitary forms which acquired their fundamental structural and behaviouristic characters in response to an environment like that of other non-social insects. But the development of permanent social living has created a new and extraordinary environment, continuous multi-millenial response to which has produced certain new characters that have come to modify, overlay and sometimes to mask the original inherited physiological, behaviouristic and even the morphological, pre-social endowment. Social life, as might easily be shown did time permit, must profoundly influence the very fundamental organic activities of reproduction, nutrition and protection, for all of these have to be regulated in a very different manner in a social as compared with a non-social species. Doubts are sometimes expressed as to whether the human individual shows any inherited effects of social living, since even language has to be learned anew by every child. But human society is a very recent affair compared with the most recent insect societies. In the latter there can be no question that the effects of social activities have become hereditary, that many of their instinctive, physiological and structural peculiarities have acquired some kind of a representation in the germ-plasm. It will be advisable, therefore, in tracing the evolution of the various social insects in subsequent lectures to enter upon a more detailed description of these socially induced peculiarities.

Another more general problem is suggested by the insect society, or colony as a whole, which as I have shown in another place (1911) is so strikingly analogous to the Metazoan body regarded as a colony of cells, or indeed to any living organism as a whole, that the same very general laws must be involved. But the biologist, with his present methods is powerless to offer any solution of the living organism as a whole. He cannot appeal to the entelechy or *élan vital* however suggestive and emotionally satisfying such agencies may be to the philosophers, nor does it help him to be told that a swarm of bees or a colony of ants or termites has a "superentelechy", "une âme de la ruche", or spirit of the hive, to use the terms of Réaumur and Maeterlinck, conceived as controlling the entelechies of the various individuals. This is merely another photograph, and a very dim and elusive one at that, of the problem. We can only regard the organismal character of the colony as a whole as an expression of the fact that it is not equivalent to the sum of its individuals but represents a different and at present inexplicable "emergent level" in the sense of Alexander (1920), Sellars (1922), C. Lloyd Morgan (1923), Parker (1924), Wheeler (1926), Smuts (1926) and others.

In conclusion, I may enumerate very briefly the subjects of the subsequent lectures, which are restricted to the truly social insects belonging, as we have seen, exclusively to a single large division of the Hymenoptera, the Aculeata, and the order Isoptera, or termites. In the second lecture, I shall attempt to deal with the general phylogeny of the Aculeata, in the third to sixth lectures, with the more special evolutions of the wasps, bees, ants, and termites. The seventh and eighth will be devoted to the problem of polymorphism, or pleomorphism, a phenomenon which has resulted from the incisive division of labour incidental to social life. In the ninth I propose to discuss the reciprocal activities or intercommunication of stimuli and food (trophallaxis) among the social individuals and in the tenth and eleventh lectures the various types of parasitism that have developed out of this reciprocity, both among the social insects themselves and between them and alien insects. In the concluding lecture I shall indicate the probable course of future development in insect societies or their fate on a planet, the natural balance of whose faunas and floras is being rapidly disturbed by a much younger and more powerful social animal—man.

II

THE ORIGIN OF THE TEREBRANTIA AND ACULEATA

WITH the single exception of the termites, all the groups of really social insects recognized in the preceding lecture, are confined not only to a single order, the Hymenoptera, but also to a single sub-order, the Aculeata. And since these social organisms have undoubtedly arisen from solitary species still represented by more or less closely allied forms in the existing fauna, we are confronted with the problem of the phylogenetic origin of the order Hymenoptera as a whole and of the Aculeata and each of its eleven social types in particular. So much interesting material bearing on this problem has been accumulating during recent years that even for its brief consideration this and the three following lectures will be required. We may perhaps reach nothing more conclusive or satisfactory than has resulted from other similar phylogenetic disquisitions but we shall encounter many singular and suggestive phenomena which, I trust, will lose none of their intrinsic value by being interpreted as representing significant stages in a very long and complicated evolutionary process. I believe, moreover, that such a review as the one I am undertaking is really necessary for the purpose of discriminating between the ancient morphological and behaviouristic characters, which the social insects still retain as a heritage from their solitary ancestors, and those which they have more recently acquired in response to their peculiar communal environment.

The Hymenoptera are a singular and sharply characterized order of immense extent. According to

Handlirsch 70,000 species have already been described and the number of existing species has been estimated at about 200,000. There are nearly 100 families embracing many hundred genera. The order is, therefore, next to the Coleoptera, the largest in the class Hexapoda. That such an order should have been very unevenly investigated notwithstanding the host of talented entomologists who for more than a century have devoted their lives to the task, will astonish no one who realizes its magnitude, intricacy and difficulty.

It did not escape that remarkable taxonomic genius, Latreille, that the order Hymenoptera may be most naturally divided into three sub-orders, which he called Phytophaga, Terebrantia and Aculeata, the first comprising the sawflies and horntails, the second the parasitic and gall-producing species and the third the wasps, bees and ants. In their essential outlines these groups have been retained down to the present day, though masked by the introduction of new names and an increasing and well-founded conviction that the Terebrantia and Aculeata are really much more closely related to each other than is either to the Phytophaga. Other entomologists have called this latter sub-order Sessiliventres, Symphyta, or Chalastogastra and have used the terms Petioliventres, Apocrita, or Clistogastra for the Terebrantia *plus* the Aculeata. Handlirsch (1924), in Schröders "Handbuch der Entomologie" is, I believe to be commended for having recently returned to the Latreillean trinary division and terminology. The Austrian entomologist, however, has added a fourth sub-order, the "Pelecinoidea", for the single American genus Pelecinus, which has always been regarded merely as an aberrant Terebrant. This is not so commendable, and the same may be said of Rohwer's (1917) and Bradley's (Comstock, 1924) recent attempt to make a sub-order "Idiogastra" out of the old family Oryssidæ, which obviously belongs among the Phytophaga. Certain families of anomalous character, comprising the Trigonalidæ, Chrysididæ, Cleptidæ, Bethylidæ and Rhopalosomatidæ, have been shifted back and forth between the Terebrantia and Aculeata. All authors agree that the three sub-orders are to be arranged in an ascending series, the Phytophaga being in structure and habits the most primitive and most nearly ancestral, the Aculeata the most recent and most highly specialized group.

Let us leave for the moment the interrelationships of the three sub-orders and turn to the question of the phylogenetic origin of the Hymenoptera as a natural group. Till very recently the position of this order, as determined by the morphology of the species and dearth of significant fossil material has been one of isolation among the other orders. While most of the latter could be traced back through the paleontological record with more or less probability either to Palæodictyopteroid or to Protoblattoid ancestors, the Hymenoptera were left hanging in a void. The older entomologists and Ashmead (1896) more recently were inclined to ally the Hymenoptera with the other higher Metabola (Lepidoptera, Diptera, etc.), but Handlirsch (1908) in his great work on the fossil insects endeavoured to derive them from the Protoblattoids and placed them next to the Coleoptera, which were supposed to have had a similar ancestry. His arguments, however, have not been generally accepted. He reviewed all the fossil evidence, which consisted of some fifteen species of Pseudosirex from the Jurassic (Malm of Solenhofen Bavaria and Lower Purbeck of England), a single Terebrant (*Ephialtites jurassicus* Meunier) from the Malm of Cataluña, Spain and a great number of species of the most various families from the Tertiary of Europe and North America. The species of Pseudosirex are very similar to recent Siricidæ and are merely interesting as showing the great antiquity of this already highly specialized family of wood-boring Phytophaga. The Ephialtites is undoubtedly a true Terebrant, but is so poorly preserved that it might be referred to any one of several existing families. Handlirsch regarded it as an

Ichneumonid, but Professor C. T. Brues has suggested to me that it is more probably a Chalcidid (Callimome ?) or a Braconid. In 1925, Martynov described four Hymenoptera from the Jurassic beds of Turkestan, viz., *Anexyela gracilis*, a saw-fly, *Paroryssus extensus*, the type of an extinct family (Paroryssidæ) related to the Oryssidæ, *Mesaulacinus oviformis*, an Aulacid, and *Mesohelorus muchini*, a Helorid Proctotrypoid of very modern aspect. These forms show a high development of the Phytophaga and Terebrantia at a period long antedating the Tertiary, but they furnish no clue to the derivation of the Hymenoptera from more primitive insects.

Tillyard in his important papers on the Panorpoid complex (1918) dissents from Handlirsch's views and is inclined to derive the Hymenoptera from Mecopteroid forms. This is clear from the following statement: "My own researches upon the Neuropteroidea and Panorpoidea, as far as they have gone, have convinced me that these two sub-classes have a great deal in common. They have also revealed the possibility of the Hymenoptera and even the Coleoptera having a closer relationship with the Panorpoidea than is generally suspected. Not only do many signs point to the Mecoptera (Panorpatæ) as being a central Order round which all the rest of the Holometabola may be more or less closely grouped, but the Paleontological evidence also points unmistakably in the same direction". He also calls attention to the fact that fossil Mecoptera "of the genus Permochorista very near the existing Australian genus Tæniochorista exist in the Permian of New Castle, New South Wales. No other holometabolous insects are known from Palæozoic strata. The other Mecoptera must have arisen well before Permian times."

Some light has now been shed on the probable origin of the Hymenoptera by Tillyard's recent studies (1924) of the Permian fossils collected by Yale University in the Wellington Shale, near Elmo, Kansas. Some 4,000 specimens of insects have been taken from this soft

PL. I.

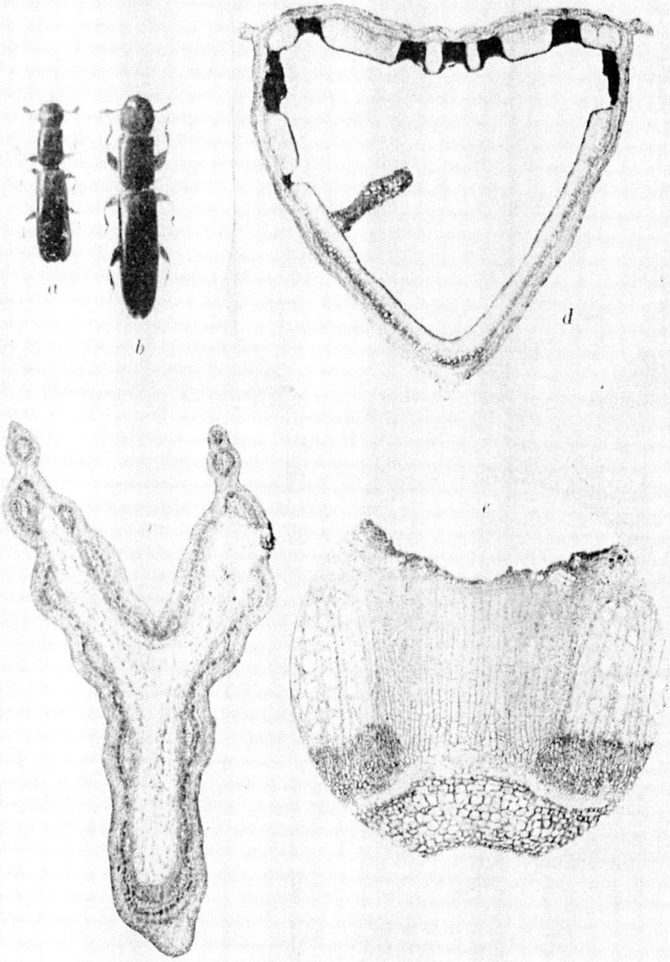

FIG. 1. a. *Eunausibius wheeleri* Schwarz and Barber. (× 10); b. *Coccidotrophus socialis* Schwarz and Barber. (× 10); c. Cross section of base of very young petiole of adult *Tachigalia*, showing pith still in the cavity. d. Cross section of base of large petiole of *Tachigalia* inhabited by the ant *Pseudomyrma damnosa*. The dark areas in the wall are nutritive parenchyma, which is not disturbed by the ants, but nourishes their coccids. e. Cross section through one of the strands of nutritive parenchyma showing the cells with their homogeneous, amber coloured contents. (Photographs by Prof. I. W. Bailey.) (*See* p. 17.)

[*face p.* 28

Plate II.

FIG. 2. *Permohymen schucherti* Tillyard, complete specimen as preserved in the Lower Permian Beds of Kansas. Length of forewing, slightly over half-an-inch. (After R. J. Tillyard.)

FIG. 3. *Protohymen permianus* Tillyard, complete forewing, much enlarged, from a specimen found in the Lower Permian Beds of Kansas. (After R. J. Tillyard.)

[*face p.* 29

argillaceous limestone which represents the deposit of a marly fresh-water lake (Dunbar, 1924, Tillyard 1924*b*, 1926*b*). Though the specimens consist very largely of wings, they are often so well preserved that even the colour-pattern is visible. Among the material, which has thus far yielded a new order, Protodermaptera, a peculiar Palæodictyopteron, *Dunbaria fasciatipennis* Tilly, many Ephemerids, small cockroaches, large Protodonata and true Odonata, Tillyard has discovered the wings of three species and genera of a new order which he calls the Protohymenoptera. The three species, *Protohymen permianus* (Fig. 3), *Permohymen schucherti* (Fig. 2) and *Asthenohymen dunbari*, are assigned to two sub-families. More recently, Tillyard (1926*a*) has published further observations on these insects, including a description of some fragments of the head and thorax of *P. schucherti* and the wings of six additional species of Asthenohymen. Protohymen is the most primitive form, with a wing-venation readily comparable with that of the Xyelidæ and Siricidæ among existing Phytophaga, and resembling that of the Mecoptera and Neuroptera but without affinities to the Protoblattoids, or primitive cockroaches. The two pairs of wings are much alike and without tornus or termen, the posterior wings being but slightly shorter and broader than the anterior pair and not connected with them in flight. The wing-membrane is stiff and glassy as in existing Hymenoptera. Protohymen and Permohymen have a distinct pterostigma but this structure is lacking in Asthenohymen. The wings described by Tillyard, therefore, seem to be so clearly intermediate between those of the Mecoptera and present-day Hymenoptera that Handlirsch's hypothesis of a Protoblattoid origin of the latter order will have to be abandoned and we shall have to assign it a place at the head of the Hexapod series among the modern members of the Panorpoid complex, the Lepidoptera, Diptera and Mecoptera (Panorpatæ). This assignment would also agree better with the facts of ontogeny and avoid Handlirsch's awkward postulation of an independent development of complete metamorphosis in the Hymenoptera.

Returning now to the evolutionary process within the Hymenopteran order, we may profitably confine our remarks to a very brief account of the three sub-orders and their interrelations. Handlirsch is undoubtedly right in stating that the existing forms are all so highly specialized in particular directions that phylogenetic derivations become very speculative. The first group to be considered, the Phytophaga, or Symphyta, is, as previously stated, unanimously regarded as the most primitive. Similar agreement prevails in regard to the relationships of the families within this sub-order. Thus the leaf-eating Pamphiliidæ and Xyelidæ may be properly regarded as the most ancient not only of all existing Phytophaga but of all existing Hymenoptera. From the Xyelidæ, one series of developments seems to have led to the Tenthredinidæ and Cimbicidæ, another from the Pamphiliidæ to the wood-eating Siricidæ, Xiphydriidæ and Cephidæ and to the peculiar family Oryssidæ, which appears to be an offshoot from the Siricidæ. This course of development, shown in the accompanying table of Yuasa (1922), is supported by the study of both adults and larvæ. The larvæ of the Pamphiliidæ are regarded by Handlirsch as the most primitive among Hymenoptera.[1] They are cylindrical, with three pairs of well-developed thoracic legs and a pair of cerci, but without abdominal prolegs. Yuasa (1922), however, regards an eruciform larva wth ten pairs of abdominal prolegs as the archetype. This is approached by the Xyelidæ and Tenthredinidæ. In the wood-boring Cephidæ (Middleton 1927), Xiphydriidæ and Siricidæ (Yuasa) the abdominal prolegs have disappeared and the thoracic legs are much reduced. The Cephid larva, however, retains eyes and cerci. Finally in the Oryssidæ all the thoracic and abdominal appendages have been lost and the larva is vermiform

[1] For good accounts of the life-histories of Pamphiliids see H. Schmidt (1910) and H. C. Severin (1920).

like that of the Terebrantia. This condition is explained by the fact that the larva of Oryssus, as Burke (1917) has recently shown, is parasitic on the larvæ of various Buprestid beetles. (See also Rohwer and Cushman, 1917 and Comstock, 1924.)

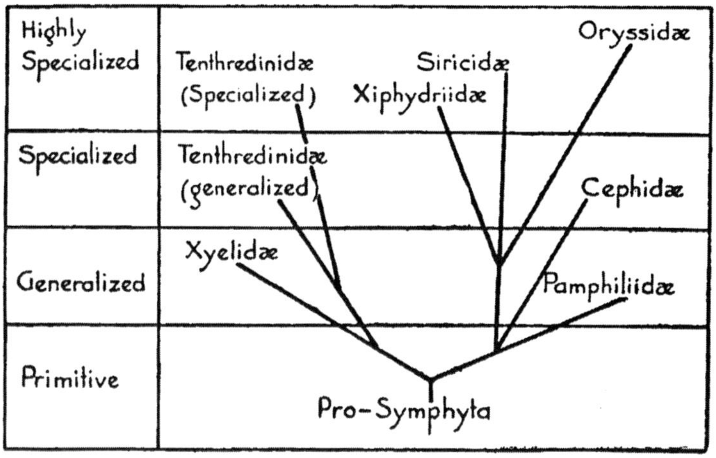

PHYLOGENETIC TREE INDICATING THE PROBABLE AFFINITIES OF THE VARIOUS FAMILIES OF THE TENTHREDINOIDEA. (After Yuasa.)

There appear sporadically among the Phytophaga certain ethological peculiarities which become more strongly accentuated and more widely distributed among the Terebrantia and Aculeata. I have just alluded to the parasitism of Oryssus, the only known case in the sub-order and clearly foreshadowing the universal occurrence of the phenomenon in the Terebrantia. Though all Phytophaga are vegetarian as larvæ, the adults of some of the sawflies, especially of the genera, Tenthredo and Tenthredella (Mràzek, 1909, Rohwer, 1913*a*, and others) are decidedly entomophagous. Parthenogenesis, moreover, a phenomenon of great frequency among the Terebrants and Aculeates, seems to be well-established in the Tenthredinidæ. The males of many species are much scarcer than the females. According to MacGillivray (1914), von Siebold found that the males of only one-fifth of the German species were known, and Cameron was able to recognize the males of only one-third of the British species. When parthenogenesis occurs it is apparently mixed or indeterminate, since in some sawflies the unfertilized eggs are known to produce only males (arrhenotocous), in others only females (thelytocous), and in still others both sexes (ampherotocous). So far as known, and with some exceptions, the unfertilized eggs of Terebrants and Aculeates produce only males. Gregarious and sub-social habits are frequent in the Tenthredinidæ, especially during the larval and pupal stages, but for lack of time cannot be adequately treated in this lecture.[1]

While the Clistogastra, embracing the Terebrantia and Aculeata, are readily distinguished from the Phytophaga by the petiolate abdomen, more or less reduced wing-venation, highly vermiform larva and differences of habits, it is not so easy to draw a hard and fast line between the Terebrantia and Aculeata. One of the best distinctions lies in the fact that the ovipositor in the Terebrantia has separate apophyses and is used for depositing the eggs in or on other insects or in plant tissues whereas in the Aculeata it has the apophyses fused and is modified to form a sting which is used as an organ of attack or defence and not for oviposition. There are exceptions, however, since the Sapygidæ, which are placed among the Aculeata, use the ovipositor like the Terebrants, and the Trigonalidæ, which are assigned by most hymenopterists to the latter suborder, have the ovipositor so greatly reduced that it can be of no use in oviposition. There are also a number of Aculeata (Chrysididæ, many ants and some bees) in which the sting is similarly vestigial.

[1] For references to some of the literature on this topic see my "Social Life Among the Insects," 1923, p. 290-292. The papers by Curtis (1845), F. Smith (1866) and Ducke (1916) also contain interesting accounts and figures of the social cocoons of Brazilian sawflies of the genus Dielocerus. The singular social larvæ observed by Mann and myself in British Guiana and Bolivia (Wheeler and Mann, 1923) may also belong to this genus.

Hymenopterists agree in regard to the phylogenetic derivation of the Clistogastra from primitive Phytophaga and in the belief that Terebrantia are more intimately related to the Phytophaga than are the Aculeata, but disagree as to which family of Terebrantia is the most primitive and as to whether the Aculeata arose from Terebrantia or had an independent origin among the Phytophaga. Handlirsch (1908) regarded the Ichneumonidæ as the most primitive group of Terebrantia and as having given rise to the Aculeata. Brues (1921) rejects this view so far as it concerns the origin of the Terebrants and regards the Stephanidæ as the family which links up most naturally with the Phytophaga. He says: " The Stephanidæ are structurally primitive and strikingly like the Oryssidæ in the peculiarly horned head which had been remarked on before the habits of the Oryssids were known. On account of the presence of a costal cell in the wing, the polymorphic family Evaniidæ is necessarily also more primitive than the Ichneumonidæ or Braconidæ and although one sub-family, the Fœninæ, resemble the Stephanidæ, as has been already noted by Bradley. Some Braconidæ, the Stephaniscinæ, Spathiinæ and Hormiinæ, are much like Stephanids, so much so that it is difficult to believe that they are not directly derived from them. One other family, the Capitoniidæ, recently segregated from the Braconidæ, appears to be very definitely related to the more generalized Evaniidæ (Aulacinæ). Omitting in this brief consideration several less pertinent families, and ignoring other recently segregated ones, we have left only the Ichneumonidæ, related possibly through the Alysiidæ to the Braconidæ. Structurally this relation seems plausible, but as the Alysiids attack almost exclusively the highly specialized Diptera it is very difficult to regard them as closely related to the ancestors of the Ichneumonidæ, so highly diversified in habits and structure. The latter then are not so easily derived and may go back to Evaniid-like forms ". Brues also points out the singular association of the primitive Terebrant families with wood-boring insects. This was seen in the Oryssidæ among the Phytophaga and is conspicuous in the most generalized group of Evaniidæ, the most primitive Braconidæ, many of the structurally primitive Ichneumonidæ and the Capitoniidæ.

With the exception of the Jurassic Mesaulacinus, Mesohelorus and Ephialtites previously mentioned, all the known fossil Terebrantia are of Tertiary age. There are few species known from the Eocene, but many, representing numerous genera and families have been recorded from the Baltic Amber of Lower Oligocene age and from the shales of Oeningen, Radoboj and Florissant, which are assigned to the Miocene. The amber and Florissant fossils were critically studied by Brues in a series of papers (1906, 1908*a*, 1910*a*, 1910*b*, 1923*b*). His list of the known Teritiary Terebrantia published in 1910 comprises 212 species distributed among 101 genera and twenty-one families, a sufficient amount of material to give a fair picture of the extent and diversity of the sub-order during early and mid-tertiary times. Although the species all differ from those now living, they nevertheless belong for the most part to well-known recent genera, and all the known Tertiary families are still extant. Some of the extinct genera, however, are peculiar or unique. Kinsey (1919) has described three species of the primitive Cynipid genus Aulacidia from the Baltic amber and Florissant Miocene, and Cockerell (1915, 1916*a*, 1916*b*, 1920, 1921*a*, 1921*b*) has described a number of other Terebrants from the Miocene of Florissant, from the Upper Oligocene of the Isle of Wight and from Burmese amber, and a few from still earlier formations, the Green River Eocene. These last he refers to two genera of Ichneumonidæ and one of Braconidæ. But all the Tertiary genera and species described up to the present time are quite as highly specialized in structure and were presumably also as specialized in habits as their living congeners and therefore throw no light on the phylogeny of existing Terebrant and Aculeate families.

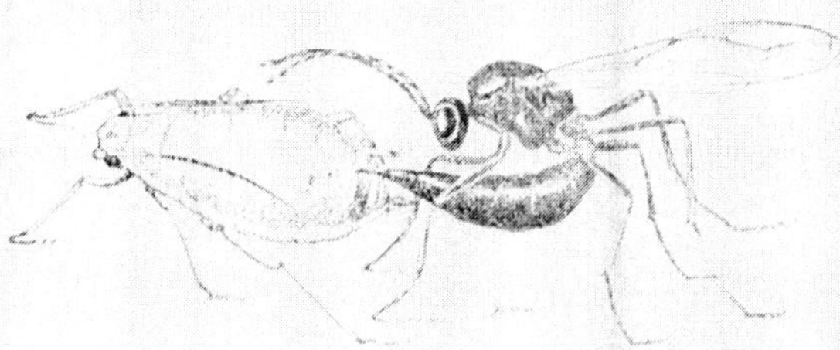

FIG. 4. A Terebrant, *Aphidius testaceipes*, ovipositing in the body of the spring grain-aphis, *Toxoptera graminum*. (After F. M. Webster.)

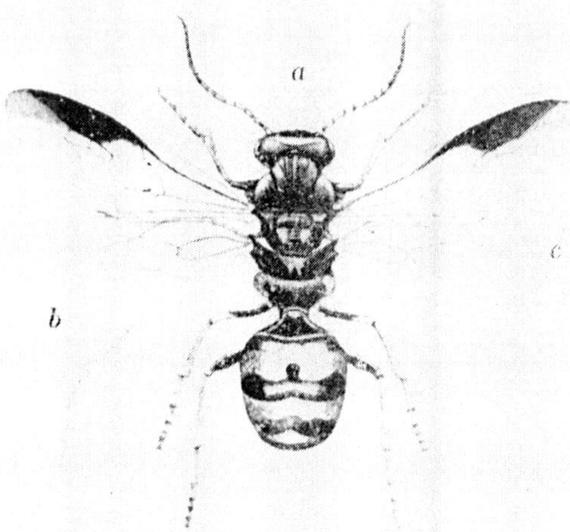

FIG. 5. a. Female of a North American Trigonalid, *Bareogonalos canadensis*; b. antenna of same; c. antenna of male. (After W. A. Schulz.)

[*face p.* 34

Plate IV.

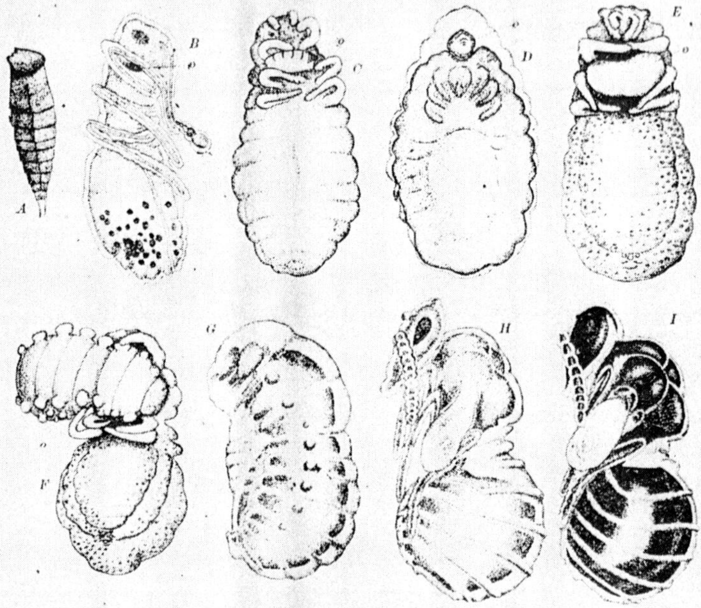

FIG. 6. Development of *Orasema viridis*. A. Planidium, or First larval stage of *Orasema*. B. Pupal worker of *Pheidole instabilis* with phanidium (o) attached to side of neck; C. Female *Pheidole* pupa with somewhat older planidium (o) attached in sternal region; D. Female *Pheidole* pupa with planidium (o) in same stage as in preceding figure, attached behind head; E. Female *Pheidole* pupa (Phthisogyne) with older Orasema larva (o) in sternal region; F. *Orasema* larva (o) beginning to pupate, with vesiculate knobs on its surface; G. *Orasema* pupa fallen from its host and developing within the vesiculate skin; H. Fully formed pupa; I. Pupa pigmented and ready for eclosion. (*See* p. 39.)

[*face p.* 35

Since, with the exception of the gall-flies (Cynipidæ) and certain plant-infesting Chalcidoids, all the Terebrantia oviposit on or in other insects (Fig. 4) and as larvæ feed on their tissues, the name " Parasitica " is often applied to the sub-order. The matter calls for brief consideration, because it has often been pointed out that the word " parasite ", as applied to the Terebrants, to many Diptera and to some other insects which have similar habits, is inappropriate. These insects differ from parasites properly so called in three respects, First, they attack other insects and true parasites with the possible exception of certain Crustacea and Infusoria, do not attack members of their own phyla; second, they nearly always eventually kill their hosts, and third, they live with their hosts only during the larval or larval and pupal stages and as adults show no " degenerative " effects of the association. They are, therefore, really refined predators. Attempts have been made to embody these distinctions in a term. " Pseudoparasitism " might be acceptable, were it not that it has been used by Blanchard (1889) in another sense.[1] Riley, in 1893, clearly recognized the need of a different term when he called the entomophagous Hymenoptera and Diptera "fatal parasites", to distinguish them from the " parasites proper " and " inquilinous parasites ". O. M. Reuter (1913, p. 53) introduced the term " parasitoids " for these peculiar predators, and I (1923) and more recently Root (1924) have employed this term. It is probable, however, that it will not be generally adopted and that " parasite " will continue in vogue. If the distinction is clearly understood there can be no harm in such usage, but it should be noted that it has naturally created confusion in phylogenetic discussions, and particularly in connection with Handlirsch's derivation of the Aculeata from Terebrant ancestors. The phylogenist

[1] " By 'pseudoparasites' we mean animals or plants which are found accidentally on the surface or in the interior of the human or animal body, whether they be normally parasites of other species or normally lead an independent life."

is naturally so convinced of the high adaptive specialization of all parasites that he avoids using them as hypothetical ancestors, but this objection does not hold in the present instance, if we regard the Terebrantia as a peculiar class of predators, or parasitoids, especially as the lower Aculeata have the same habits. This has been clearly stated by Picard in his valuable work on the insects of the fig-tree (1919, p. 167). He says: " It is difficult to see how the mode of life of an externally feeding Terebrant differs from that of certain predators, such as the Scolias. Both approach their prey and without moving it, attach to it an egg which will produce a larva similarly attached. The opinion of Lepelletier, who took great pains to prove that the Scolias could not be regarded as parasities, no longer seems to us to be of much interest. A Polysphincta larva on a spider is quite comparable to that of the Pompilids and especially to those Pompilids which enter the nest of a spider and oviposit on it without dragging it forth and burying it. It is true that the predators paralyse their victims, but the Terebrants also possess a poison gland and many of them immobilize the larvæ which they attack, at least feebly and momentarily ".

I shall not be able to discuss at length the many extraordinary modifications exhibited by the ontogeny of the Terebrants, such as polyembryony, which has been so carefully studied within recent years in France, Italy and the United States. The more frequent phenomena of hypermetamorphosis, however, call for a partial consideration in connection with a peculiar and very primitive family, the Trigonalidæ, concerning the affinities of which hymenopterists have shown the greatest diversity of opinion. Some (Westwood, Ashmead, Schimedeknecht) have regarded these insects as primitive Aculeates, while others (Börner, Handlirsch, Bradley, etc.) regard them as primitive Terebrants. Such different allocations would imply that the family should be placed near the point of bifurcation of both the Terebrants and Aculeates from the ancient Phytophageous stem of the Hymenopterous order.

W. A. Schulz, who monographed the Trigonalidæ in 1907, describes the family as containing forty-two species, seventeen genera and five sub-families, distributed in various parts of the world. The species, however, are all very rare and local and apparently on the verge of extinction. They are rather small, often vividly spotted or banded, with a very archaic appearance (Fig. 5), showing a bewildering mixture of Phytophagous, Terebrant and Aculeate traits. The multiarticulate antennæ and the number of abdominal segments are Terebrant but the head recalls the sawflies and the general form of the body the Bethylidæ and the shape of the abdomen the Mutillidæ. In some species (Bareogonalinæ and Nomadininæ) the anterior portion of the head is prolonged and snout-like, like that of the Mecoptera. Cerci are present, but the ovipositor is reduced to a minute vestige. Börner (1919) found the maxillæ to possess well-developed paraglossal appendages and to be so much more primitive in structure than in other Terebrantia that he created for the family a special section, the "Archiglossata". The wing-venation is very primitive, and in usually possessing three closed cubital cells is much like that of Aculeates. The tarsi have well-developed plantar vesicles, structures which, according to Börner, are lacking in Terebrants but of frequent occurrence among Phytophaga and Aculeata.

The habits of Trigonalids are very imperfectly known and, as I shall endeavour to show, have been incorrectly interpreted. The insects are usually seen flying near the ground and alighting on the low vegetation in places where social wasps are common. This and the following data have led authors to regard them as parasitic in the nests of Vespids: A single specimen of the South American *Seminota depressa* was shaken from a nest of *Polistes canadensis* in the British Museum, and a specimen of *S. mejicana* from a nest of *Paracharlergus apicalis* in the Museum of Paris. G. W. Taylor (1898) took twenty-three males and four females of the North American *Bareogonalos canadensis* (Fig. 5) resting on grass and weeds within a few feet of the entrances to *Vespa occidentalis* nests, and inferred that the parasites must have developed in these structures. Schulz (1905) described the male *Nomadina cisandina* from a specimen taken by O. Dreher in a nest of *Polybia dimidiata* at São Paulo, Brazil. From these facts Schulz concluded that the Trigonalids must enter the wasp nests and oviposit on the brood, and even interpreted the peculiar dilated joints of the palpi as special adaptations for finding a way about in the dark recesses of the nests.

Two other observations, however, show that the Trigonalidæ occasionally have very different hosts. As early as 1891, Riley and Howard bred the North American *Tapinogonalos pulchella* from a puparium of a Tachnid fly (*Exorista lobeliæ*) which was living in the caterpillar of *Acronycta lobeliæ*, and in 1909, Bischoff reared a specimen of the North American *Sycogaster pullata* from the cocoon of an Ichneumonid, *Ophion macrurus*, which had parasitized a caterpillar of *Telea polyphemus*. In both these cases, therefore, the Trigonalids were behaving as hyperparasites. Apart from the fact that the pupæ of these insects are naked, i.e., not enclosed in cocoons, the foregoing are all the data we possess concerning their habits. Bugnion (1910) has published an account of the anatomy of the European *Pseudogonalos hahni*, to which I shall return presently, and Cockerell (1917) has described a fossil Trigonalys (*T. pervetus*) from the Burmese amber, which is supposed to be of Miocene age.

Now I believe that some light may be thrown by a brief review of what is known of the development of the Terebrant families Eucharidæ and Perilampidæ not only on the Trigonalidæ but also on certain obscure points in the ontogenetic development of the Hymenoptera in general. Some eighteen years ago (1905*a*) I found in the nests of a Texan ant (*Pheidole instabilis*)

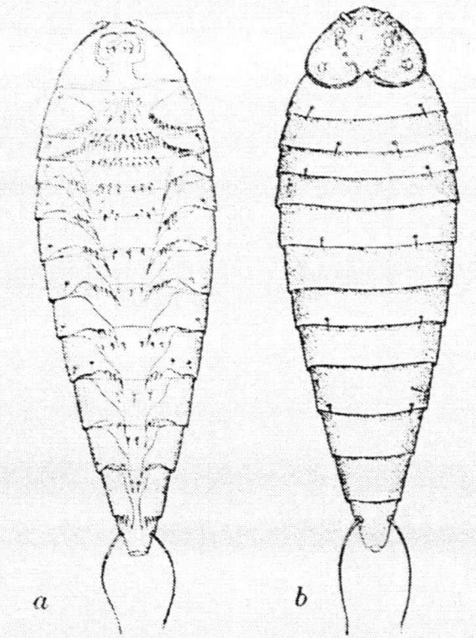

FIG. 7. Planidium of *Perilampus hyalinus*. a. Ventral view; b. Dorsal view, x about 350. (After H. S. Smith.)

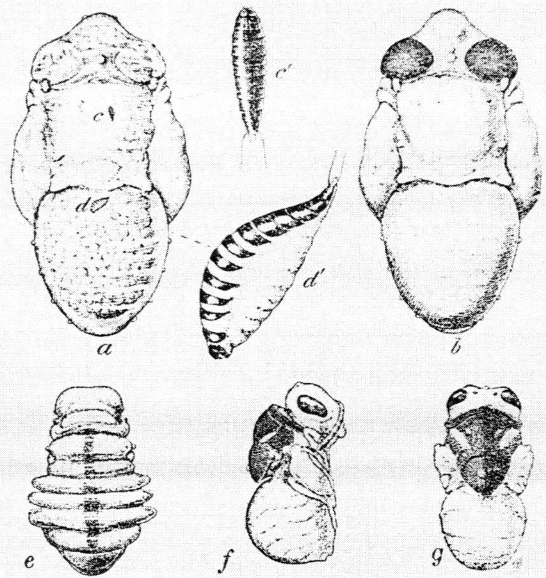

FIG. 8. Development of *Perilampus hyalinus*. a. Pupa of *Varichaeta aldrichi* parasitized by *hyalinus*. b. unparasitized pupa of *V. aldrichi*. c. planidium of *P. hyalinus* before feeding. c′. same more enlarged. d. planidium after feeding. d′. same more enlarged. e. Mature pupa of *P. hyalinus*. f. Lateral view of pupa just before eclosion. g. Dorsal view of same. (After H. S. Smith.)

[face p. 38

Plate VI.

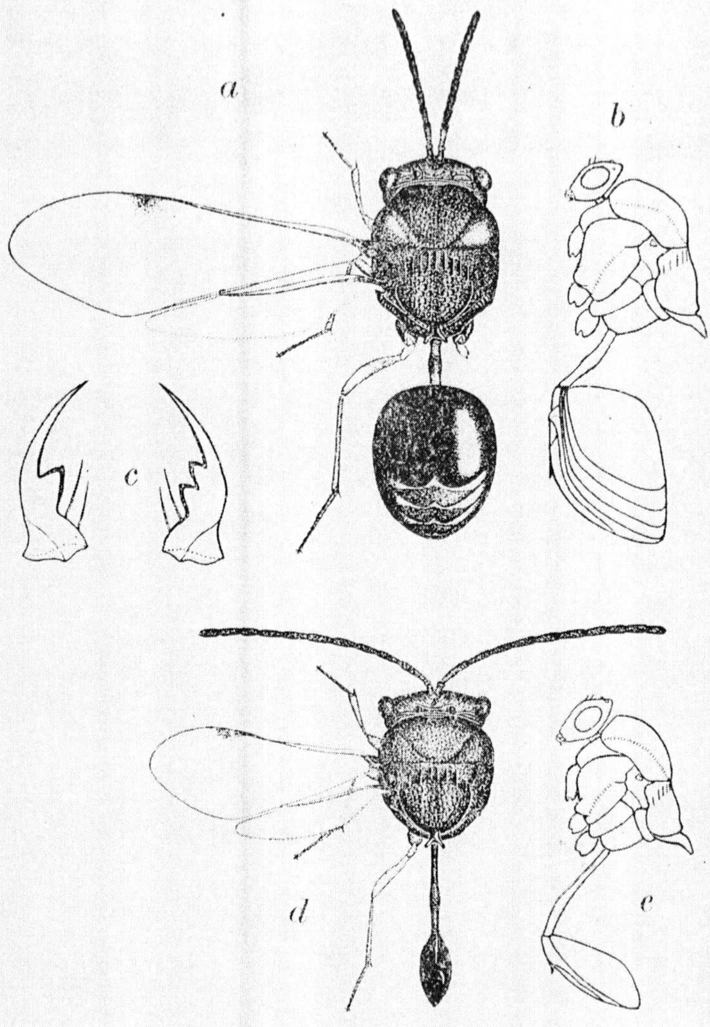

FIG. 9. *Schizaspidia tenuicornis* Ashm. a. Adult female, dorsal view. (× 11); b. Lateral view. c. Mandibles. d. Adult male, dorsal view. (× 11); e. Lateral view. (After C. P. Clausen.) (*See* p. 42.)

[*face p. 3*

a metallic green Eucharid (*Orasema viridis*), which, judging from dissections of the female, produces an enormous number of very minute eggs. I did not see the eggs after oviposition but only the minute, active first-stage larvæ (Fig. 6A), which are about 0.16 mm. long, dark-coloured, with heavily chitinized dorsal surface, well-developed head, thirteen post-cephalic segments and a pair of appendages (cerci) at the posterior end. These larvæ, which I subsequently called "planidia", attach themselves to the necks of the ant larvæ and pupæ (Fig. 6B to F) and after feeding and moulting, pass into a stage comparable with the usual, plump vermiform Hymenopteran larva. After attaining their full growth they pupate *in situ* (Fig. 6G), without spinning cocoons, drop from the host and soon afterwards emerge as adult Orasemas, which are carried about and fed by the ants but eventually escape from the nest to the outside world.

A notable advance in our knowledge of the planidium and its development was later made by H. S. Smith in his studies of Perilampus (1912, 1916, 1917). In his first paper he showed that the planidium of *P. hyalinus* (Fig. 7) is sometimes found crawling over the surface of small web-worm caterpillars (*Hyphantria textor*) and that it bores its way into the caterpillar and does not undergo any further change till it encounters inside the host the larva of one of its common parasites, i.e., either the Tachinid *Varichæta aldrichi* (Fig. 8) or the Ichneumonid *Limnerium validum*. In case the planidium encounters a Varichæta maggot it enters it, even if it is very small, and remains quiescent till the maggot forms its puparium. If a Limnerium larva is encountered the planidium enters it, and remains inactive till the following spring. "Varichæta pupates in the fall so the planidium makes its exit at this season of the year and hibernates externally. *Limnerium validum* hibernates as a larva and pupates in the spring, hence the planidium in this case remains internal until spring, when it emerges by means of its efficient mandibles and spined head, completing its development in precisely the same manner as those individuals which happen to have chosen Varichæta as their host". The further development of the Perilampus planidium on the parasites of the web-worm is very similar to that of Orasema, since the minute parasite after feeding externally produces a vermiform larva before pupation (Fig. 8*e*). Thus *Perilampus hyalinus* behaves as a superparasite. Both it and other species of the genus have been bred from several other caterpillar parasites, which are listed by Smith, but there are records of Perilampus bred from other insects (Chrysopa, *Anthonomus grandis* and the Sarcophagid parasites of grasshoppers). In 1915, W. R. Thompson published an account of the internal anatomy of an unidentified planidium which he found on Noctuid caterpillars feeding on witch hazel (*Hamamelis virginica*) at Ithaca, New York, and in 1922, Miss Norma Ford described a planidium of a Perilampus which she found attached to a meadow grasshopper (*Conocephalus fasciatus*). In this case the true host is probably the larva of a Sarcophaga, species of which have been shown by Swenk (1911), Kelley (1914), and others, to be parasitic in grasshoppers.

In his papers of 1916 and 1917, Smith also solved the problem of the place in which the Perilampus female lays her eggs. After breeding *P. chrysopæ* from Chrysopa cocoons he succeeded in observing the oviposition of the parasite. The eggs are laid singly on the leaves near Aphis colonies infested by Chrysopa larvæ. The female deposits about fifty of her minute eggs in a single day. "The egg is very faintly attached to the surface of the leaf at one end. Hatching takes place in seven to ten days and the first larval stage is of the planidium type described in the above mentioned paper (1912). For several days previous to hatching the dark-coloured planidium can be plainly discerned through the transparent egg-shell. The planidium is active immediately upon hatching, crawling rapidly about, but soon it attaches itself to the leaf by means of the caudal sucker and stands out at right angles to the surface. In this position it remains for days at a time, motionless, excepting when some insect comes within its reach, when it suddenly becomes frantically active, reaching and swaying back and forth in its attempt to attach itself to the prospective host (the Chrysopa larva). If the latter should unfortunately come too near, the planidium attaches itself with lightning-like quickness to a hair or bristle of the host. It then leisurely crawls down the hair to the host's body and attaches itself by its mouth hooks. Quite often the planidia are found attached to the egg-stalk of Chrysopa, assuming a position at right angles to the axis of the stalk. This shows an interesting instinct in the planidium, since it may and actually does, as the writer has observed, waylay the young Chrysopa larva as it leaves the egg and crawls down the stalk. The prescience of the mother Chrysopa in placing her eggs at the end of a long egg-stalk to overcome the cannibalistic propensities of her progeny is in this case their undoing, since the planidium attached to the Chrysopa egg-stalk is sure to reach its proper host, while those upon the surface of a leaf are quite as likely to attach themselves to an Aphid or other insect". Smulyan (1916) has, in fact, found a Perilampus planidium attached to the antenna of an Aphid. In tracing the further development of *P. chrysopæ*, Smith found that the planidium clings to the Chrysopa larva without feeding till the latter is ready to pupate, though it may be occasionally removed from the host by ecdysis. After the Chrysopa larva has spun its cocoon and pupated, the planidium begins to feed and develops after the manner of *P. hyalinus* previously described.

Further interesting and important observations have also been made within recent years on the Eucharidæ, which like Orasema are all ant-parasites. Forel in 1890 found an adult *Eucharis myrmeciæ* in a cocoon of an Australian bulldog ant (*Myrmecia forficata*). More recently Brues (1919) has described the planidium and later larval and pupal stages of *Psilogaster fasciiventris* taken by Tillyard from cocoons of another bulldog ant, *Myrmecia gulosa*, and G. C. and E. H. Wheeler (1924) have described *Schizaspidia polyrhachicida* from the cocoons of a Philippine ant, *Polyrhachis* (*Myrmhopla*) *dives*. But the most extensive observations, which go far towards filling the gaps in my observations on Orasema, have been made by Clausen (1923) on *Schizaspidia tenuicornis* (Figs. 9 and 10), which parasitizes the common Japanese carpenter ant, *Camponotus japonicus*. During late August, *S. tenuicornis* was found to thrust its minute eggs in masses into the buds of various trees (*Morus alba*, *Castanea sativa*, *Cladrastis amurensis* and *floribunda*) (Fig. 10*a*). The number of eggs deposited by a single female was found to range from 940 to 1230, with an average of 1050. Several females often ovipost in a single bud and so many buds may be utilized that a single shrub was estimated to bear 4,320,000 eggs! There is a great mortality among the eggs, owing to an eventual falling away of the bud-scales and exposure of the eggs to humidity, mould, etc. The planidium develops in the egg during the winter and hatches during July. It is scarcely more than 0.1 mm. long and capable of a rather limited looping movement by means of its sucking mouth and a group of stout spines which replace the cerci of other planidia at the posterior end of the body (Fig. 10*b* to *d*). The larvæ make their way out of the buds and move about on the twigs and leaves. Here they come in contact with the Camponotus workers attending the plant lice and visiting the fruit of the trees and attach themselves to the hairs on the tarsi of the ants (Fig. 10*m*). They are thus carried into the ant nests and pass from their carriers to the ant brood. "They manage, largely by chance, to become attached to the ant-larvæ and after moving about over the body for a time settle dorsally on one of the sutures between the head and first thoracic segment, or between the first two of the latter. The mandibles are then embedded firmly in the derm of

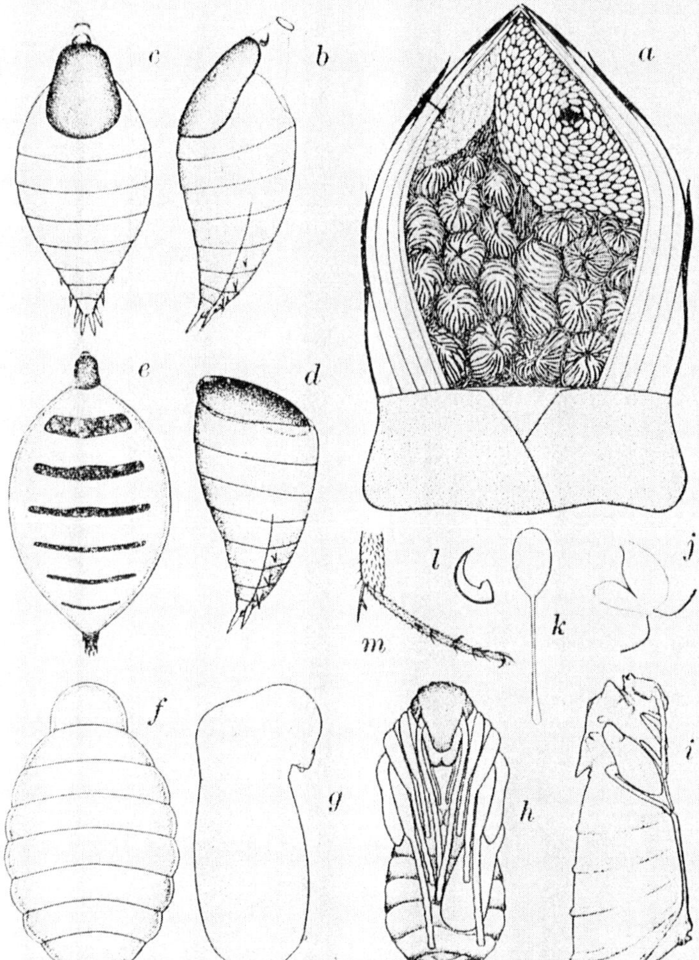

FIG. 10. Life history of *Schizaspidia tenuicornis* Ashm. (After C. P. Clausen.) a. Mulberry bud with scales removed to show two egg-masses *in situ* (x 20); b. First stage larva, lateral view (x 500); c. First stage larva, dorsal view (x 500); d. First stage larva, lateral view, with head retracted (x 500); e. Advanced first stage larva, dorsal view (x 200); f. Second stage larva, dorsal view (x 85); g. Third stage larva, lateral view (x 10); h. Third stage larva in situ on male pupa of Camponotus (x 9); i. Female pupa, lateral view (x 10); j. Sagittal section (diagrammatic) of the mouthparts of the third stage larva. k. Ovarian egg (x 100); l. Mandibles of first stage larva; m. Three first stage larvæ adhering to the leg of a Camponotus worker (x 50). (After C. P. Clausen.)

[*face p.* 42

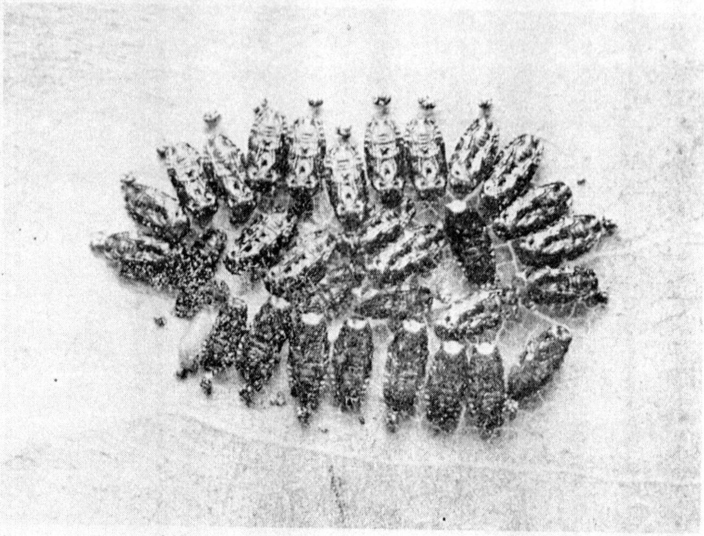

FIG. 11. Pupæ of *Comedo* sp. on the lower surface of a leaf, showing their concentric arrangement and the masses of fæces deposited near their posterior ends. One of the larvæ (to the left, below) has not yet pupated. (Photograph by Prof. C. T. Brues.) (*See* p. 54.)

FIG. 13. *Sphex procerus* carrying caterpillar of sphinx moth to her burrow. (Photograph by Prof. Carl Hartman.) (*See* p. 52.)

[*face p.* 43

the host and no further movement takes place". Two stages follow, during which the planidium feeds and becomes vermiform (Fig. 10*f* to *h*), and after the host has spun its cocoon and become a semipupa and eventually a pupa, the parasite manages to regain its hold after the host's ecdysis. The attachment to the ant pupa is almost invariably on the metathorax just beneath the legs or wing-pads (Fig. 10*h*). Development now becomes greatly accelerated, and feeding is completed within a few days after the second moult. Pupation occurs within the host cocoon and the adults, after emergence in the nest, are not, like Orasema, fed and cared for by the ants, but forthwith escape to the outside and mate in the open, the males hovering about the nests and seizing the females as soon as they emerge into the light and before they can take flight. The life of the adult *S. tenuicornis* (Fig. 9) is extremely short as shown in Clausen's table, which is here reproduced:

Egg stage .. .. approximately	11 months.
First larval stage	20 days.
Second larval stage	3 days.
Third larval stage	4 days.
Pupal stage	6 days.
Adult (within cocoon)	1 day.
Adult (outside nest)	2 to 3 days.

In my paper of 1907 I took it for granted that the female *Orasema viridis* must lay her eggs in the ant-nest, but Smith and Clausen have given reasons for supposing that this Eucharid probably lays on the leaves of bushes, like Perilampus, or in buds, like Schizaspidia. Attention should be called, however, to the fact that Orasema is more of an ant-guest than Schizaspidia since the Pheidole workers actually feed, fondle and carry it about. Moreover, during November, 1923, Dr. W. M. Mann came upon another species of Orasema in the nests of a small Pheidole in Lower California and saw five of the males actually copulating with female pupæ in the nest chambers! This observation would seem to indicate, though it does not, of course, prove that the eggs of Orasema are laid in the ant-nests.

After this review of the complicated development of the Perilampidæ and Eucharidæ as ascertained by American entomologists, we may return to the Trigonalidæ. It will be noticed that the very rare and sporadic occurrence of these insects both in the adult and larval stages, and the occurrence of the latter both as primary parasites and hyperparasites in very diverse hosts points to a development with a minute, active, first larval stage like that of the Perilampids and Eucharids. This becomes practically certain when we turn to Bugnion's anatomical observations on the female of the European *Pseudogonalos hahni*. He says: " The female possesses neither ovipositor nor sting. All that can be seen are a few small chitinous pieces situated at the orifice of the genital canal and probably serving to hold the penis during the copulatory act ". And his description of each ovary shows that it consists of as many as 300 to 400 slender ovarioles, each containing a series of very minute eggs, only 0.18 mm. long. Counting five eggs to an ovariole, as represented in his figure, a single female would produce between 3,000 to 4,000 ova! Bugnion seems not to have noticed the significance of this extraordinary fecundity, which is much greater than that of the Japanese Schizaspidia described by Clausen. It can only mean, when we consider the great rarity of the Trigonalidæ in all their stages, that there must be an enormous wastage of eggs and this can only be due to a mode of oviposition and development like that described for the Perilampidæ and Eucharidæ. The minute eggs are probably laid on the leaves of bushes or weeds and the emerging larvæ are probably much like planidia and attach themselves to passing wasps or caterpillars, thereby gaining access to the brood of the former and the Hymenopterous or Dipterous parasites of the latter.

Certainly the first larval stage of the Trigonalids when found will be of great interest, since it may well be, if we may judge from the archaic habitus of the adult, of a more primitive and more generalized type than the planidia of the Chalcidoids above described. Perhaps it actually has thoracic legs and would, therefore, be a true triungulin. The striking resemblance both in structure and behaviour of the Hymenopterous planidium to the triungulin larvæ of the Meloid and Rhipiphorid Coleoptera and the Strepsiptera has been remarked by several observers.[1]

There are, furthermore, both among the Terebrantia and Diptera a number of first larval forms which resemble planidia and others, like the cyclopoid and eucoiliform larvæ of certain Platygastridæ and Cynipoidea, which seem to represent peculiar modifications of the planidial type. Richardson (1913) and Mrs. E. H. Wheeler (1923) have distinguished at least ten or a dozen different first larval types among the Terebrantia and H. L. Parker (1924) has published an elaborate paper on the larval forms of Chalcids. Among the Diptera planidium-like larvæ occur among the Acroceridæ, and active first stage larvæ which behave like planidia occur among the Nemestrinidæ, leaf-ovipositing Tachinidæ and Bombyliidæ.[2]

[1] For recent accounts of the larvæ of American Meloidæ and lists of literature see the papers of Williams and Hungerford (1914), Milliken (1921), Parker and Böving (1924), and Brues (1924a).

[2] The Acroceridæ and Nemestrinidæ are particularly interesting in connection with the Trigonalidæ, because the two Dipterous families are similarly rare, sporadic and archaic insects, which lay a great number of minute eggs at a distance from the hosts, but in their general environment. King (1916) has shown that the female *Pterodontia flavipes* flies up and down in front of a tree-trunk and shoots her eggs at the bark to which they stick. The active planidium-like larvæ attach themselves to passing spiders, bore into their bodies and then become plump, vermiform larvæ. The Nemestrinids lay their very slender eggs in abandoned beetle burrows in fence posts or rails, and the just-hatched larvæ stand erect and allow themselves to be blown by the wind over the fields, where they apparently attach themselves to Scarabæid beetles (Rhizotrogus) and eventually become parasites in the pupæ of their progeny. (See Adam Handlirsch 1882, Brauer 1883*a*, and 1883*b*, and Bruch 1917*a*). The Bombyliidæ are much more abundant insects than the Acrocerids and Nemestrinids because the female lays her eggs very near the larval hosts. (See Shelford 1913, Fabre, Riley, Chapman, etc.) Clausen, King and Teranishi (1927) have recently published an account of two interesting Tachinid fly parasites

Types of development with active triungulin or campodeiform larval stages are known to occur also in certain Neuroptera, e.g., Mantispa (Brauer, 1869) and Coleoptera, e.g., the Carabid *Lebia scapularis* (Silvestri, 1905), which is parasitic on the elm-leaf-beetle and several Staphylinids of the genus Aleochara, which parasitize the puparia of various Muscids (Wadsworth 1915, Scott 1920, Kemner 1926). In all these cases entomologists speak of hypermetamorphosis, because there are at least two distinct larval forms, a minute, active and abstemious first stage larva, which seeks out the host and a lethargic, gluttonous and obese second stage larva which develops after securing contact with the latter. Brauer (1869) and more recently Heymons (1907) have objected to this use of "hypermetamorphosis" on the ground that it should be applied only to cases like the Meloidæ, originally described by Fabre, i.e., to cases in which a quiescent larval instar (pseudochrysalis or coarctate larva) supervenes after the trophic larval stage and before the true pupa. It is probable, however, that entomologists will continue to use the term as heretofore for cases of successive larval di- or polymorphism. Brauer's and Heymons' contention that such cases cannot be sharply differentiated from those of ordinary metamorphosis would carry more weight if biological phenomena in general lent themselves more easily to precise logical discrimination.

More important than the academic question of the use of terms is the question as to whether we are to regard such types of development as those of the Perilampidæ, Eucharidæ and Trigonalidæ (presumably) as more primitive than the ordinary Hymenopterous type of development or as having been secondarily acquired.

---

of the beetle *Popillia japonica*, which show a striking contrast in the number of eggs laid and the first larval stages. *Centeter cinerea*, which oviposits on the thorax of the beetle, produces only 62 eggs and has no specialized first larval stage, whereas *Prosena siberita* produces 800 eggs, which hatch in the uterus of the mother. The young larvæ are active and somewhat like planidia. They are laid on the ground into which they bore in search of their hosts, the grub of the Popillia.

Some authors have taken the latter view. Brues (1919*b*) expresses this very clearly when he says, " it is evident that the planidium and triungulin have been interpolated in the larval development just as the whole larval stage has been added to the life-cycle of the metabolous insects. Their presence, especially in Hymenoptera, is quite secondary ". I entertain serious doubts concerning the accuracy of this inference. Cases like Mantispa, Aleochara, Lebia, the Meloidæ, Acroceridæ, etc., seem to me to indicate very clearly that the obese vermiform larval stage has been intercalacted secondarily and that the triungulin or planidium, so like the typical and admittedly primitive campodeiform larva of most Megaloptera, Neuroptera and Adephagous Coleoptera, is really an archaic, though often considerably modified stage which has been suppressed in most Terebrantia and in all Aculeata, with the possible exception of the active first-stage larva of the Chrysididæ. The young larvæ of Pamphiliidæ among the Phytophaga, in possessing only thoracic legs and cerci, may be readily compared with the triungulin and the gap between such a larva and the planidium may be bridged, perhaps, when the first stage larva of the Trigonalidæ is discovered. The eruciform larva of modern Mecoptera, Trichoptera, Lepidoptera and Tenthredinidæ is evidently derived from the campodeiform, and there is general agreement that the vermiform may arise either from the eruciform or the campodeiform larva.

If I am not mistaken, therefore, the remote ancestral Terebrants sought out their prey, or hosts as active larvæ, since their mothers had not yet acquired the habit of laying their eggs directly on or in the larval food. These ancient forms were what I have called " atrophaptic " or " dystrophaptic " insects. The occurrence even in some Ichneumonids and Braconids of a rather active first larval stage remotely like a planidium, e.g., in such species as *Calliephialtes messor*, according to Cushman (1913), and *Ephedrus incompletus*, according to Mrs. E. H. Wheeler (1923), is significant in this connection. The future study of the very early development of the more primitive families such as the Stephanidæ, Evaniidæ, and Megalyridæ may have some interesting disclosures in store for us.[1]

Did time permit it would be interesting to trace the increasing attention bestowed by the female Terebrant on her offspring from the atrophaptic and dystrophaptic to the eutrophaptic condition. Even in the last there are two phases, one in which the female makes no nest but seeks out the host or prey and lays the egg on or in it, and another in which the female makes a nest and provisions it with prey or food for the young. The former method is characteristic of most Terebrants and of some Aculeates. The female may even enter the water to find the prey or host. The known cases of this singular behaviour, exhibited by several different groups of Terebrentia have been briefly reviewed by Brues (1921, p. 151). "In several well-known cases, the behaviour of the adult parasites has become so profoundly modified that the females not only enter the water in search of their hosts, but they may be, occasionally at least, accompanied by the males. The first observation of this sort was made nearly a century ago by Francis Walker (1836) on Agriotypus, and the well-known observer Sir John Lubbock (1863) later gave an account of the habits of two Chalcis flies in which he describes the actual process of swimming. One species, the Mymarid *Cataphractus cinctus* makes use of its ciliated, paddle-shaped wings for this purpose, while the other, a Trichogrammid (*Prestwichia aquatica*) propels itself by means of the legs. Numerous other contributions, notably those of von Siebold (1858), W. Müller (1889), Marchal (1900), Rousseau (1907), Heymons (1908), Schulz (1907*b*, 1910*a*, 1910*b*), and Matheson and Crosby (1912) have added much of interest, not only in bringing to light aquatic

[1] I find that Genieys (1924) has recently described the larva of an European Evaniid, *Zeuxevania splendidula*, which feeds on the eggs of a cockroach (*Loboptera decipiens*). Even the first larval stage proves to be a typical maggot like that of most other Terebrants.

members of several families, but in determining some of the host species on which they prey. In many cases the adaptation to aquatic life is not so perfect as the cases just mentioned, although several other species are known to swim readily, using either the legs or wings, which usually show modifications adapted to such behaviour ".[1]

The final, eutrophaptic stage in which the female brings the prey or food to the young in a previously prepared cell or nest occurs only among the Aculeata and only among the more highly specialized members of the sub-order. We shall deal with these conditions in the three following lectures.

[1] To the list of forms mentioned by Brues may be added the aquatic Braconid, *Giardinaia urinator*, discovered by De Stefani Perez (1902).

III

THE EVOLUTION OF WASPS

In the preceding lecture I called attention to the fact that the Terebrant, or parasitic and the Aculeate, or stinging Hymenoptera are so closely related that most modern authors unite them in a single sub-order, the Clistogastra, and that the Aculeates are usually supposed to be derived from Terebrant ancestors though others would derive both groups from common or closely allied hypothetical forbears among primitive Phytophaga. Whichever view we accept and notwithstanding the fact that the Terebrants use the ovipositor for egg-laying whereas it is converted into a sting in the Aculeates and, with the exception of the Sapygidæ, Chrysididæ and Dryinidæ, is not employed in oviposition, we must admit that the Terebrants have already developed certain characteristics which occur with more or less modification and elaboration among the higher Aculeates. Among these peculiarities attention may be called to the following:

(1) Even among Terebrants there exists a sporadic tendency to a development of polymorphism in one or both of the sexes. Thus in the Braconid *Sycosoter lavagnei*, according to Lichtenstein and Picard (1917) and Picard (1919), there are two forms of each sex, one winged, the other apterous. Two forms of females are also well-known in certain Chalcidids, e.g., in Harmolita (Webster and Reeves, 1909) and Cynipids, and among the fig-insects of the Chalcidoid tribe Idarnini (Phylotrypesis) Grandi (1921, 1923) has revealed the existence of a peculiar polymorphism very much like that of certain

worker ants, though confined to the male sex. In *Pezomachus flavocinctus* Strickland (1912) has described a peculiar polymorphism of the male, which may be winged, subapterous or apterous. These cases of the loss of wings in one or both sexes and especially in the female and considerable variation in the size of the head and mandibles in individuals of one of the sexes among Terebrants somewhat diminish our surprise at their more frequent occurrence among certain groups of Aculeates.

(2) In certain Terebrants the ovipositor may act like the Aculeate sting and introduce into the prey a poison which induces at least a temporary paralysis of the musculature. This has been observed by Silvestri (1907) in the Chalcidid *Elasmus flabellatus*, which completely paralyses the larva of the Tineid *Prays oleellus* in which it oviposits. Doten (1911) found that *Microbracon juglandis* completely paralyses the caterpillars of *Ephestia kühniella*, so that the latter remain fresh for from two to four weeks, and Picard (1919) has observed a similar but feebler paralysis of the larvæ of *Hypoborus ficus* by *Sycosoter lavagnei*. The paralysis of blowfly larvæ by *Alysia manducator* has been carefully studied by Altson (1920) and Myers (1927).

(3) It is well-known that some Terebrants lay their eggs within the body while others lay them on the surface of the prey, but that the latter method alone is employed by the Aculeates, with the exception of the Dryinidæ. Several authors and more recently Picard (1919) have called attention to the fact that external oviposition is employed among the Terebrants

when the prey is concealed in a plant or earthen cavity, or in a cocoon or puparium and that internal oviposition is the rule when the host is exposed, or free-living. Among the exceptions we may mention Polysphincta, which parasitises spiders (Lichtenstein and Rabaud, 1922) and the Chalcid *Schizonotus sieboldii* which, according to Cushman (1917) lays its eggs on the exposed pupæ of a Chrysomelid beetle (*Melasoma interruptum*). The same author calls attention to the Eulophid *Euplectrus comstocki* which oviposits on the cotton worm. Silvestri (1911) has described this habit also in the European *Euplectrus bicolor* which parasitises the caterpillars of *Plusia gamma* and *Mamestra brassicæ*. The general rule of oviposition on the outside of the prey when it happens to be in a cavity applies also to the Aculeata, but the Rhopalosomatidæ and some Pompilidæ are known to attach their eggs to free-living insects.

(4) A phenomenon frequently noticed in Aculeates that paralyse their prey (Figs. 13 and 14) is their malaxation or feeding on portions of it. This is very natural because the adult Aculeate is then merely continuing to eat portions of an insect of the kind on which it was reared as a larva. The same tendency is apparent in many Terebrants but in them usually takes the form of feeding on the juices exuding from punctures made by the ovipositor in the egg, larva, pupa, or adult of the host species and immediately after oviposition or after the ovipositor has been used merely as a food-procuring instrument. This singular behaviour was first observed by Marchal (1905, 1909) in *Tetrastichus xanthomelænæ* which oviposits in the eggs

of the elm-leaf beetle (*Galerucella xanthomelæna*) and in *Aphelinus mytilaspidis*, which oviposits in a scale (*Aspidiotus ostreæformis*). It has also been observed by several American entomologists in *Tetrastichus asparagi* which oviposits in the eggs of the asparagus-beetle (*Crioceris asparagi*); by Saunders in *Aphelinus fuscipennis*, which parasitises the scale *Aspidiotus rapax* (Howard 1908); by Silvestri (1910) in *Tetrastichus xanthomelænæ;* by Doton (1911) in *Pteromalus puparum* and a species of *Meraporus*, as parasites of the pupæ of the cabbage-butterfly (*Pieris rapæ*) and in *Microbracon juglandis* as a parasite of the meal-moth (*Esphestia kühniella*); by Roubaud (1917) in *Nasonia brevicornis* which oviposits in the puparia of Muscids; by Lichtenstein (1921) in the Chalcid *Habrocytus cionicida* which lays its eggs in the pupæ of a weevil (*Cionus thapsi*); by Trouvelot (1921) in the Braconid *Habrobracon johansseni*, by Whiting (1921*b*) in *H. brevicornis* which attacks the meal-worm; by Caffrey (1921) in *Anastatus semiflavidus*, an egg-parasite of the moth *Hermileuca oliviæ*; by Balfour Browne (1922) and Picard (1923) in the Chalcid *Melittobia acasta* which lays its eggs on the larvæ and pupæ of a great variety of Hymenoptera, Diptera and Coleoptera, and by Voukassovitch (1924) in the Ichneumonid *Dicælotus erythrostoma*, a parasite of *Polychrosis botrana*. The most singular of these cases is the Habrocytus described by Lichtenstein, because the Chalcid runs its ovipositor through the cocoon of the beetle and the space between the cocoon and the pupa and into the latter. It leaves the organ in this position for half an hour while a secretion hardens around it and forms a delicate tube through which the parasite sucks the blood of the prey as soon as the ovipositor is withdrawn. It will be seen that in all the cases cited the mandibles are not used in feeding because the prey is too large and too tough-skinned, whereas in the larger and more powerful Aculeata the mandibles may be readily employed for a similar purpose. There are, however, small Aculeates (Bethylidæ) which, as I shall show in the sequel, use the sting like the Terebrantia for securing food.

(5) Perhaps we may detect the germs of a feeble sociability in those Terebrants which develop in numbers in the same host and when mature, pupate in compact masses of cocoons. In one genus of Chalcididæ, Cratotechus, or Comedo (Fig. 11), as it is now called (Howard 1891), the pupæ are naked and are attached to a leaf side by side in a regular ellipse enclosing some of the individuals, an arrangement which seems to presuppose a peculiar "Fühlung", to use a German expression, among the larvæ. It must be admitted, however, that sociability is much less pronounced among larval Terebrants than among the more ancient and primitive Phytophaga and that there are many instances of internecine destruction among the young larvæ when the prey is too small to nourish more than a single parasite.

While the various peculiarities cited under these five heads may be regarded as representing primitive and very ancient ethological tendencies which reappear as convergent, or parallel developments in more pronounced and more specialized forms among the Aculeates, the majority of the latter depart from the Terebrants, first, in having learned to excavate cavities in the soil or wood or to build cells of earth or vegetable matter or even from inspissated secretions of their own bodies, and second, to store in such structures insect or plant food for the use of their young and themselves. The slow evolution of these two peculiarities can be readily traced within the Aculeate sub-order and especially among the social wasps, and bees, but the subject is so extensive and has been so often treated in popular works, that I shall give it only such brief attention as my time will permit.

Concerning the division of the Aculeata into superfamilies and their sequence and interrelations there is at the present time so much difference of opinion among hymenopterists that the student of ethology is left in a quandary. Of the two most recent classifications, that of Handlirsch (1924) and that of Bradley, Rohwer and Bequaert (Comstock, 1924), the former is the more conservative and will probably appeal more strongly to European workers, whereas the latter, which embodies some of the views of Ashmead, will be more acceptable to the American hymenopterists. I am myself inclined to prefer Handlirsch's classification so far as it retains many of the old Latreillean groups, though some of the latter obviously require modification. The following simple arrangement will be adequate for the purposes I have in view:

SUB-ORDER ACULEATA

(VESPOIDEA OR VESPIFORMIA)

Superfamily	Bethyloidea
Family	Dicrogeniidæ
,,	Bethylidæ
,,	Dryinidæ
,,	Embolemidæ
,,	Sclerogibbidæ
,,	Rhopalosomatidæ

Superfamily	Tubulifera
Family	Cleptidæ
,,	Chrysididæ
Superfamily	Heterogyna
Family	Sapygidæ
,,	Plumariidæ
,,	Anthoboscidæ
,,	Scoliidæ
,,	Tiphiidæ (incl. Methocinæ and Myrmosinæ)
,,	Mutillidæ
,,	Thynnidæ
,,	Formicidæ
Superfamily	Pompiloidea
Family	Pompilidæ
Superfamily	Diploptera
Family	Vespidæ

(SPHECOIDEA, OR SPHECIFORMIA)

Superfamily	Fossores
Family	Ampulicidæ
,,	Sphecidæ
Superfamily	Anthophila
Family	Apidæ

The precise position of the Pompilidæ is a matter of dispute, since they have been placed by Börner and Handlirsch and many hymenopterists before them among the Fossores, and by Ashmead, Rohwer, Bradley, Banks and others, among the Vespoidea. Until their affinities have been more accurately determined, I can see no harm in regarding them as an independent superfamily. Paleontology sheds no light on the origin of the various families in the foregoing table. Most of them, like the families of the Terebrantia, are represented in the early and middle Tertiary by genera identical with or closely allied to those now living, and morphology and ethology show very clearly that all the families of extant Aculeata are highly specialized and that most of them cannot be derived from one another. There are, nevertheless, certain patterns in development, structure and behaviour which indicate certain definite evolutionary tendencies. These seem to be clearly foreshadowed in the group Bethyloidea, which is placed at the beginning of the sub-order and deserves special consideration on account of its numerous and often elusive morphological relationships to all the other superfamilies.

Although Haliday (1835) regarded Bethylus as a Fossor, it and the allied genera were placed among the Proctotrypoidea until 1902, when Ashmead transferred them to his Vespoidea. There is now no doubt that they are a primitive, though peculiarly and diversely specialized, cosmopolitan group of Aculeates. Fully 1,200 species, mostly of small size, and more than 100 genera have been described and distributed among some six families. The most heterogenous affinities of the various Bethyloids to other Aculeates have been noticed by specialists. Kieffer, who monographed the Bethylidæ (1908) and Dryinidæ (1907), regarded the former as "most closely related to the genera Meria and Tiphia among the Scoliidæ," and the genus Mesitius as attaching them to the Tubulifera (Chrysididæ). In 1909 he described a genus Parachrysis, which can be assigned to either the Bethylidæ or the Chrysididæ. Westwood (1881) noticed the resemblance of the Bethylid genus Apenesia to the Mutillidæ. Perkins (1905) saw in the Bethylidæ relationships to the old Fossorial series on the one hand and to the Proctotrypoidea on the other. Brues (1910*b*) notes the relationship of the Bethylidæ, Dryinidæ, and especially of the genera Epyris and Pristocera to the Ampulicidæ and of the Sclerogibbid genus Mystrocnemis to the Tiphiidæ and Ashmead's family Cosilidæ (Anthoboscidæ). He believes that the Embolemidæ are an ancient group related to the forms from which the Proctotrypoids are descended. Bridwell (1917*a*) regards the relationships of the Bethylidæ to the Proctotrypoidea as purely superficial but those of the Scoliidæ and Chrysididæ as "eminently natural".

Börner (1919) places both the Bethylidæ and Chrysididæ among the Terebrants but regards them as being connected by the Cleptidæ. Brues (1922, 1923*c*) noted the resemblance of the genus Algoella (Apteropompilus), which seems to be a Bethylid, to the Anthoboscidæ (Sierolomorpha) and the Mutillidæ, and described from the Baltic amber a peculiar genus, Palæobethylus, which strongly resembles the Ampulicidæ, especially Rhinopsis. Handlirsch (1924) places the Bethyloidea as the first superfamily in the Aculeata but Bradley (Comstock 1924) intercalates the family Bethylidæ between the Formicidæ and the Vespidæ and the family Dryinidæ among the Sphecoidea after the Ampulicidæ.

The family Rhopalosomatidæ deserves discussion on account of its dubious position in the classification. It consists of only three known species and genera, *Rhopalosoma poeyi*, occurring from the Antilles to Maryland and Missouri, *Paniscomima erlangeriana* in South Africa, and *Hymenochimæra abnormis* in India. Hood (1913) discovered the larva of *Rh. poeyi* attached to the third abdominal segment of an active jumping tree-cricket (*Orocharis saltator*) on Plummer's Island, near Washington, D.C. The adult Rhopalosoma is nocturnal and closely resembles in colour and the texture of its integument the nocturnal Vespids of the genus Apoica and the nocturnal males of certain ants (Eciton, Diacamma, Lobopelta, etc.), Westwood (1874), who first carefully considered the relationships of the insect, said that it was "*quoad affinitates animum excrucians*", but thought it might be a Vespid, Nylander regarded it as an ant, Haliday as a Fossor, Fred. Smith as an Ophionine Ichneumonid, Cresson as a Braconid, Sharp as allied to the Scoliidæ and Sphecidæ. Ashmead (1902-1903) erected a special family for its accommodation and placed it between his Cosilidæ and the Thynnidæ. Morley (1910) is satisfied that it can be placed nowhere among the Terebrantia, but is an Aculeate allied to the Scoliidæ, with the venation of an Ichneumonid. Rohwer (1913*b*) believes that "perhaps the Rhopalosomidæ and the Dryinidæ had a common origin, as the larva would indicate". Turner and Waterston (1917) maintain that the male genitalia of the Rhopalosomatidæ "show a very distant relationship to those of the Eumenidæ and differ strongly from those of all other Hymenoptera. Thus, of the many attempts to assign a place to the group, that of Westwood, who connects it with the Vespidæ, is the most correct. It is almost certainly derived from the same stock as the Eumenidæ, but strongly modified by nocturnal and probably also by parasitic habits. The wide range of the family points to considerable antiquity, though it seems to be absent from Australia". Handlirsch (1924) places Rhopalosoma in the Heterogyna, between the Sapygidæ and Anthoboscinæ, Bradley (Comstock, 1924) between the Vespidæ and Bethylidæ. Brues has suggested in conversation that it may, perhaps, belong near the Bethylidæ, a position which would be indicated also by the Dryinid affinities suspected by Rohwer.

More interesting and more significant for our purposes than the citations of divergent opinions concerning the taxonomic relationships of the Bethyloids are the habits of these insects. Omitting the Rhopalosomatidæ, it appears from the published data that there are at least three different methods or types of dealing with the prey, of oviposition and development in the superfamily. These may be called the Bethyline, Sclerodermatine and Dryinid types respectively.

(1) The Bethyline type has been observed in the genera Bethylus (Perisemus), Pristocera, Epyris and Holepyris. Until recently very little was known about these insects, apart from the very fragmentary observations of Haliday (1835), Ferton (1901-21) and Nielsen (1903). Haliday saw a female *Bethylus fuscicornis* drag a paralysed Tineid caterpillar over the ground to a hollow reed, enter the reed to make a reconnaissance, return and pull the caterpillar into the cavity. Ferton found in a hollow briar four green larvæ of the same species of Bethylus devouring a green caterpillar. With them was an adult Bethylus female, evidently the mother of

the larvæ, though he regarded her as one of their sisters. Nielsen, on splitting a piece of briar in which a Pemphredon had once nested, saw, a few inches from the entrance, a Tortrix caterpillar which a *Bethylus fulvicornis* forthwith carried further into the cavity where he found a second caterpillar that had been almost devoured by four Bethylus larvæ. He noticed that the first caterpillar had been paralysed and concluded that *B. fulvicornis* is in a more primitive stage than the true fossorial wasps, because it failed to make a nest and because the caterpillars were lying loose in the burrow without being separated from one another by partitions, whereas all true fossorial wasps that live in stalks make cells. Bethylus differs from the Fossores, moreover, in laying several eggs on each caterpillar, a habit common among the Terebrants. Bridwell (1917), near Cape Town, South Africa, observed an Epyris female dragging a Tenebrionid larva over the ground. But the most complete observations were made by F. X. Williams (1918) in Hawaii on *E. extraneus* (Fig. 12). This small black wasp, with the habitus of a Tiphia, captures and stings till it is nearly immobile the much larger larva of a Tenebrionid (*Conocephalum seriatum* Fig. 12B), seizes it, apparently by a palpus, and carries it over her back into some crack in the soil, where she fashions a rude cell around it and glues a single egg to the middle of the venter of its first abdominal segment (C). The hatching larva at first lies on the surface (D) but later assumes an erect position at right angles to the longitudinal axis of the victim (E). When full grown the larva forsakes the remains of the prey and spins a cocoon (G). Notes indicating very similar habits in *Holepyris hawaiensis* have been published by Bridwell (1917*b*, 1919) and by Hyslop (1916) and Hayes (1927) on the North American *Pristocera armifera*. Hyslop found the single egg of the latter species in the ground attached to the ventral surface of an Elaterid larva, *Limonius agonus*, and Hayes found it attached to the larva of another Elaterid, *Æolus elegans*.

Plate IX.

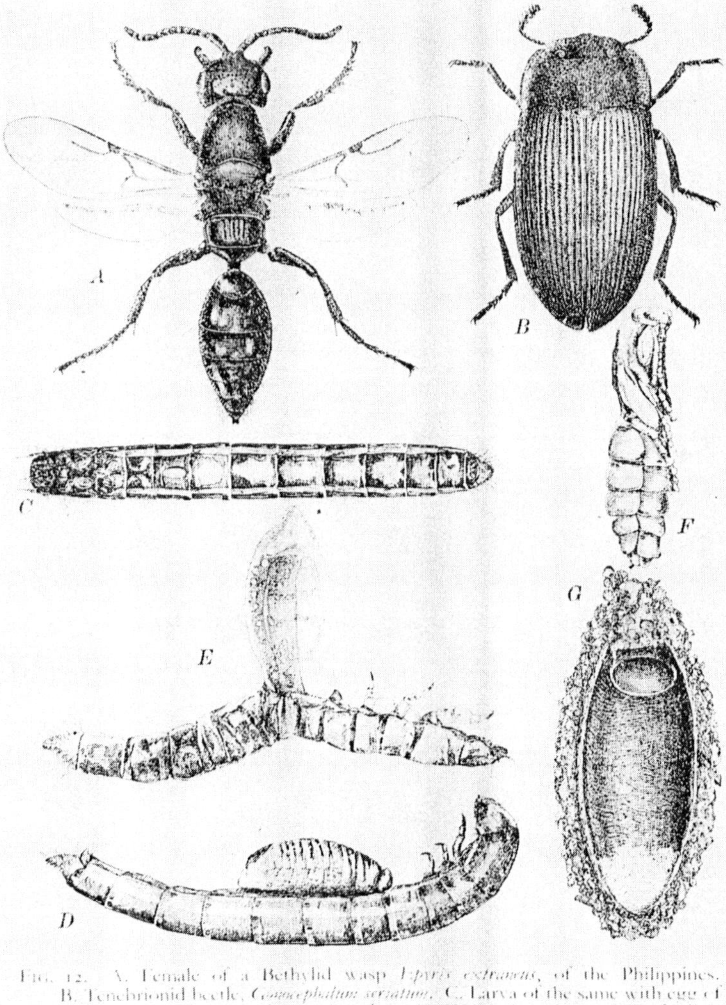

FIG. 12. A. Female of a Bethylid wasp *Epyris extraneus*, of the Philippines. B. Tenebrionid beetle, *Gonocephalum seriatum*. C. Larva of the same with egg of *E. extraneus* on middle of ventral surface. D. Young *E. extraneus* larva feeding on the larva of *G. seriatum*. E. Later stage of same. F. Pupa of *E. extraneus*. G. Cocoon of same. (After F. X. Williams.)

[face p. 60

FIG. 14. Burrow of *Sphex procerus* in section, showing filling of debris in the tunnel and the paralyzed Sphinx moth caterpillar in the cell, with the egg glued[1] to its side. (Photograph by Prof. Carl Hartman.) (*See* p. 52.)

A B C D

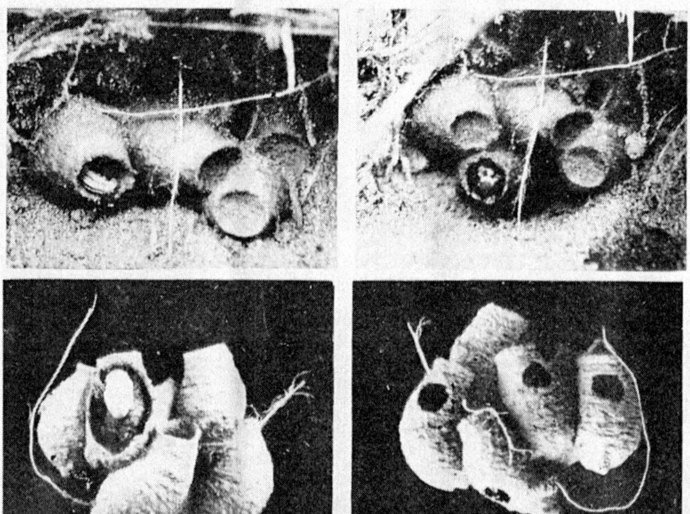

FIG. 15. Four stages in the mud nest of *Odynerus dorsalis*. A. Showing one cell open and being stored with small caterpillars. B. Nest on the following day, showing wasp resting in a new cell made on the previous afternoon. C. Nest with one cell opened to show the wasp larva feeding on caterpillars. D. Same nest, showing holes made by the escaping wasps. (After Prof. Carl Hartman.) (*See* p. 69.)

[*face p.* 61

(2) The Sclerodermatine type of development has been observed in the genera Cephalonomia, Lælius, Goniozus, Scleroderma, Neoscleroderma, Anoxus, Apenesia, Sierola and Parasierola. In these cases the host, which is a Coleopteron, a Lepidopteron or possibly in some cases an ant larva, is sought out by the Bethylid and several eggs are laid on it after it has been paralysed by repeated stinging. The hatching larvæ consume the prey and after spinning cocoons pupate near its remains. The most complete of the earlier observations of this type of behaviour are those of Busck on *Lælius trogodermatis*, as recorded in Howard's "Insect Book" (1905). The female Lælius attacks and paralyses the larva of Dermestes, bites off the long hairs on its ventral surface and there attaches one to six eggs. The entire life-cycle requires only thirty to thirty-five days. The winter is passed in the pupal stage, the cocoons at that season being firmer and darker in colour than those of the summer generations. Busck found that though the Lælius will paralyse every Dermestid larva it encounters, it does not lay on all of them, and that though the larva is paralysed it may pass feces or even moult. When this occurs the eggs attached to its cuticle perish. He also found that the female lays more than twice her bulk of eggs and that she is sometimes parthenogenetic but then produces only male offspring. Brief accounts of the habits of species of Goniozus or of their prey have been published by Audouin (1842), by Popenoe and Marlatt (1889), Ashmead (1893), Picard (1919) and others, and of Parascleroderma and Cephalonomia by Picard. Ashmead in his monograph of the North American Proctotrypidæ (1893) and Kieffer in his monograph of the Bethylidæ (1908) have published a number of host records. Much more extensive observations on Scleroderma, Sierola, Parasierola, and Cephalonomia were made by Bridwell (1917*c*, 1919, 1920) in the Hawaiian Islands where these interesting insects happen to be so numerous that Fullaway (1920) described in a single paper, 171 new species of the genus Sierola alone. Several host records

of Hawaiian Bethylidæ have also been published by Swezey (1915).

Scleroderma immigrans, a species introduced into Hawaii from the Philippines, was studied in considerable detail by Bridwell (1920) and more recently I have been able to rear a species from our Southern States, *S. macrogaster*. Its habits are so similar to those of *immigrans* that I may describe it as a paradigm of the genus especially as all the species seem to have essentially the same habits. As was noticed by Westwood (1881) and Bridwell in other species of Scleroderma, *macrogaster* has two forms of males and two forms of females, winged and apterous, but the winged female and apterous male are rather rare. The dimorphism of the sexes is, therefore, very similar to that observed by Picard in *Sycosoter lavagnei*. The behaviour of both forms of each sex is the same. A batch of cocoons of *S. macrogaster* from Texas was given me by Bridwell with a statement that I should probably be able to rear the species on the larvæ of some Clytine beetle. The original host was unknown to me, but I found, just as Bridwell had shown in the case of the Sclerodermas of Hawaii, that I could rear the Texan species on a variety of larvæ and pupæ. I succeeded best with larvæ of the hickory borer (*Cyllene pictus*) and the larvæ and pupæ of another Cerambycid, *Liopus cinereus*, fairly well with other beetles (pupæ of *Pissodes strobi*, larvæ of Cleridæ, pupæ of *Thymalus fulgidus*) and larvæ of Ichneumonidæ, less satisfactorily with the soft larvæ and pupæ of ants (*Camponotus americanus* and *Lasius americanus*). When the female Scleroderma, which is only 2.5 to 3 mm. long, is placed in a small vial with an active Cyllene larva, which is many thousand times larger, she at first avoids it but eventually climbs on to it, grips its cuticle with her mandibles and stings it, often at first in the region of the mandibular muscles, and she keeps on biting and stinging it, as she moves from place to place, as long as the muscles beneath the skin gripped by her mandibles show any signs of contraction, till the larva is completely paralysed. This may require from one to four days' depending on the size of the victim. There is no regular sequence or position for the punctures. Bridwell interprets the stinging of the mandibular muscles first as an adaptation to prevent the prey from crushing the Scleroderma, but this very rarely occurs even when the prey is first attacked in some other spot. After the prey has been rendered motionless the Scleroderma feeds for several days by tweaking the larva at various points and imbibing the blood which exudes in very minute droplets from the previously made sting punctures. Her abdomen now rapidly enlarges with the growth of her ovaries and she soon begins to lay numerous glistening white eggs, which are rather large in proportion to her size, on the surface of the prey and most frequently in the intersegmental constrictions. The prey, though alive, remains perfectly motionless and neither moults nor undergoes any further development. If she has been given a pupa, however, development may not be inhibited though its abdomen remains paralysed and in rare cases the skin may harden so that the hatching Scleroderma larvæ die and the imago of the host, though moribund, actually emerges. The eggs laid on a larva or young pupa produce minute larvæ which at first lie on the surface but later become spindle-shaped and erect, so that the host bristles with them like a porcupine. The older larvæ acquire the colour of the juices of the prey ; those feeding on the pink larvæ or pupæ of Liopus becoming red. They are always spotted with white, owing to the large masses of urate crystals in their fat bodies. The mother Scleroderma remains with the larvæ, often stands over them and may sometimes lick them, holding them meanwhile in her fore feet. She also continues occasionally to drink the host's blood, which exudes about the deeply inserted heads of her larval offspring. Although she will sometimes eat her eggs I have never seen her attack one of her larvæ. The devouring of some of the eggs seems to be due to a tendency to regulate their number according to the volume of the prey. When the larvæ are mature they fall away from its shrivelled and exhausted remains and spin snow-white cocoons in a cluster. Pupation covers a period of fourteen to thirty days. The males emerge first from their cocoons, at once eat their way into the female cocoons and fecundate the pupæ. They also mate readily with mature females, sometimes copulating with the same individual five to eight times after brief intervals. The same female may also mate with several males in succession. So great is the ardour of the latter that they often attempt to mate with one another. The mother being a long-lived insect may mate with one of her sons and will readily paralyse another beetle larva, rear another brood and mate again with one of her grandsons. The males die very soon after mating. If virgin females are isolated and given prey, they paralyse it readily but their oviposition is conspicuously delayed. The eggs and larvæ, however, develop normally but produce only males.

(3) The third type of behaviour is seen in the Dryinidæ which are well-known parasites of Homoptera (Fulgoridæ, Jassidæ, Membracidæ) and ants. The female Dryinids, with the exception of those of the small genus Aphelopus, have singular chelate fore tarsi with which they hold the host motionless while they are thrusting an egg into its body. The resulting larva eventually breaks out of the side of the abdomen in a peculiar sack, which in some forms seems to consist of the larval exuviæ, in others of the evaginated hypodermis of the host, and after reaching maturity detaches itself and spins a cocoon on a leaf or more rarely enters the ground to pupate. The behaviour of the Dryinids has been described by a number of observers—Perris (1857), Mik (1882), Swezey (1903), Perkins (1905), Kieffer (1907), Jeannel (1913), Keilin and Thompson (1915) and others. Giard (1889*a*, 1889*b*), also described the larva of *Aphelopus melaleucus* on *Typhlocyba rosæ* and noticed that infected individuals of this Homopteron suffered from " parasitic castration ". Matausch (1909, 1911) described similar changes

produced in American Membracids by Aphelopus larvæ, and Kornhauser (1919) has published an important study of the effects of *A. theliæ* on our large and common Membracid *Thelia bimaculata.* My time will not permit of a consideration of the peculiar action of the parasite on the male and female genitalia of the host, but attention should be called to the fact that *A. theliæ*, according to Kornhauser, is polyembryonic. It lays in the nymphal Membracid a single egg, which produces forty to sixty of the parasites. After they reach maturity inside the host, they escape from it and pupate and pass the winter in the ground. *A. theliæ* is also interesting in connection with the cases of ovipositor-feeding described in the first part of this lecture. After the Aphelopus female has selected a Thelia nymph, she "grasps the caudal part of the nymph's abdomen between her mandibles, and, holding firmly with her legs to the abdomen of the host, tries to thrust her sword-like ovipositor cephalad through the intersegmental membranes of any two abdominal terga. The nymph struggles as the ovipositor pierces, and from the anal tube exudes a drop of liquid. This the Aphelopus grasps in her mandibles and is gone in a second". Kornhauser also noticed on other occasions that the Aphelopus after ovipositing generally mounts the abdomen of the nymph to secure a drop of excrement (honey-dew).

The preceding descriptions show that behaviour among the Bethyloidea is very heterogeneous, that the Dryinids behave essentially like Terebrants, that some of the Bethylids resemble the Ampulicids, Sphecids and Pompilis, which hunt, paralysed and drag their prey to a place of safety, and that others resemble the Scoliids, Tiphiids and Thynnids in seeking out and paralysing the prey in its own retreat. Scleroderma, furthermore, seems to exhibit a generalized pattern of behaviour combining that of Terebrants, Tiphiids and to a slight extent that of the Formicidæ. The Sphecid and Tiphiid methods, however, are not very sharply separated, since some of the recorded observations show that Scoliids and Tiphiids may occasionally pick up the prey on the surface of the soil, transport it a short distance and then bury it. Thus, according to Bridwell, Gosse long ago observed an *Elis 4-notata* carrying a large lamellicorn beetle larva in its mouth, and Forbes (1908) and Davis (1919) found that when Lachnosterna larvæ are placed in the soil near a female Tiphia she will sting and paralyse and then proceed to inter them, though under normal circumstances she seeks her prey under ground and oviposits on it *in situ*. Ferton (1911) observed a small Myzine (*M. andrei*) in North Africa hunting a Terebrionid larva of the genus Tentyria over the sand, stinging it once and forthwith burying it beneath the surface.

The Sclerodermas are also of unusual interest because they exhibit all the essential rudiments of social behaviour since the long-lived mother insect lives on the same food as the young, shows an interest in her larvæ and even licks them and is capable of bringing up several successive broods. When the mother and her female offspring, or several females from different broods are confined in the same vial with a Cyllene larva, instead of the rivalry one might expect, there is co-operation among them in paralysing the prey and all lay their eggs on it in common. We could, in fact, derive the social behaviour of wasps, bees and ants from such conditions as we find in Scleroderma. I do not, of course, believe that these social insects are derived from Bethylid ancestors. Although the existing Bethylids are undoubtedly very ancient, they are in many respects highly specialized and there is no reason to suppose that they have given rise to any of the other extant families of Aculeata.

The social wasps all belong to a single family, the Vespidæ, which alone constitutes the superfamily Diploptera, so called because the wings are folded longitudinally when at rest, a sharply defined and very compact group. Most of the species are solitary, and, as Ducke (1910, 1914) has pointed out, the social species possess no morphological characters that can be recognized as correlated with their mode of life or as separating them as a group from the solitary forms. For this reason the ethological peculiarities and especially the methods of nidification acquire considerable taxonomic significance. The older authorities distinguished two families, the Masaridæ and the Vespidæ, the former being solitary and strictly anthophilous, provisioning their nests with honey and pollen, the latter either solitary or social and more or less anthophilous as adults, but usually providing their larvæ with insect food. It was known that some social species, e.g., *Nectarina lecheguana* regularly, and a few species (Polistes, *Polybia occidentalis*) occasionally, store honey in their nests. The solitary species of the European and African genera Masaris, Ceramius and Celonites, according to H. de Saussure (1853), Giraud (1871), Ferton (1901, etc.) and Brauns (1910-1911) do, indeed, store honey in the cells of their mud nests, but according to Davidson (1913) the species of the allied genus Pseudomasaris of the South Western United States, like the Eumeninæ, provide their young with paralysed caterpillars. Brauns' observations on *Ceramius lichtensteini* in the dry Karroo Plains of South Africa are of peculiar interest, because they show that this insect has not only adapted itself to living in very arid situations but has also become sub-social. The nesting habits are described as follows: "I have often excavated the nests of *C. lichtensteini*. The gallery descends perpendicularly and in a straight or curved line, according to the nature of the soil, often to a depth of four or five feet when in soft earth. The wasp hollows out small, verticillate cavities along this main gallery and builds oval clay cells in them. The cells are like those of *Sceliphron spirifex* and allied forms, but are roundish oval. It is certain that the wasp feeds her larvæ with flower-honey till they are full-grown. I have never found stored pollen or honey masses in the cells with the larvæ, even when they were small. As soon as the larva has completed its growth the wasp closes the cell. In order to pupate the larva constructs within the clay cell a parchment-like cocoon, which is in general flask-shaped and round, and has a strongly flattened bottom, like a small medicine bottle. In such cells the larvæ may live for many years, depending on the periodical droughts, without pupating. At the present time I have a number of these cells that have been kept dry for three years and still contain untransformed living larvæ".

Bequaert (1918), in his recent revision of the solitary and social wasps of the Congo, and Bradley (1922) in his monograph of the Masarine wasps, have abandoned the family Masaridæ and regard the Diploptera as constituting a single family, the Vespidæ, comprising nine closely inter-related sub-families. These may be arranged in the following sequence:

VESPIDÆ.

Solitary or at most Subsocial:	Social:
Euparaginæ.	Stenogastrinæ.
Masaridinæ.	Ropalidiinæ.
Eumeninæ.	Polybiinæ.
Zethinæ.	Polistinæ.
	Vespinæ.

There is no significant specialization in bodily structure as we pass from the first to the last of these sub-families, but there is a very definite gradual development and specialization of social propensities, showing that evolution among the Vespidæ has been very largely on the functional (physiological and ethological, i.e., instinctive) planes.[1]

Nothing is known of the habits of the Euparaginæ,

[1] See Ducke (1914, p. 319): "I have already emphasized the fact that (contrary to Ashmead!) the Vespinæ and the Eumenidinæ are merely subfamilies of the family Vespidæ, since both even at the present time are connected by very distinct transitional forms and since we know of no morphological character that would distinguish them completely. Both of these subfamilies are founded, therefore, solely on the presence of females with abortive genital apparatus (so-called workers) in the Vespinæ, whereas such individuals do not exist in the Eumenidinæ, but at most fertile females which occasionally build a common nest."

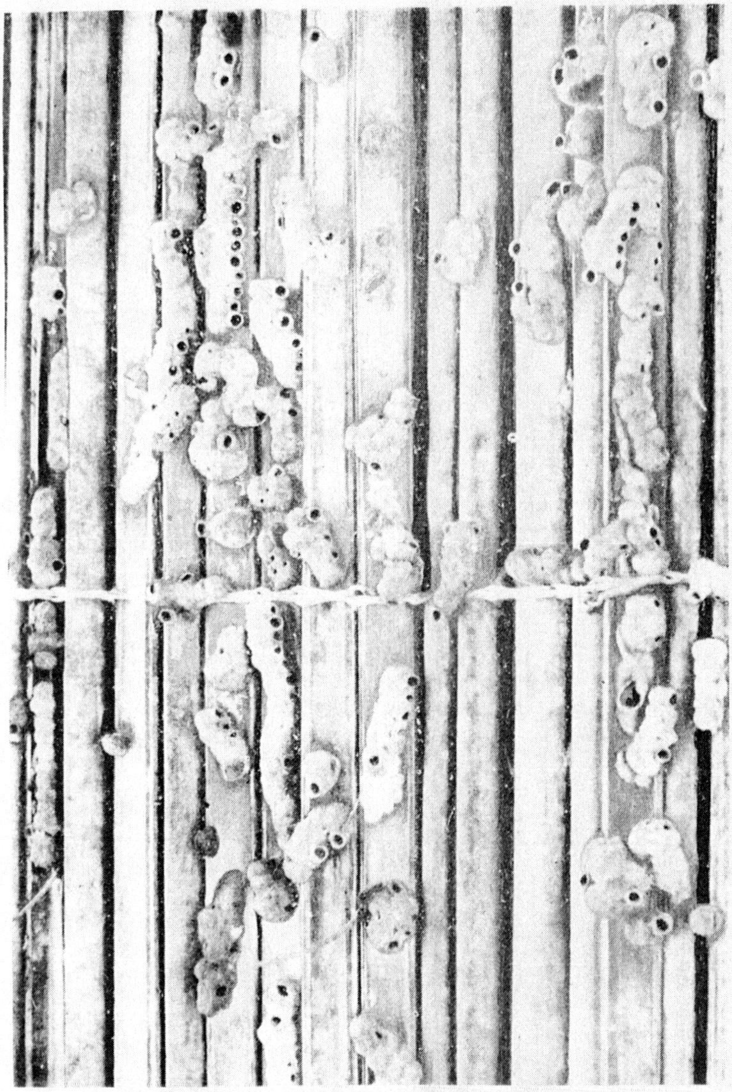

FIG. 16. Mud nests of *Synagris cornuta* on the thatching of a native hut in the Congo. "Some of these nests show very distinctly the short neck with its slightly widened opening curved to the side and downwards. Such a chimney is built at the entrance of the cells containing eggs or larvæ still nursed by a female." (After Bequaert from a photograph by H. O. Lang.)

[*face p.* 68

Plate XII.

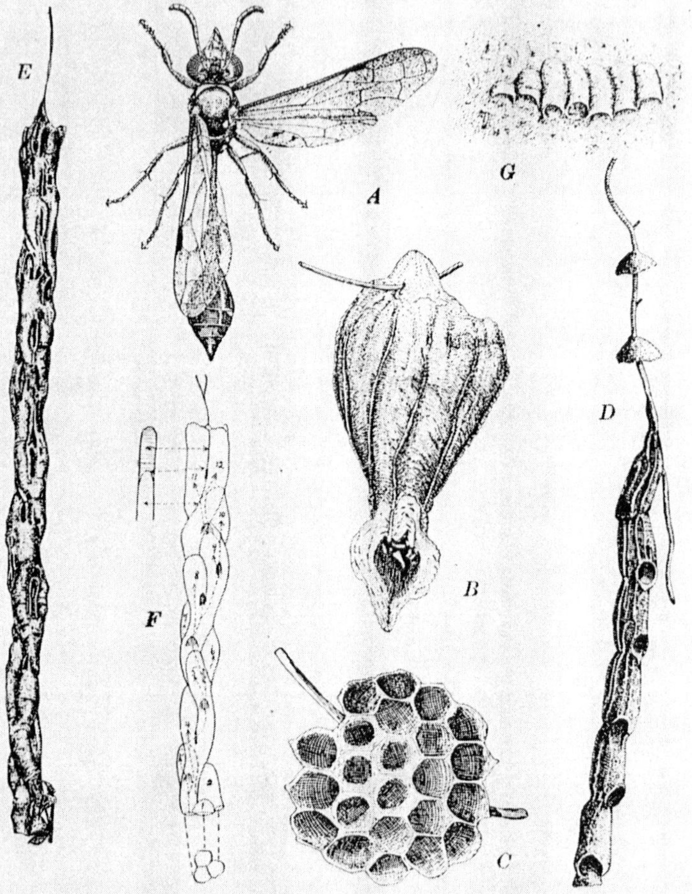

FIG. 17. Nests of Stenogastrine wasps from the Philippines. A. *Stenogaster micans* var. *luzonensis*, female. B. Completed nest of same. C. Nest with only the basal portion completed. D. Nest of *Stenogaster* sp., with umbrella-like "guards." E. Nest of *S. depressigaster*. F. Diagram of same showing arrangement of cells and passage-ways. The numbers indicate the cells. The tops of the passage-ways are shown in two planes by series of parallel lines. G. Nest of *S. varipictus* on the bark of a tree. (After F. X. Williams.)

[*face p.* 69]

a group consisting of only three rare species from the South Western United States and closely related to the Masaridinæ. The primitive and central group of the whole series is obviously the Eumeninæ, which comprise by far the greatest number of species, many of which, belonging to the genera Odynerus (Fig. 15) and Eumenes, are familiar to all students of Hymenoptera in temperate regions. They either excavate cells in the ground, occupy the cavities of plant stems, or construct elegant clay cells in exposed situations. They all lay the egg first and then for the prospective larva bring in the provisions which consist of paralysed caterpillars, thus reversing the usual sequence of prey and egg in most other solitary wasps. It is from this common body of Eumenine wasps that the five sub-families of social species have been descended, according to Ducke (1914), along at least five or six independent lines. As indicated in his diagram some of the social genera are traceable to Eumenes-like, others to Odynerus-like ancestors. But even among existing species of the sub-family Eumeninæ there are some very significant advances in the direction of social life. Roubaud found that certain African species of Odynerus and Synagris (Fig. 16) instead of adopting the usual method of mass-provisioning, i.e., of placing the complete requisite amount of paralysed caterpillars for the larva with the egg and closing the cell, actually adopt the method of progressive provisioning of the growing larva from day to day and thus exhibit a distinct and significant advance towards social behaviour. The latter condition seems to arise naturally out of a seasonal scarcity of the prey, as Roubaud (1908, 1910*b*, 1916) has shown in the case of *Synagris spiniventris* and *calida*.

Socialization has advanced somewhat further in the Zethinæ, which have been studied recently by Ducke (1914), Howes (1917) and F. X. Williams (1919*b*). H. de Saussure (1852, 1875) had shown that the cells of the South American *Zethus* (*Didymogastra*) *romandinus* consist not of clay as in the Eumeninæ but of a mixture of wood fibre and a gummy material, and according to Forbes (1885) the cells of the East Indian *Zethus cyanopterus* are made of minute pieces of leaves. An account of this same species, studied by Williams in the Philippines, was published in my " Social Life among the Insects (1923) ". He noticed that the female feeds her larva from day to day with small caterpillars which she has partly eaten. She faithfully guards the larva and while it is small and there is ample room, sleeps in the cell. She closes the latter as soon as the larva is full-grown and proceeds to build another. Ducke's account is so interesting that the greater portion of it may be quoted. The nest of *Zethusculus lobulatus* which he frequently encountered in Pará, "consists of nearly cylindrical cells built on lianas, the cells opening partly on one side, partly on the other and consisting of irregular, roundish particles of leaves apparently held together by a gum-like substance. These cells have a peculiar scabby appearance and are at first green but in a few days become dark brown and very hard As soon as the cell has been built to half its height, the wasp lays an egg on its floor and continues the construction. The young larva hatches by the time the cell is completed and is fed at first with very small and later with larger caterpillar-like insect larvæ, which have been paralysed by the wasp. The cell is closed by the mother over the full-grown larva with the same material used in making the cell, but with a greater admixture of the resinous substance. The nests are nearly always built by several, often by as many as a dozen, associated wasps. On one occasion I saw two females a short distance apart, each building cells which were later united to form a single common structure by the addition of intermediate cells. In the larger nests females and males are found, and at least a part of the new generation remains in the nest, which is then enlarged by the young females. The building female passes the nights and rainy days in the open cell on the floor of which the larva lies and is thus protected by the mother wasp's

FIG. 18. A. Nest of *Polybioides melaina* of the Congo. B. The same partly destroyed showing the pendent combs, which have cells on both sides. (After J. Bequaert, from photographs by H. O. Lang.)

[*face p.* 70

Plate XIV.

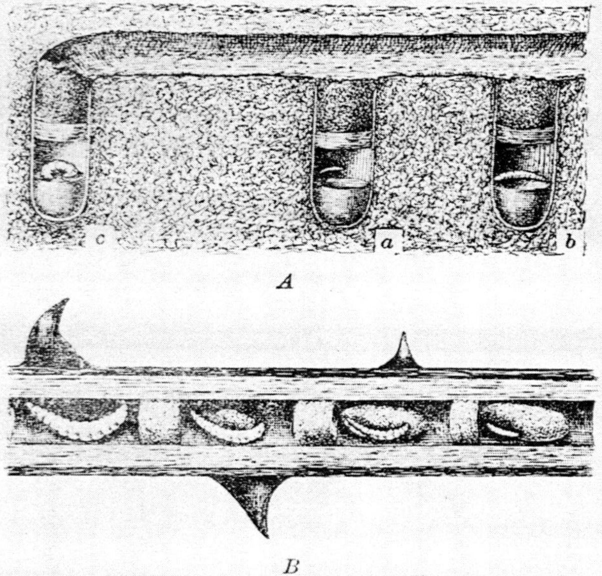

FIG. 19. Nests of Solitary bees. A. Nest of *Colletes succinctus* in the ground. (After Valery Mayet.) a. Cell provisioned and supplied with an egg. b. Cell with young larva. c. With older larva. B. Nest of a small carpenter bee (*Ceratina cucurbitina*) in a hollow Rubus stem; showing egg, three larvæ of different stages and bee bread in three of the cells. (After Dufour and

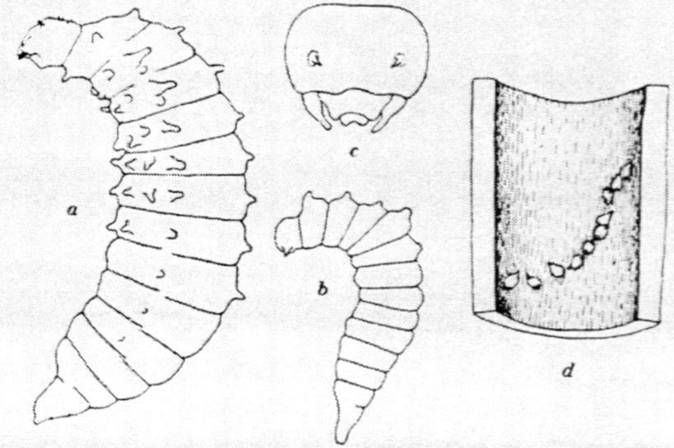

FIG. 20. a. Adult larva of *Allodape pringlei* Cam. b. young larva (4 mm. long). c. head of adult larva from above. d. interior of a hollow stem of Rubus, with the eggs fastened by one pole in a row. (After H. Brauns.) (See p. 85.)

[*face p.* 71

body. Her abdomen is directed downward while her face and antennæ project from the cell. The males assume a similar position in the empty cells."

The accounts of Ducke and Williams show that at least some of the Zethinæ are, to all intents and purposes, social wasps. They are evidently as far advanced as some of the first subfamily of social Vespidæ, the Oriental Stenogastrinæ (Fig. 17A). Williams' rather fragmentary account of four Philippine species of this group, which exhibits extraordinary diversity in nidification (Fig. 17B to G), indicates that they must be very primitive. Three of the species use wood-pulp, or carton in nest construction and have therefore advanced in this respect to the stage of the four other sub-families of social wasps, but one species (*Stenogaster varipictus*), like many Eumeninæ, makes its cells at least largely of earth (G.).

The Polybiinæ (Fig. 18) are a large and heterogeneous, tropical group, comprising a much greater number of genera (23) and species than any other subfamily of social wasps and ranging from socially very primitive forms like the African Belonogaster to highly specialized forms like the neo-tropical Chartergus and Nectarina. Great differences are also apparent in the architecture of the nest, which in the more primitive genera consists of a single naked comb of hexagonal cells attached to some support by a peduncle, and in the more advanced forms of a single comb or of several combs super-imposed on one another and inclosed in an envelope with an opening for ingress and egress. The combs are in some cases pedunculate (stelocyttarous), in others attached directly to the support or to the envelope (phragmocyttarous). In nearly all cases the nest is made entirely of paper, but in a few American species of Polybia clay may be employed. Except in *Polybia atra*, it is always above ground and attached to the branches or leaves of trees, or to the under side of some shelter (roofs, banks, etc.). According to Roubaud (1916), in primitive forms like Belonogaster, as a rule, a single fecundated female starts the nest by building a single pedunculate cell and then adding others in circles concentrically to its periphery as the comb grows, but not infrequently the foundress may be joined by other females before the work has progressed very far. Each larva is fed with pellets of malaxated caterpillars till it is full-grown, when it spins a convex cap over the orifice of its cell and pupates. The emerging females are all like the mother in possessing well-developed ovaries and in being capable of fecundation. In other words, all the females of the colony are physiologically equal, and even such differences in stature as they may exhibit have no relation to fertility. The colonies are small, the nests having usually only about fifty to sixty cells, rarely as many as 200 or 300. In larger colonies there is a certain rude division of labour since the older females devote themselves to egg-laying, the younger to foraging for food and nest materials and the recently emerged individuals to feeding the larvæ and caring for the nest. The males, too, remain on the comb, but behave like parasites and exact food whenever it is brought in by the foraging females. Belonogaster is described as a polygynous wasp, because each of its colonies contains a number of fecundated females. When the colony has reached its full development the females leave in small companies and found new nests either singly or together. This phenomenon is known as "swarming" and occurs only in the wasps of the tropics where it seems to be an adaptation to the favourable climatic conditions. In the higher South American genera of Polybiinæ, however, the females are not all alike but are differentiated into true females, or queens, i.e., individuals with well-developed ovaries and capable of fecundation and workers, i.e., females with imperfectly developed ovaries and therefore sterile or capable only of laying unfertilized, male-producing eggs. Many of these wasps, according to R. von Ihering (1896, 1903) and Ducke (1910, 1914) are polygynous and regularly form new colonies and nests by sending off swarms of workers with one or two dozen queens. The colonies often become extremely populous and comprise hundreds or even thousands of individuals. Some of the species (Nectarina, Polybia) have a habit of storing a considerable amount of honey in their combs, while others are known to capture, kill and store within the nest envelope, and even in the combs, quantities of male and female termites or male ants as a supply of food to be drawn on when needed.

The Ropalidiinæ are a small group of only three genera of paleotropical wasps, the best known of which is Ropalidia. These are primitive forms which build a single naked comb like that of Belonogaster and feed their young with pellets of malaxated insects. The colonies are small and polygynous, but according to Roubaud (1916), true workers can be distinguished though they are few in number compared with the true females. Swarming seems to occur in some species.

The Polistinæ are represented by only two genera. One of these, Polistes, is cosmopolitan and, like Ropalidia and Belonogaster, makes a single, naked comb, suspended by a central or eccentric peduncle to the underside of some shelter. As there are several common species in Europe and the United States, the habits of the genus are well-known. The nest is usually established and in its incipient stages constructed by a single female, or queen. A certain number of her offspring are workers though they seem often to lay male-producing eggs. True females are rather numerous in the colonies of some tropical species, which may therefore be regarded as polygynous, and some of the latter may, perhaps, swarm. In temperate regions, however, the Polistes colony is an annual development and usually not very populous. The young females are fecundated in the late summer and pass the winter hidden away under bark or in the crevices of walls, whence they emerge in the spring to found new colonies. Several of the species, even in temperate regions, occasionally store small quantities of honey in their combs.

Like the Polistinæ, the sub-family Vespinæ includes

only two genera, Vespa and Provespa. The species of the former, the only genus besides Polistes that occurs in the north temperate zone, are the largest and most typical of social wasps. So far as known the species are strictly monogynous. The nest, founded by a single large fertile female, consists at first of a small pendent comb, like that of Polistes, but while there are still only a few cells a more or less spherical envelope is built around it. The eggs first laid produce workers, which are much smaller than the mother and incapable of fecundation. They remain with their parent, enlarge the comb and envelope and, to accommodate the rapidly increasing brood, build additional combs in a series from above downward, each new comb being supported by one or more peduncles attached to the comb above it (stelocyttarous). At first large numbers of workers are produced, but later in the summer males and females appear. Owing to the greater size of the females, the cells in which they are reared are considerably larger than the worker cells. After the mating of the males and females the colony perishes, with the exception of the fecundated females, which hibernate like the females of Polistes and during the following spring found new colonies. In the Vespinæ, therefore, a very distinct worker caste has been developed, though its members occasionally and perhaps regularly lay male-producing eggs. The species of Vespa are usually divided into two groups, one with long, the other with very short cheeks. In Europe and North America the long-cheeked forms, as a rule, build aerial nests above ground, the short-cheeked forms in cavities which they excavate in the ground. The colonies are often very populous by the end of the summer (3,000 to 5,000 individuals).

To some of the peculiarities of the social Vespidæ, such as their feeding habits, castes and rudimentary polymorphism which are merely touched upon in the preceding paragraphs, I shall have occasion to return in future lectures in connection with similar phenomena in the other social insects.

IV

THE EVOLUTION OF BEES

THE Vespidæ briefly considered in the last lecture are a circumscribed group of Aculeates comprising a considerable number of solitary and a comparatively small number of mostly tropical social species. Nevertheless, the various sub-families as they stand represent a gradual evolution of social habits from very primitive types like Synagris, the Zethinæ and Stenogastrinæ to the distinctly specialized Vespa. The course of development is shown in the gradual emergence and differentiation of a worker caste, advancing complication in nest-architecture, an increase in the population of the colony and an increase in the stature of the fertile female or queen, as shown in the genus Vespa. The colonies of some Polistinæ, and especially of the Polybiinæ and Vespinæ, have certainly attained to a status of very definite social integration, but we must admit that it is not of a very high order. The differentiation of the worker caste is very feeble in the Polybiinæ, and the colonies of the Vespinæ and most Polistinæ are merely annual growths, a condition which not only sets very definite bounds to the size of their population but must greatly retard or even inhibit further progress in social development. Probably such progress among the Vespidæ is henceforth possible only among some of the more primitive genera of Polybiinæ in the American or African tropics.

If we turn our attention now to the second family of Aculeates which have developed social habits, the bees, or Apidæ—the Anthophila, or Mellifera of Latreille—we find ourselves confronted with a more formidable task. The very size of the group—some 20,000 species, as estimated by Friese (1923)—is staggering, and the fact that the habits of only a few of the genera and those mostly European, have been carefully studied, renders ethological generalizations premature and tentative. At the present time no other family of Aculeates displays anything like this magnificent efflorescence of species on our planet, although it is surpassed by the Formicidæ in number of individuals. The great majority of Melliferan species are confined to two belts, corresponding to the warmer portions of the north and south temperate zones. The tropical bee fauna, especially in the Old World, is rather meagre and the number of species steadily diminishes from the warmer temperate zones towards the poles.

Our difficulties are increased by the present plight of the taxonomists, which arises in turn from obvious ethological conditions. The whole family Apidæ consists of species which either actively collect nectar and pollen and are therefore intimately associated with the flora, or of species which have abandoned this habit to become parasitic on other more or less closely related species. The first group has developed remarkable modifications of organs and structures, including the tongue, the hind legs and the pilosity of the body and appendages for collecting a highly specialized food, whereas the second group shows a degeneration or loss of these very modifications. Other organs, notably the wings and their venation, are extraordinarily uniform in both groups. Nothing could be more unfavourable for the taxonomist, because clearly adaptive structures do not as a rule furnish good characters for the definition of sub-families, tribes and genera, and the absence of similar characters in the parasitic species, which in many instances are actually derived phylogenetically from the genera of their hosts (see Wheeler 1919*a*) and should therefore be placed near them in any natural classification is, to say the least, disconcerting. Add to this the further consideration that there are several cosmopolitan or widely distributed genera of bees comprising great numbers of very closely related species (Megachile, Halictus, Andrena, Anthophora, Prosopis, Sphecodes, Colletes, etc.), and many other peculiar or local genera which are either monotypic or consist of a few species with unstable or elusive characters, and we can understand why no one has been able to construct a classification of the Apid sub-families and tribes that anyone else will accept. Such attempts are, in fact, premature, because the great bee-faunas of all the continents, except Europe, are inadequately known even to the taxonomists. This is notably true of the faunas of Australia and South America where there are many primitive types that may ultimately throw considerable light on the relationships of genera now confined to other parts of the world.

There has long been complete unanimity of opinion in regard to the extremely close structural affinities of the bees to the Sphecids or Fossores.[1] These affinities are so close that in Bradley's classification (Comstock, 1924) the Apidæ are treated merely as one of the families of the super-family Sphecoidea and that even expert hymenopterists have occasionally described bees as Fossores. Thus the type species of Neolarra (*pruinosa*) was regarded as a wasp allied to the Bembicinæ or Larrinæ by Ashmead (1890) and Fox (1893), but Baker (1896) later recognized it as a bee allied to the genus Phileremulus. The Chilean bee genus Lipanthus (=Psænythia) was also originally placed among the Fossores. The pronounced Fossorial habitus of many bees, especially of the less hairy parasitic genera (Sphecodes, Nomada, Epeolus, etc.), or of species like Prosopis, which swallow

[1] The only author, to my knowledge, who has rejected this opinion is Börner (1919). He places the bees, because they lack the strigil on the hind legs, near the Vespidæ. Obviously the absence of this organ cannot be regarded as a proof that the Apidæ are not closely related to or derivable from Sphecoid forms. Handlirsch's suggestion (1924) that the singular tibial and metatarsal structures of the bee's hind legs may represent a highly modified strigil adapted to collecting pollen from the surface of the body, would if substantiated also dispose of Börner's contention.

the nectar and pollen and have therefore failed to develop the usual external collecting apparatus, is very striking. Hence hymenopterists have been led to derive the bees from Fossores and to suppose that they must have had entomophagous ancestors. In the latter half of the past century, during the romantic period of Darwinian speculation, H. Müller, well-known for his fine work on the relations of bees to flowers, considered this hypothesis in two very thoughtful papers (1872, 1881). The adult Sphecids are, as a rule, expert flower-visitors and Müller surmised that their training in the search for insect prey and in excavating or constructing cells, supplemented by an increase in bodily stature, had enhanced their ability in finding and extracting the nectar of flowers. He adds: " But all these advances together are insignificant compared with the enormous progress in flower exploitation which we witness in the family of bees. The bees, originating in the family of Fossores, and natively endowed with caution and discrimination, with tireless energy and with a skill in creeping into and out of cavities gradually perfected and inherited in connection with the care of their brood, have learned to confine themselves to provisioning their young with floral foods exclusively. Naturally they could not fail enormously to increase their eagerness in visiting and their cleverness in exploiting the flowers ".

The derivation of the bees from the Fossores has been repeatedly considered by Friese, and most fully in his recent book (1923). He maintains that the Apidæ had a diphyletic or possibly even a triphyletic origin, and presents three ancestral trees of the family, one constructed by himself, and based on the wing venation, one by Tosi (1895) based on the anatomical structure of the gizzard, or proventriculus, and one by Langhofer (1897), based on the structure of the tongue. All three of these schemata were independently conceived and all agree in deriving the bees from two independent hypothetical Fossorial stocks, one of which gave rise to

Prosopis, Colletes, etc., the other to Sphecodes, Halictus, etc.[1] Friese indicates a third possible ancestral source of the bees among the Vespidæ (Odynerus) for the peculiar Australian genus Hylæodes, which, so far as known, has given rise to no other forms. This supposed polyphyletic origin of the Apidæ is certainly suggestive of a possible masked or concealed convergence, or parallelism of development due to anthophilous habits among only remotely related Sphecid or even Vespid groups.

Although views like those of H. Müller and Friese have also been held by von Alten (1910) and others, one might go even further and maintain that the Apidæ are not derived from any of the Sphecidæ as that family is now constituted, but that the Apidæ themselves are merely the taxonomic equivalent of an anthophilous sub-family, or if diphyletic of two anthophilous sub-families of the same rank as any of the extant entomophagous families, e.g., the Bembicinæ, Crabroninæ, Sphecinæ, etc. There is no satisfactory evidence that the bees ever fed their young with dead or paralysed insects. If the Aculeata are descended from the Phytophaga directly instead of through the Terebrants, the vegetarianism of the bees, for aught we know to the contrary, may be primitive. This position is actually taken by Roubaud (1918) when he says: "Without doubt the Apidæ, which were exclusively adapted to a melliferous diet, never used their sting for the capture of prey. Very probably they represent forms that have remained exclusively vegetarian and as such have retained feeding habits most like those of the ancestral type. The adaptation to a carnivorous diet must have occurred secondarily in certain originally vegetarian forms as a result of the employment of the sting against insects of small stature. But the facts which might enable us to establish a real affiliation in these habits are still lacking."

[1] Friese regards the species of Sphecodes as very primitive, predatory bees and not as parasites. His discussion again opens up the whole controversy in regard to the behaviour of Sphecodes, which seemed to have been settled by the researches of Nielson, Ferton, Breitenbach and others (see Wheeler 1919*a*) and is interesting though by no means convincing. We eagerly await the publication of the second part of Stöckhert's paper (1923), which will probably throw much light on the matter.

Two sets of facts might seem to lend some support to Roubaud's contention. First, the existing Sphecidæ comprise a number of sub-families or, according to Kohl (1896) and Handlirsch (1924), groups of genera, all of which, so far as known, are too highly specialized both structurally and ethologically to serve as Apid ancestors, and second, the paleontological evidence indicates that throughout the Tertiary at least, the relations of the Sphecidæ and Apidæ were essentially those obtaining at the present time. Cockerell (1909*a*, 1917) has described from the Baltic amber (Lower Oligocene) Sphecids of the well known genera Crabro and Pison and also numerous highly specialized bees (Electrapis, Protobombus, Chalcobombus, Sophrobombus, Ctenoplectrella and Glyptapis), from the shales of Florissant (Miocene) Fossores of the sub-families Crabroninæ, Pemphredoninæ, Philanthinæ, Nyssoninæ, and Sphecinæ, contemporaneous with such highly specialized genera of solitary bees as Ceratina, Anthidium, Dianthidium, Heriades, Andrena, Calyptapis (near Melissodes), Libellulapis and Lithandrena (near Andrena), and from the shales of Oeningen (Miocene) bees of the modern genera Lithurgus, Xylocopa and Andrena. We shall therefore have to seek the ancestors of both bees and Sphecids in the Cretaceous or probably even in precretaceous formations.[1] It must be admitted,

[1] Owing to the fact that the bees could only have developed in symbiosis with the flowering plants the paleobotanist might be expected to throw some light on the most probable period of Apid evolution. The only recent authoritative statement I have seen on this matter is that of Wieland (1924), who says that "probably the opening of the Cretaceous time saw in all the world fully 40,000 species of non-flowering plants, and many thousand seed-bearing plants representative of all the modern orders. Both monocotyls, including the typical palms, and dicotyls, or net-veined types, are modern in form and widely distributed, although the known species of these higher types do not become numerous until the mid-Cretaceous. Since then, the main course of plant evolution can be summed up as a rapid disappearance of naked seeded types, or gymnosperms, to 600 species of relatively limited distribution, with the steady increase of the flowering plants or the angiosperms to their present numbers, about 120,000 species. E. W. Berry insists that there has been much recent evolution within the dicotyls. But it would not be absurd to hypothesize for the dicotyls of early Cretaceous time fully half as many species as now exist".

however, that the older view of a change in the ancestral bees from entomophagy to anthophily is not improbable when we consider the case of the Masaridinæ among the Vespidæ.[1] I called attention in my last lecture to the fact that the primitive genus Pseudomasaris in the South Western United States stores insect food for its young, but that the less primitive genera Masaris, Celonites and Ceramius in Europe and Africa are, according to all accounts, strictly anthophilous, like the bees. It is, of course, obvious, as Roubaud (1918) implies, that the character of the food is imposed on the offspring by the adult insect, and that the adult Sphecid, unlike most Masaridinæ and the bees, really has a double diet, since it feeds both on nectar and insects. In most cases the insect diet is represented merely by the malaxation and tasting of the paralysed prey, and may be interpreted as a vestigial act reminiscent of the Sphecid's own larval diet.

When we turn to the reproductive behaviour which has led to the development of societies we find a most extraordinary parallelism between the family Apidæ as a whole and that of the Vespidæ as described in my previous lecture. The progress from the solitary condition, shown in more than 95 per cent. of the species, to the conditions in the most highly socialized form, the honey bee, is, so to speak, a repetition of the various wasp *motifs* set in a different key. Everyone of the thousands of species of solitary bees has its own peculiarities of behaviour, but the differences are usually so insignificant that the habits as a whole are very monotonous. With the exception of the parasitic bees, which have been secondarily evolved from non-parasitic forms, all the solitary bees make their nests either in

[1] Handlirsch (1924, p. 781) clearly maintains this view when he says: " It is very probable that the Apides go back to older primitive Sphegids, the strigils on the hind legs of which were perhaps converted into a complicated brushing apparatus. In these forms the vegetarian regimen is certainly not to be derived from the primitive phytophagy of the Symphyta, but from the mode of life of zoophagous forms, which cared for their brood."

the ground (Fig. 19A) or in the cavities of plants (Fig. 19B), in crevices of walls, etc., or construct earthen or resin cells. Some species line their nest cavities with pieces of leaves or petals of flowers, with plant-hairs or particles of wood, or with films of secretion which resemble dried collodion or gold-beater's skin, the "baudruche" of French authors. Most of these materials, as will be noticed, are derived from plants. The nest usually consists of several cylindrical or elliptical cells arranged in a linear series or more rarely in a compact cluster, and as soon as a cell has been completed, it is provisioned with a ball or loaf-shaped mass of pollen or of pollen soaked with honey and called "bee-bread", an egg is laid on its surface and the cell is closed. We have here again the typical mass-provisioning of the solitary wasps, very similar to that of the Eumeninæ and Sphecidæ, except that vegetable instead of animal substances are provided for the young. Nevertheless, the pollen and honey are ideal foods, since the former is rich in proteins and oils and the latter in sugar and water, and both contain sufficient amounts of various salts for the growth of the larvæ. As in the care of the solitary wasps the mother bee dies before her progeny emerge.

In many solitary Sphecids (Bembix, Stizus, Philanthus, etc.) which nest in the earth, the females establish their burrows in close proximity to one another, though each cares for her own brood and resents the intrusion of her neighbours. The very same phenomenon is repeated among the bees of many genera (Panurgus, Eucera, Osmia, Andrena, Spinoliella, Melitoma, Pachycentris, Anthophora, etc.), and has been noticed by many observers. In some of the cases several females use a common entrance to the cells, but there is nothing to show that such "parœcism" is a decisive step towards social life. It may, of course, be an expression of a vague associative or gregarious tendency, but it may also be the result of the attraction exercised by a circumscribed and peculiar environment on a number of female bees of the same species. The species of Euglossa, often mentioned in the literature and known to construct clusters of ellipsoidal cells of dung, earth, and resin, are in all probability cases of the same kind of association. Latreille (1809), who first described this genus of large and beautiful neo-tropical insects and noticed the development of the corbula on the hind tibiæ, a structure which occurs in no other bees, except the higher social species, placed the group near the Bombinæ, and he, Lucas (1878) and others were, therefore, led to suppose that the Euglossas might be social. Lepeletier (1827) however, who failed to find the wax-combs on the hind legs, dissented from this opinion. Möbius (1856) even claimed to have detected a waxen lining to the cells of *E. surinamensis*. Within more recent years, Schulz (1902), Ducke (1902, 1903, 1905), R. von Ihering (1904), and Schrottky (1922), who have observed the Euglossas in South America, agree that they are not social, though some of the species, e.g., *E. nigrita*, construct large nests of more than 200 cells. In such cases the nest is made by several females but each cares only for her own brood, though a common entrance is constructed. The existence of wax in the cells has not been confirmed, but Friese (1923, p. 76) states that wax pectens are present on the hind basitarsi of Euglossa, Eumorpha and Eulæma.

Apart from a few imperfectly described and therefore dubious cases, all the known social bees belong to five groups comprising certain species of the genera Allodape and Halictus and all the species of the three highly specialized sub-families Bombinæ, Meliponinæ, and Apinæ. The three latter have long been known to be eminently social and their behaviour has been so often described that I shall later touch on only a few aspects of it. Suffice it to say, for the present, that we have no evidence that the social habits of these sub-families have had a common origin. There is, moreover, no phylogenetic connection between Allodape and Halictus, which belong to widely separated groups of bees, the former to the Tarsilegidæ, the latter to the Podilegidæ, and neither is directly ancestral to the Bombinæ, Meliponinæ, and Apinæ. We may therefore assume, at least till proof to the contrary is forthcoming, that social habits have appeared on at least five different occasions in the family Apidæ, just as we found five independent sources of social development among the Diploptera, or Vespidæ. Owing to the unusual interest attaching to Allodape and Halictus I shall devote a large part of this lecture to a discussion of their behaviour.

The South African bees of the genus Allodape exhibit an extraordinary transition from a solitary to a social mode of life but of quite a different type from that of Halictus and the three higher sub-families. The species are small and rather closely related to our European and North American carpenter bees of the genus Ceratina, but their larvæ are peculiar in possessing pairs of appendages with which they hold on to their food. Some notes on the habits of Allodape kindly sent me by Dr. Hans Brauns were published in my "Social Life Among the Insects" and in the French edition of the present work. Since he has lately (1926) published a detailed account of these bees, I substitute several paragraphs from his very interesting paper and reproduce two of his figures.

"The old authors considered the species of the genus *Allodape* to be mostly parasites; and from the time of Lepeletier St. Fargeau we find them quoted in hymenopterous literature as being parasites or messmates. This mistake is not easily explained, because the species of *Allodape* bear on the legs a distinct collecting apparatus of hair-brushes. They are in fact true social bees, although their colonies are as primitive as can be. There is no building of cells, no shrouding of the brood whatsoever. The female excavates a tunnel into the dry stem of iridaceous plants, Iris, Rosa, Aloe, Rubus, amaryllid plants, etc., or any other plant with a soft pith; it removes the latter entirely, and the nest is then ready. I believe that in most cases labour is not even involved. They as often as not use the stems previously hollowed out by *Ceratina* bees, or perhaps *Diploptera* wasps. Some of the smaller species ulilize the galleries made by Coleoptera in dead logs or poles, or again the long thorns of mimosa trees. In the plains of the Karroo, where suitable plants are not as common as in other parts of South Africa, some of the small species even nest in the ground, the steep sides of dry rivers and " sluits ", and even in the walls of houses. So, for instance, does the widely distributed *Allodape pringlei*, Cam. (Fig. 20), but this nesting in the ground is comparatively rare.

" The colony is founded by the female alone at any season of the year. It is, of course, clearly established that at higher altitudes the seasonal temperature and the want of flowers must impede or hinder the continual gathering of food, but there is no hibernation. A warm sunny day finds the bee out of doors, even in the middle of the winter season. Certain plants flower only in the cold or winter season, as, for instance, the Ericaceæ, some Aloes, Lycium, and others. Something is always to be had. After excavating the stem to a convenient depth, or putting the last touch to the ready-made gallery, according to her taste, the female deposits the egg at the more or less rounded bottom of the hollow part. The egg is of large size in comparison with the size of the bee, and of kidney shape. As soon as the larva is hatched the mother commences feeding it. The egg or eggs that will hatch last are always to be found at the bottom of the cavity ; above them are to be found the younger larvæ, then the more advanced ones, and towards the entrance are the pupæ. Sometimes, however, this disposition will be found to be disturbed. The mother in carrying food to the young larva at the lower part of the cavity must perforce constantly pass the pupæ and larger larvæ, and may therefore slightly disturb the above-mentioned order, but this occurs rarely.

" A special worker class, such as we find in the genera *Apis*, *Bombus*, *Trigona*, and *Melipona*—that is to say, a more or less crippled female, is not to be found in

Allodape. All females are fertile, and, as far as known, take a share in breeding and feeding the larvæ. In all probability the young females, after fertilisation, constantly found new nests. But they return for a considerable time to their original home and there feed the foster brother- and sister-larvæ. The males return to the nests as well, and the latter when opened in the evening are found to be crowded with a population of both sexes. The old mother bee is easily recognized through her tattered wings and other signs of hard work and old age, but otherwise there is little difference in size between herself and her female progeny.

"There are three types of larvæ in the genus *Allodape*, corresponding to three methods of feeding. In the first case the larva is provided on the thoracic part of the body with a pair of movable and contractile fleshy protuberances which we will call, in accordance with their functions, arms. The larva of *A. ceratinoides*, Grib. (Fig. 21), is here figured as a conspicuous example of this form. These arms enable the larva to retain firmly the small morsel of food, deposited between them by the mother bee, who feeds in this way a number of larvæ at one visit. They close these arm-like processes round the morsel of food, and are thus able to retain their hold of it until it is devoured. On opening the nest shortly after the entrance of the mother bee one finds a number of larvæ supplied in this manner with food particles, and busy feeding. Nearly all species of large size, in fact the great majority of the South African species, belong to the section the larvæ of which are of this type.

"The second type of larva is illustrated by that of *A. pringlei*, Cameron (Fig. 20). It is easily seen from this figure that it is altogether different from the first. The egg is no longer lying loose at the bottom of the cavity, but a row of eggs (up to ten or more in number) is fastened, perpendicularly, probably by an agglutinous substance, by the lower polar end to the walls of the hollowed stems, and arranged in a more or less spiral

Plate XV.

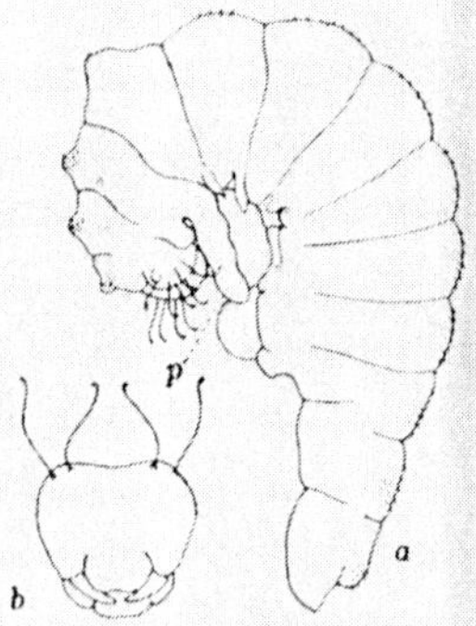

Fig. 21. a. Adult larva of *Allodape ceratinoides* Grib. b. Head. c. Left pseudopodion of same. (After H. Brauns.)

Fig. 22. An unusually large comb of the East Indian honey bee, *Apis dorsata*, from a photograph taken by Dr. J. W. Chapman in Negros Oriental, Philippine Islands. (See p. 102.)

[*Face p.* 86

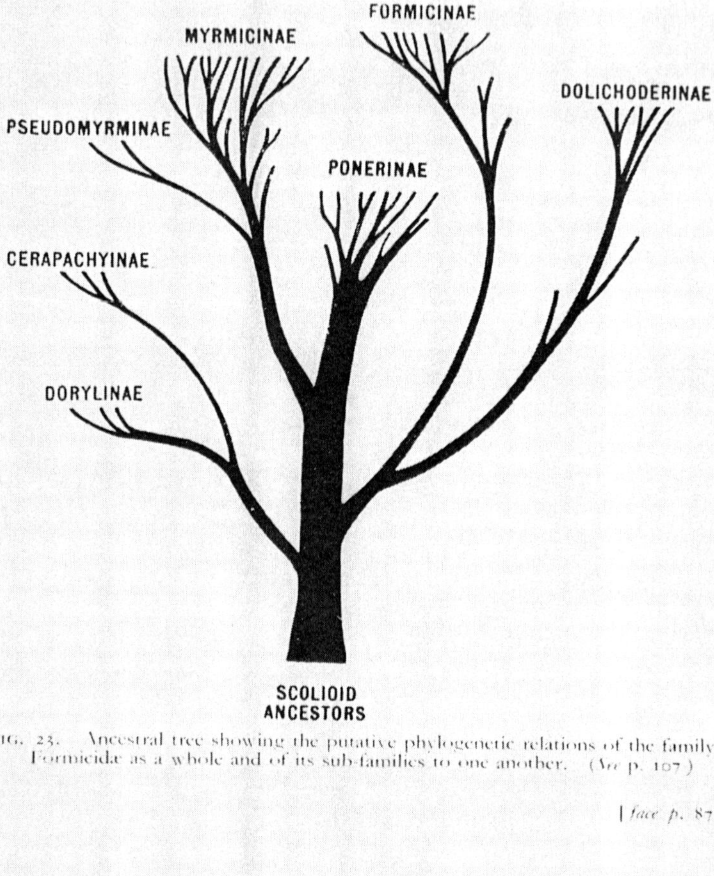

FIG. 23.—Ancestral tree showing the putative phylogenetic relations of the family Formicidæ as a whole and of its sub-families to one another. (*See* p. 107.)

[*face p.* 87

curve (Fig. 20*d*). The larvæ (*a*) are provided with a great number of short, fleshy, and contractile appendages disposed round the segments. Friese uses the term Pseudopodion for these appendages. By the aid of these fleshy protuberances the larvæ are able to keep their position in the same level of the lumen of the gallery. By pressing these proturberances against the rugosities of the walls, and also against the bodies of their neighbours, they maintain the position of the body in the gallery in such a way that the apical end of the larvæ and the heads of the same are each at the same level. It follows that the hatching of the larvæ must take place, approximately, at the same time. It is probable that the bee lays and fastens the row of eggs in rapid succession. The feeding of the larvæ is accomplished in the following manner. The mother bee deposits the food in the space left between the heads of the larvæ so that they can all feed at the same time from the same food supply. The pupæ and the larvæ after the last casting of skin are found near the entrance, and the agglomeration of larvæ still feeding are below them, nearer to the bottom.

" In principle, the manner in which the larvæ are fed in the two typical instances here given is really the same. The larvæ are not fed directly by the mother bee, nor by regurgitation, but they are fitted with integumentary organs for holding the food near the buccal parts, and also in one case for clinging to the walls of the cavity. They use the food supply in a way similar to that which that eminent observer of ants and their life-history, Professor W. M. Wheeler, has shown us so clearly to happen in the case of Ponerine Ant larvæ. It may be stated, however, that the larva casts off the integumentary appendages at the last moult leading to the pro-pupal stage, when it has already assumed the maggot-like appearance of the usual bee larva.

" The third method of feeding the larva is, in comparison with the two already described, the most interesting and the most important. I have only recently observed it in the case of a small species making its nest in the Karroo plains in the dry stems of a native asparagus. There is no doubt about the result of my observations since I have come across about a dozen colonies.

"In this species the mother bee carries in food for one larva only at a time. As in the first type, she first lays the egg, which in comparison with her own size is enormous, being almost equal to one-third of her own bulk. After the oviposition, she brings in enough food to last for the larva until it reaches the pupa stage. The volume of this pollen mass is about the same as we find in cases of solitary bees, as for instance in any of the partitions of the nest of a *Ceratina* species of the same size. It is quite sufficient for one larva. If we open the nest we find one larva clinging to a pollen mass, another to the next one, and so on. On the top of all is a fresh egg, and above it a pollen mass which the bee is about to complete. When several larvæ are present the one on the bottom is the most advanced. I found the colonies provided in this manner in all stages of growth. The external appearance of the larva is also different from that of the two first-described types. The larva clings closely and broadly with its very flat ventral surface to the food mass. Only two comparatively small and inconspicuous fleshy protuberances, the anterior larger than the posterior, are to be seen on each side of the larva. It will be necessary to observe more closely the habits of this species. More will probably be found providing their larvæ in this manner.

"There is only one difference separating the economy of this bee from that of the general type of solitary bees, namely that in this instance there is no partition between two adjacent larvæ and their store of food.

"This point is of great importance, and should be thoroughly grasped. All other solitary bees provide first the food, then lay the egg on it, and close the cell or partition, proceeding after this with another partition resting on the closed one. In the case of this *Allodape* the process is the same, except that the egg is first laid, then a sufficiency of provender carried in for the future larva; another egg is laid above the first, food brought in for the same; and so on. But there is no partition separating the larvæ and their food as in the case of all other solitary bees; and the gallery is not closed after the last egg has been provisioned with the required food.

"The reason for this method of oviposition may perhaps be explained as follows: As mentioned before, the egg is very voluminous in comparison to the size of the mother bee. It is hardly credible that a small bee should lay eggs one-third her own size in rapid succession. Such eggs take time to mature in so small an ovarium, and cannot therefore be laid at short intervals. Owing to this, the bee has time enough to bring in sufficient food to feed each larva till the pupal stage is reached.

"In the first type of oviposition and food storage, it is true, as already stated, that the egg is also very large, but we do not know the time required for the completion of oviposition. In the second type, where numerous eggs are fastened to the walls of the gallery, the size of these eggs is small, and accords with the size of the mother bee. From the description of their position it is evident that they must be laid in rapid succession in order to allow all the hatched larvæ to feed from the same food supply, and at the same time".

Brauns' observations on Allodape are of considerable interest, because they reveal within the limits of a single genus a series of stages beginning with a mass-provisioning of the young, like that of the solitary bees and wasps, and ending with a stage of progressive provisioning. And not only has the latter led to an acquaintance of the mother with her offspring but to a temporary affiliation of the offspring with the mother. It would seem that these conditions must have had their inception, as Brauns suggests, in so simple a matter as the omission of the series of partitions which all other solitary bees construct between their provisioned cells. In this connection it may be noted that in the gastrilegous bees of the genus Lithurgus, which are related to Megachile and excavate digitiform tunnels in woods, there are indications of an approach on the one hand to the conditions in Allodape and on the other to those of the higher social bees. Gutbier (1914) found that the palearctic *L. fuscipennis* lays several eggs on the same mass of bee-bread and that this is devoured in common by the hatching larvæ. On reaching its full growth and before pupation, however, each larva isolates itself by means of partitions of excrement. In an Australian species, *L. dentipes*, known to occur also in the Carolines and other Pacific Islands, Ludwig (1904) and Friese (1905, 1923) observed that same of the digitiform cavities of the nest are used as storehouses for pollen.

Halictus is the largest of bee genera, of world-wide distribution and comprising about 1,000 described species, a great number of which differ by such insignificant and elusive characters that they are the despair of taxonomists. The habits of only a small number have been studied, mostly with fragmentary and often conflicting results. The European species have been investigated by the following observers: Walkenaer (1817), Eversmann (1846), F. Smith (1855), Breitenbach (1878), Fabre (1879-80, 1903), Verhöff (1891, 1892, 1897), Friese (1891, 1923), J. Pérez (1895), Aurivillius (1896), Ferton (1898), von Buttel-Reepen (1903), Semichon (1912), Fahringer (1914), Armbruster (1916), Legewie (1922) and Stöckhert (1923). In the United States these very common bees have attracted very few students. The only important papers I can recall are one by J. B. Smith (1901) on an unidentified Halictus and *H.* (*Augochlora*) *humeralis*, one on *H.* (*Chloralictus*) *pruinosus* and its parasites by Melander and Brues (1903) and some notes by Rau (1926) on the nesting habits of *H.* (*Chloralictus*) *zephyrus*. R. von Ihering (1904) has published a description of the nests of *Augochlora gramminea* in Brazil, and Lüderwaldt (1911) a few notes on the cells of the Brazilian *Neocorynura erinnys*. More recently Herbst (1922) has given us some valuable data on the habits of three of the Chilean sub-genera, Augochlora, Agapostemon and Corynura.

The Halicti, so far as known, nest in the soil or more rarely in rotten wood, in aggregations which may comprise hundreds of nests. Each nest is started as a narrow, more or less perpendicular gallery, usually less than two feet in length, but in the case of *H.* (*Augochlora*) *humeralis* reaching a depth of six feet. The ellipsoidal cells are excavated singly or in groups along the gallery. In some species (*H. sexcinctus*) they are combined in the form of a rude comb, which in *H. quadricinctus* is surrounded and partly isolated by a capity. The peculiar structures of the latter species were first seen by Eversmann, but those of both species have been more recently described by Verhoeff and others. Many Halicti, as Fabre observed, line their gallery and cells with a thin layer of a secretion which hardens as a hydrofuge varnish. For many years there has been a smouldering controversy over the number, sexual composition and parthenogenesis of the annual generations of these bees. Since there would be no advantage in entering into a full discussion of the various opinions that have been entertained by different authors, I shall select for brief consideration only the interpretations of Fabre and Armbruster and especially of Stöckhert, through whose recent very careful investigations the whole matter assumes a new and intensely interesting aspect.

Since the days of Frederick Smith melittologists have agreed that an Halictus brood develops during the summer and produces males and females in the fall and that after mating and the death of the males, the females hibernate in the maternal burrows or their vicinity and during the following spring renovate the old nest or excavate new nests and begin to provide for their brood. Fabre, who studied mainly *H. calceatus* (=*cylindricus*) and to some extent *scabiosæ* (=*sexcinctus*) and *fodiens*, maintained that the overwintered fecundated females produce only female offspring and that these produce parthenogenetically the fall generation consisting of both sexes. He therefore believed that there are two annual generations, an autumnovernal, with fecundated females, and a parthenogenetic summer generation. His contentions that the latter consists only of parthenogenetic females and that the two sexes of the fall generation mate in the nest, prove to be erroneous, but as we shall see, his belief that the overwintered females survive for many months and act as janitresses of the nests was well-founded. But this is neither the sole nor the most important function of the old females.

Armbruster undertook to determine the number and character of the annual generations by a statistical study of the flight-records of German Halicti and reached the conclusion that the males and females mate in the fall but that, at least in Northern Europe, the females, as well as the males, die after provisioning their cells. Hence the brood alone passes the winter and produces nothing but females the following spring. These produce without fecundation a summer generation of females which in turn become the parthenogenetic mothers of the fall generation of males and females. According to Armbruster, therefore, there are three annual generations, since the fall and spring females are not the same individuals, as Fabre and others have maintained, and there are two parthenogenetic generations, the first producing only females (thelytocous), the second both sexes (ampherotocous). Armbruster believes that the same conditions obtain also in Sphecodes, the parasite of Halictus, and goes so far as to speak of a peculiar " Halictus-type " of generation, unique among bees, since the unfertilized eggs do not conform to the Dzierzon rule but produce females in the summer and both sexes in the fall.

Stöckhert somewhat scornfully rejects these contentions as presumptuous, erroneous and based merely on "paper statistics". His own observations were carried out during seven years in Upper Bavaria, for the most part on *H. malachurus, maculatus, sexcinctus, immarginatus*, and *puncticollis*, which happen to represent most of the European groups of the genus. The observations relating to reproduction and parthenogensis were carefully controlled by Zander through dissections of the female reproductive organs. It appears that the number of annual generations differs in different species of Halictus. Thus some northern forms, e.g., *H. lineolatus*, have only one generation. like many species of Andrena, and others *H.* (*Lucasius*) *clavipes* and *H. soror* have two, each consisting of both sexes, also like certain Andrenas. Moreover some Halicti with long flying periods may have a third generation intercalated in the autumn, especially when the advent of winter is delayed (*H. morio, puncticollis, villosulus*). But whether there are one, two, or three generations, the males die in the fall, though rarely in some South European species a few may survive and fly during the following spring, whereas the females always hibernate.

Although Stöckhert shows that both Fabre and Armbruster were wrong in their account of one or more parthenogenetic generations, there is in certain Halicti a type of reproduction even more interesting on account of the light it sheds on the evolution of social habits. *H. malachurus* may be taken as a paradigm. This bee nests in hard, clayey soil, often in congregations of more than a thousand burrows. The overwintered females sometimes appear as early as the middle of March and begin to clean and varnish the burrows and to visit the flowers of willows and dandelions. Although several females have been hibernating peacefully in the same burrow, their number is now reduced to three, two or normally only one, owing to the struggles among them for possession of the burrow. The ousted individuals have to move to unoccupied nests or to dig new ones. The brood-cells are built directly off from the main gallery, provisioned and provided with eggs. If more than one female remains in the nest, these daughters of the same mother nevertheless build separate cell-clusters. By the middle of June the brood emerges and consists almost exclusively of females. They are smaller than their mother and differently sculptured and belong to a form which taxonomists have described as *H. longulus*. The first individuals to appear are unusually small. There are no *longulus* males. The mother still survives and her *longulus* daughters remain with her and begin to construct brood cells. This work is carried on mainly at night, as Fabre observed in *H. calceatus*. The eggs for these cells are laid by the mother and not by the *longulus* daughters, which, however, collect the provisions consisting mainly of the pollen of Hieracium and Leontodon. The old mother remains at home and guards the nest entrance in the manner observed by Fabre and others. The *longulus* females forage till about the last of September, although each individual probably lives only four to six weeks. Their number increases till in August there are about a dozen in each nest. Females of the *malachurus* type begin to appear about the beginning of August, also from eggs laid by the old mother. They are very, drowsy and lethargic, compared with the very active, bustling *longulus* females and collect no pollen though they visit the flowers for nectar. The males, which have been appearing and increasing in numbers in the meantime, and are also the offspring of the old mother Halictus, pay no attention to the *longulus* females, but eagerly pair on the surface of the nest with the young *malachurus* females when they leave the burrow for the first time on some sunny day. There is therefore a kind of abortive marriage flight. Dissection of the *longulus* females shows that they are never fecundated. The old mother, after surviving the preceding winter and producing the summer brood of *longulus* and fall brood of *malachurus*, though much worn and with frayed wings, lives till the end of the season. Zander's dissections showed that her ovaries may contain ripe eggs and her spermatheca an abundance of sperm till the end of August. Somewhat later she loses her power of flight, crawls away and dies. The *longulus* females and the males then also perish, but the young fecundated *malachurus* females go into hibernation in their mother's nest.

The habits of *H. maculatus* were found by Stöckhert to be very much like those of *malachurus*. The summer females are also smaller than their mother though they have the same sculpture and have not been described by taxonomists under a different name. In *H. immarginatus* the summer females, however, differ in size, punctuation and rugosity from their parent and are known as *pauxillus*. The summer form of *H. puncticollis*, though of the same size as the mother, is sufficiently different to have been named *villosulus* by the systematists.

In this brief account of Stöckhert's observations you cannot fail to have noticed the very striking and suggestive resemblance of *H. malachurus* to Vespa and Bombus. The sterile *longulus* females correspond, of course, to the workers of Vespa and Bombus, while the old mother *malachurus* represents the overwintering wasp and humble bee queen. In fact, the social organization of *H. malachurus* and the allied species, apart from the number of co-operating individuals, is almost as highly developed as that of Bombus. The resemblance is further enhanced by the fact that in some of the allied species of Halictus (*sexcinctus* and *quadricinctus*) the brood cells have the form of a rude comb and in *H. scabiosæ*, according to Fahringer (1914), some of the brood-cells may be used for storing pollen. You will notice also that there is no infraction of the Dzierzon rule in Halictus, no parthenogenetic origin of females, and that Fabre's and Armbruster's claims to the contrary must have been due to a misinterpretation of the facts. Stöckhert's observations, to which I shall have occasion to return in another connection, indicate that there may be many surprises in store for us when the life-histories of these seemingly monotonous and uninteresting bees have been subjected to more careful scrutiny and experimentation on other continents, especially in South America and Australia.

The higher social bees, the Bombinæ, Meliponinæ and Apinæ, representing the most highly specialized forms among the podilegous division of the family Apidæ, present so many knotty problems to the phylogenist that a whole course of lectures would hardly suffice for their adequate consideration. It is often assumed that these three sub-families must be rather closely related, partly because their morphological structure is similar and partly because they all secrete wax and employ it in building their combs and brood-cells, but their habits and the genesis of their societies are so diverse that we are led to suspect that we may be dealing with three convergent developments from as many independent ancestral sources rather than with three diverging branches of a common stock. The colonies of such forms as *Halictus malachurus*, which, as we have seen, are very similar to those of the Bombinæ, are also very suggestive in this connection. My time is so limited that I shall have to confine myself to a few notes on the paleography, secretion of wax, the degree of polymorphism and the methods of colony formation in the three sub-families.

Social bees have been described from the Baltic amber (Lower Oligocene), Sicilian amber (Miocene) and various shales of the Upper Oligocene and Miocene of Europe, and an attempt has been made by von Buttel-Reepen (1906, 1915) to interpret some of these fossils as actual ancestors of our modern *Apis mellifica*. But the species placed by the early entomologists in the modern genera Bombus, Melipona, Trigona and Apis on recent examination by Cockerell (1909*a*, 1909*b*), a specialist in bees, have been shown to belong for the most part to quite different and extinct genera. Of the species of Bombinæ described from the Baltic amber, *Bombus pusillus* Menge, is only three millimetres long and cannot therefore be a Bombus; *B. carbonarius* Menge and *Bombusoides mengei* Motschulsky are, according to Cockerell, quite unrecognizable. Of the five species of Bombus described from the Upper Oligocene and Miocene, Cockerell (1909*a*) has shown that two, *B. jurinei* and *abavus*, belong to the genus Xylocopa, and one cannot help entertaining some doubts about the generic identification of the others, *antiquus* Heyden, *crassipes* Novak and *grandævus* Heer. Cockerell (1909*b*) finds, however, three distinct genera allied to Bombus in the Baltic amber, namely Protobombus with two, Chalcobombus with two and Sophrobombus with a single species. Protobombus is described as being more closely allied to Bombus than to Apis and possibly in the direct line of descent to the former genus. The other genera represent side lines not leading to any of our modern bees. We may conclude, perhaps, that the Bombinæ were a larger and more diversified group during the Tertiary than they are at the present time, since, if we omit the parasitic Psythirus, there is now only a single genus.

Species of Melipona and Trigona have been cited from the Baltic amber by Brischke and Burmeister (Handlirsch 1908), but Cockerell regards the generic references as quite worthless. Nevertheless, two Meliponine bees belonging to an extinct genus, *Meliponorytes succini* and *sicula*, have been adequately described and figured by Tosi (1896) from the more recent Sicilian amber. They are of unusual interest as showing that the sub-family, now confined to the Neotropical, Ethiopian and Indo-australian regions, was once represented in Europe. Authentic Meliponinæ may therefore occur in the Baltic amber, which contains so many insect genera that are at present confined to Australia and the East Indies.

The two fossil species of Apis adduced by von Buttel-Reepen (1906) in support of his view that the honey-bee must have originated in Germany, have met a sad fate. *Apis adamitica* Heer, from the Miocene of Oeningen, proves to be a Lithurgus, i.e., a gastrilegous bee allied to Megachile, according to Cockerell (1909*a*), and von Buttel-Reepen's *Apis meliponoides* from the Baltic amber has been shown by the same investigator to be, not an Apis, but the type of a new genus, Electrapis. He admits that it is fairly intermediate between Apis and Bombus and that it "may be regarded as representing the ancestral stock of Apis and to some extent that of Bombus, but it is no doubt actually a side branch derived from that stock and not strictly in the line of descent leading to the modern genera". A second species, *tornquisti*, doubtfully referred to Electrapis by Cockerell, is even closer to Bombus. The *Apis proava* of Menge, also from the Baltic amber, cannot be recognized from the description and probably belongs to some other genus. Thus we are left with practically no paleontological data on which to erect a "Stammbaum" like that of von Buttel-Reepen for the Bombinæ, Meliponinæ, and Apinæ.

In my last lecture I failed, perhaps, sufficiently to emphasize the fact that there is a very striking progress in nidification among the Diplopterous wasps, starting with a construction of the nest in or of clay or earth, through the utilization of leaf-particles for the same purpose (Zethinæ) to the nidification of the five social sub-families which fabricate a veritable *papier mâché* consisting of fine vegetable fibres, agglutinated with an oral secretion . This evolution is paralleled by that of the bees. Many of the lower solitary forms nest in the soil or make earthen cells, while higher forms use a great variety of vegetable substances, sometimes even a leaf-paste, as in the European *Osmia inermis*, till finally in the three higher social sub-families the material for the brood and storage cells is produced exclusively in the form of a unique secretion, known as bees' wax. There is, therefore, in both groups a similar significant evolution from more primitive terricolous or humicolous to more aerial (dendrobious, or arboricolous) habits.

While there is no evidence that any of the solitary bees secrete wax, von Buttel-Reepen (1903) and Friese

(1923) have called attention to the fact that in freshly killed and drying female specimens of several genera (Tetralonia, Eucera, Anthophora, Pachymelus and Xylocopa) there exudes from between the abdominal segments on the dorsal side a substance which looks like wax though its chemical analysis (in Tetralonia) reveals nothing but the presence of fat. Wax, however, being also one of the lipins, or lipoids, is of a similar composition. Friese doubts whether these masses of fatty exudate can be utilized by the solitary bees, since they possess no apparatus on their hind legs for removing them from between the segments, but the matter has been given very little attention and we are not in a position to say that more definite physiological stages leading up to the wax-secretion of the higher social bees may not exist in some of the solitary species. Heselhaus (1922) has shown that the glands underlying the large velvety areas along the internal orbits in Andrena and the glands of certain abdominal segments in Osmia are structurally very similar to the true wax-glands of the honey-bees, Bombinæ and Meliponinæ as described by Dreyling (1905). The structures in Andrena and Osmia certainly do not produce a typical wax. In Andrena the secretion may perhaps be a volatile essential oil and the facial glands may therefore be scent-organs, though, as Heselhaus suggests, their proximity to the olfactory organs of the antennæ is not very favourable to such an interpretation. The regions which secrete the wax in the Bombinæ, Meliponinæ and Apinæ are known to have a different distribution on the abdomen and in the individuals of the various castes. In Bombus, wax is secreted both dorsally and ventrally by both queens and workers on the second to fourth abdominal segments, in the Meliponinæ by queens, workers and males on the second to fourth segments, but only on the dorsal side of the abdomen, and in the Apinæ exclusively by the workers and only on the ventral side of the second to fifth segments. In all cases the secretion is produced by young individuals only. The Bombinæ would seem, therefore, to represent the most primitive or generalized stage, i.e., the one which may be conceived to have given rise to the condition seen in the Meliponinæ by a loss of the glands on the ventral side of the abdomen and in the Apinæ by their loss on the dorsal side. There are also some interesting differences in the method of employing the wax in the three subfamilies, the Apinæ using it in its pure form, the Bombinæ mixed with a considerable amount of pollen and fatty substances (see Armbruster's (1914) citation of Sundwick's analyses) and the Meliponinæ with a large admixture of earth, resins (propolis) and other vegetable matter.

Since I propose to go into a lengthy discussion of the polymorphism, or caste development of the social insects in two subsequent lectures, I shall here merely touch on it in the bees. In some Halicti and in Bombus the female has developed two phases, but they exhibit only feeble structural differences like those of the social wasps, the worker being smaller than the queen, incapable of fecundation and therefore either sterile or capable of laying only male-producing ova. The differentiation is carried somewhat farther in the Meliponinæ and Apinæ and, owing to differences in habits, exhibits a significant reversal of the conditions seen in Halictus and Bombus, since the queen, though retaining and exaggerating the fecundity of the queens of these forms, shows a degenerative reduction of the pollen-collecting apparatus of the hind legs and of portions of the head (smaller size of head in Meliponinæ, reduced mouth parts in Apis).

In conclusion something may be said about the various methods of colony formation among the Bombinæ, Meliponinæ and Apinæ and the matter of monogyny ("monogamy", or "monometrosis") and polygyny ("polygamy", or "pleometrosis") touched upon in the preceding lecture. In north temperate regions the Bombus colony is an annual development initiated by a single fecundated, overwintering queen in precisely the same manner as the colonies of Vespa, but according to R. von Ihering (1903), the colonies of *B. carbonarius* and *cayennensis* in Brazil are perennial and contain several fecundated queens which establish new colonies by swarming, like many of the tropical social wasps investigated by Ducke, the von Iherings and Roubaud. According to Silvestri (1902) and H. von Ihering (1903) the Meliponine colonies are also established by swarming. There is only one mother queen in a colony and since she is unable to fly on account of the volume of her abdomen and the relatively small size and often mutilated state of her wings, the swarms of workers leave the nest with the young queens as in the case of the tropical wasps and humble bees. In the honey-bee, as is well known, there is also only a single mother queen, but she preserves her powers of flight and leaves with a swarm of workers as soon as a young queen is about to emerge in the hive. To anticipate, I may say that in the Formicidæ the great majority of species, even in the tropics, undoubtedly establish their colonies like Vespa and the Bombi of temperate regions, though colonies may also arise by swarming, i.e., by fecundated females leaving large colonies in company with workers and occasionally also by association of a few young sister queens ("primary pleometrosis").

Reserving for future consideration the aberrant cases of certain social parasites, we may therefore distinguish four types of colony formation in the social Aculeata, as follows :—

(1) The Vespa type, seen also in the northern species of Bombus and Polistes, in Halictus and most ants. The incipient colony is epitomized in a single fecundated queen.

(2) The Belonogaster type, occasionally seen also in Halictus and some ants (Lasius, Myrmecocystus (Wheeler 1917)), etc. The beginning colony consists of several fecundated, probably sister queens.

(3) The Melipona type, occurring also in many tropical Vespidæ (Polistes, Polybia, etc.)

and Bombus. The colony at first consists of a single fecundated queen and a number of workers which are her sisters.

(4) The *Apis mellifica* type, not known to occur in any other insects. The new colony consists of an old fecundated queen and a number of her worker daughters.

The fourth type may, perhaps, be a variant of the third, but it has not been satisfactorily explained, though careful investigation of the other species of Apis, *dorsata* (Fig. 22) and *florea*, of the Indomalayan region may be expected to clear up some of the difficulties. The Vespa type is monogynous whereas the Belonogaster and Melipona types are polygynous, with the difference that in Belonogaster the individuals founding the new colony are all fertile, whereas in Melipona there is a single fertile and a number of sterile individuals.

The von Iherings, Ducke and Roubaud regard the polygynous as the more primitive and the monogynous as a derived type, which has arisen in the northern wasps and bees as a response to unfavourable climatic conditions. These will not account, however, for the almost universal monogyny in ants. On the other hand, we might say that the colonies of Vespa and northern Bombi each year produce a swarm of females and workers, but that the advent of cold weather destroys the less resistant workers and permits only the dispersed fecundated queens to survive and hibernate till the following season. It seems to me that the question as to which method is the more primitive may be purely academic and that the method may depend on the mutual tolerance or animosity of the queens. This seems to be shown in the primitive societies of Halictus (Stöckhert 1923). A number of fecundated sister queens hibernate peacefully in the same maternal burrow but with the arrival of spring conflicts break out among them and the result is usually a monogynous colony. In some cases, however, two or three of the sisters may compose their differences and establish a polygynous colony. We should probably observe a similar indeterminate type of colony formation in Vespa and our northern Bombi, if the fecundated queens were in the habit of hibernating in the maternal nest, or among the ants if the queens, instead of dispersing during the marriage flight, congregated in convenient cavities. Occasionally several queens of Lasius, Formica or Myrmecocystus (Wheeler, 1917*b*) may actually do this, and von Buttel-Reepen (1905) has witnessed the beginning of a digynic colony of *Lasius niger*. For twelve months a couple of queens of this ant lived very amicably side by side and cared for their brood in common, but as soon as the first workers emerged, they violently attacked each other. On being separated, one of them, already severely wounded by her sister, was killed by the workers, so that the colony became monogynous. The instincts of the two queens thus underwent a sudden change with the maturing of the brood. In Halictus a similar change supervenes, but at a much earlier stage of colony formation, and in Belonogaster it seems not to occur.[1]

[1] Janet's observations (1904) show that in mature polygynous colonies of ants (Solenopsis and Tetramorium) kept in artificial nests, the queens are gradually killed off by the workers till only a single one remains. Since this elimination does not occur in wild colonies, Janet was probably right in suspecting that it might have been due to lack of food in the artificial nests.

V

THE EVOLUTION OF ANTS

THE third and last and from many points of view the most interesting group of social Aculeates comprises the family Formicidæ, or ants. It is essentially a tropical group and till recently our knowledge of it was so largely confined to the rather small number of species in the north temperate zone, that no comprehensive description was possible. During the past half century, however, Forel, Emery, Ern. André, Santschi, Viehmeyer, Donisthorpe, Crawley, Mann, myself, and others, building upon the foundations laid by Latreille, Frederick Smith and Gustav Mayr, have been endeavouring to complete the taxonomic edifice of the family and to gain such knowledge of the behaviour of the species as was incidentally possible. Though there still remains, of course, a great deal to be done, the data accumulated have, nevertheless, attained imposing dimensions. They are now being used in the production of compendious works, of which several are already accessible to the student. During the past twenty years general, semipopular volumes on ants have been published by Forel in French, by Escherich (1917) and Brun (1924) in German, by Emery in Italian (1915*a*) and by Donisthorpe (1915) and myself (1910*a*, 1923) in English. Forel's "Fourmis de la Suisse" (new edit. 1920) and especially his "Monde Social des Fourmis" (1921-23) cover the field so well that I can find no excuse for devoting this lecture to a survey of the structure and behaviour of ants. I shall take for granted that you have some acquaintance with these insects and shall treat, so far as my time will permit, only a few general matters such as the phylogeny of the family, its known paleontological record and its geographical distribution. In later lectures I shall enter into somewhat greater detail in regard to some other equally controversial matters.

The fact that the family Formicidæ, unlike the Vespidæ and Apidæ, consists entirely of social insects, is a serious obstacle to a determination of its phylogeny. Up to the present time 6,000 different forms, i.e., species, subspecies and varieties have been described from all parts of the world, and it is probable that our younger myrmecographers will be able to raise the total to 10,000 long before the end of the present century. Since there can be scarcely half as many forms of termites, the ants are far and away the largest group of social insects. According to Forel and Emery, the family Formicidæ comprises five subfamilies, the Ponerinæ, Dorylinæ, Myrmicinæ, Dolichoderinæ and Formicinæ (Camponotinæ). The Ponerinæ are a primitive but rather heterogeneous group (Fig. 24). The Dorylinæ comprise the famous driver ants of Africa (Fig. 39) and the army ants, or "ants of visitation" of the American tropics. The Myrmicinæ are a large and very heterogeneous assemblage of genera, which are, nevertheless, easily recognized by the two small petiolar segments of the abdomen in all the castes (Figs. 26, 48, and 72). The Dolichoderinæ (Fig. 79) and Formicinæ (Fig. 51) are both characterized by a single small petiolar segment but differ in the shape of the gizzard, anal aperture and the character of the poison glands. I have recently recognized two other subfamilies, the Cerapachyinæ (Fig. 25), formerly included among the Ponerinæ, and the Pseudomyrminæ, formerly placed among the Myrmicinæ. It seems to me that an eighth subfamily should be created for the very aberrant and specialized tribe Leptanillini, which Emery has placed among the Dorylinæ.[1] In the future the Myrmicinæ will probably be resolved into several subfamilies. The series of subfamilies which may be recognized at present, with their most general distribution are the following :—

[1] Dr. G. C. Wheeler (1928) has shown that the larva of Leptanilla is very different from that of any other ant-genus.

Dorylinæ—tropical and sub-tropical.
Cerapachyinæ—tropical and sub-tropical.
Ponerinæ—very largely tropical.
Leptanillinæ—tropical and sub-tropical.
Pseudomyrminæ—tropical.
Myrmicinæ—cosmopolitan.
Dolichoderinæ—largely tropical.
Formicinæ—cosmopolitan.

It will be seen that nearly all the North American and European ants belong to only two of the subfamilies, the Myrmicinæ and Formicinæ.

There has been considerable discussion concerning the relationship of the various subfamilies to one another, and Handlirsch (1924) has recently pronounced this discussion to be futile, since each of the five recognized by Forel and Emery contains both primitive and highly specialized forms. But a general appreciation of each subfamily, based on a careful estimate of all the genera, shows that the average differentiation of the larger groups is very unequal and that some of the subfamilies as wholes are really very primitive (Dorylinæ, Cerapachyinæ, Ponerinæ) whereas others are highly specialized (Dolichoderinæ, Formicinæ). There are, moreover, annectant groups, like the Cerapachyinæ, which link the Ponerinæ with the Dorylinæ; Rhopalomastix, which, according to Forel, seems to connect the Ponerinæ and Myrmicinæ; and Aneuretus, which seems to attach the Dolichoderinæ to the Ponerinæ. Furthermore, there are certain general morphological and ethological tendencies such as the progressive reduction of the sting in the more specialized subfamilies (some Myrmicinæ, nearly all Dolichoderinæ and all Formicinæ), the transition from terricolous to arboricolous habits, which I shall consider later, and the evolution from a pronounced entomophagous to an aphidicolous and vegetarian (spermatophagous, mycetophagous) diet, an evolution strangely analogous to that of man, as first conceived by Condorcet, from the hunting and pastoral to the agricultural stage, or Auguste Comte's three

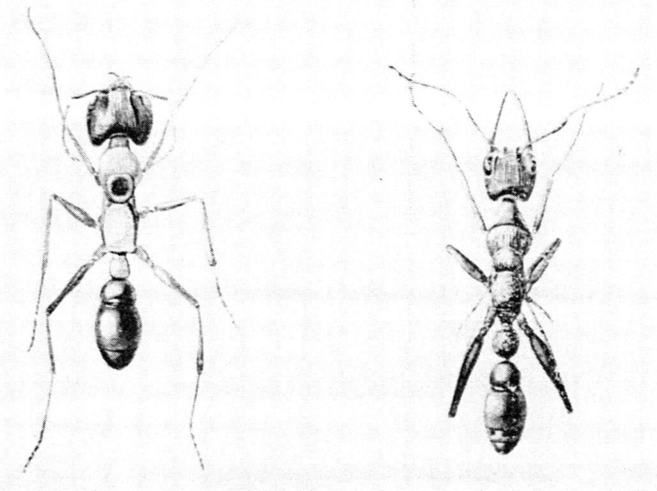

FIG. 24. Two bulldog-ants from Australia. a. *Promyrmecia aberrans* Forel, the most primitive of all known Formicidæ. b. *Pristomyrmecia mandibularis aureorufa* Forel, one of the "jumpers," or leaping ants. (After A. Forel.)

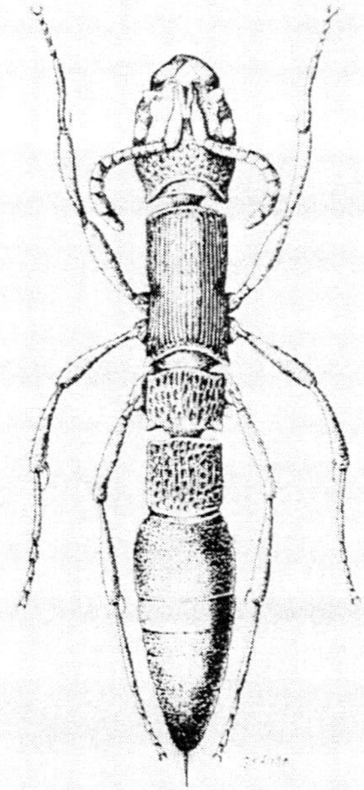

FIG. 25. Worker of a Cerapachyine ant, *Phyracaces langi* Wheeler, from the Congo.

[*face p.* 106

FIG. 26. Two singular Myrmicine ants. a. *Podomyrma ruficeps abdominalis* var. *pulchra* Forel, from Queensland. b. *Planimyrma loriai* Emery, from New Guinea. (After A. Forel.)

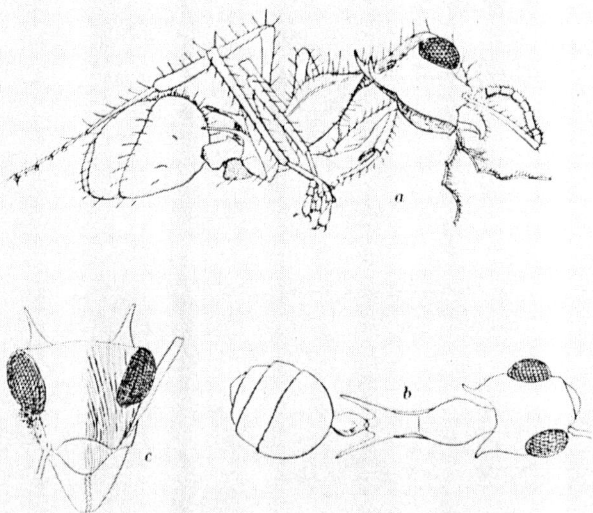

FIG. 27. *Sicilomyrmex corniger* Emery, a Formicine ant from the Sicilian Amber. a. Lateral. b. Dorsal view. c. Frontal view of head. (After C. Emery.) (*See* p. 114.)

stages of conquest, defence and industry in human history. I believe, therefore, that the discussion of the interrelationships of the Formicid subfamilies is not altogether futile. Handlirsch, who has constructed so many " Stammbäume ", may be pardoned for disliking those of other hymenopterists, but he is hardly consistent in regarding their construction as a useless occupation. They are after all only graphic expressions of conceivable phylogenetic relationships and have neither more nor less value than other schemes of epitomizing our provisional notions in obscure fields of investigation. If, therefore, contrary to the eminent Austrian entomologist's opinion, we venture to construct an ancestral tree of the Formicid sub-families, it would look something like Fig. 23.

The Ponerinæ represent the primitive stock from which the other subfamilies are conceived to have radiated, and the amount of specialization to which they have severally attained is indicated roughly by the level of their names.

There is one question which has not been explicitly discussed by myrmecologists though it is suggested by our account of the Vespidæ and Apidæ and that is: Are the various subfamilies of ants really derived from a common stock or are they polyphyletic like the various social subfamilies of wasps and bees? In the preceding diagram they are treated as monophyletic. This is the view taken by Forel in 1910 and by myself in 1920. Emery, however, has indicated a polyphyletic origin of the subfamilies in the following schema published in 1920:

Prodorylinæ (Cerapachyinæ)——Dorylinæ.

UNKNOWN HYPOTHETICAL FORMICID STOCK. — Proponerinæ {
——Euponerinæ
Amblyoponini, Ectatommini, Proceratini
— { Myrmeciini, Platythyreini } { continuing the stock with the minimum amount of differentiation.
——Myrmicinæ
}

?——Dolichoderinæ

?——Formicinæ (Camponotinæ)

Here at least four independent, parallel lines of development are suggested as arising from the hypothetical ancestral stock of the Formicidæ, which lies even further back than the Proponerinæ, the putative ancestors of the Ponerinæ. If we take Emery's point of view, therefore, we have much the same situation as in the social Vespidæ, for the hypothetical ancestors must have been very primitive and in all probability solitary or subsocial Aculeates. It is clear that if all solitary Vespids and Apids had become extinct and only the social forms had survived, we should very probably regard them as monophyletic families. On the other hand, the various Formicid subfamilies are all very closely related and, as already stated, seem even to be united by annectant groups. I deem it more probable, therefore, that the family as a whole is really monophyletic.

Our next task is to determine among the solitary Aculeates the precise locus of origin of the ants so far as this is possible with the aid of morphology, paleography and ethology. The impression conveyed by the ants as a whole is that of a sharply circumscribed and very ancient or archaic family. The archaic aspect is revealed in the many genera and species which combine obviously very old with highly specialized characters, in the pronounced sexual dimorphism, since the differences, apart from the wings, between the males and females is greater than in any other Aculeates, and in the extremely specialized worker caste, which is not only always wingless but always differs much more from the fertile female, or queen in the structure of the head, thorax and abdomen than in the social Vespidæ and Apidæ. All these differences are indications of a very long evolution, apparently much longer than that of the social wasps or bees. Obviously, the ancestors of the Formicidæ are not to be sought among the Sphecidæ, or Fossores, but on account of the thoracic structure, wing-venation, absence of strigils on the hind legs, etc., among the Vespoids, or Vespiformia. The Diploptera (Vespidæ) and Pompilidæ may also be eliminated as possible ancestors on morphological grounds. We are reduced, therefore, to the series of families (see p. 56) constituting the most ancient section of the Aculeate sub-order, the Sapygidæ, Scoliidæ, Tiphiidæ, Thynnidæ, Methocidæ, Plumariidæ, Mutillidæ, Apterogynidæ, Fedschenkiidæ, and Anthoboscidæ. The behaviour pattern of these insects, so far as known, is very uniform and of a more primitive type than that of the Pompilids, solitary Vespids and Sphecids, since they behave precisely like parasitoids and merely seek out and paralyse their insect prey and oviposit on it *in situ*, or in rare instances (Sapygidæ) act as parasites of bees and Sphecids.

Writers have not been slow in taking advantage of the opportunitiy here afforded for differences of opinion in regard to the family most closely related to the Formicidæ. Most of them, following Latreille, have settled on the Mutillidæ, but others have pointed to ant-like traits in nearly every one of the other families of Heterogyna (Börner's superfamily " Formicina ").[1]

According to Börner (1919), an ideal ant might be constructed by combining the hind wings of a male Mutilla with the maxillæ of a Myrmosid, the tibial spurs of a Methocid and the cranium of a Sapygid. He actually regards the Plumariidæ as a subfamily of Formicidæ and Plumarius as " perhaps the most ancient of ant-forms ". Now while his suggestion that the cranium of ants is more primitive than those of Scoliids, Thynnids and Mutillids is interesting, it is evident that the Sapygids and very probably the Plumariids are in many respects less like the ants than any of the other families of Heterogyna. The Sapygidæ are known to be parasitic

[1] Both Börner (1919) and Handlirsch do a good deal of lumping in this group. The latter (1924) recognizes only four families, the Sapygidæ, Rhopalosomatidæ (which I have placed in the Bethyloidea), Mutillidæ and Formicidæ, and treats the other groups above mentioned as subfamilies. His Mutillidæ, therefore, include as sub-families the Mutillinæ, Anthoboscinæ, Scoliinæ, Fedschenkiinæ, Pterombinæ, Thynninæ, Methocinæ, Myrmosinæ, Apterogyninæ, and (?) Bradobæninæ. The Plumariidæ, he believes, constitute an independent super-family, which he places at the beginning of the Aculeate series.

on solitary bees. The habits of *Sapyga quinquepunctata* as described by Fabre (1890), and those of *S. similis* as described by Nielsen (1902), show that they behave like Terebrants and use their long ovipositor-like sting for piercing the wall of the bee's cell and laying the egg near or on the egg of the host. The young Sapyga larva after devouring the latter proceeds to devour the supply of bee-bread. On account of this vegetarianism, Nielsen would place the Sapygidæ among the Apidæ, where they were once placed by Curtis (near the parasitic bees of the genus Nomada) " or rather in an independent allied family ".[1]

The singular Plumariids are known only from the male sex. There are only two genera, Plumarius (Konowiella) in Argentina and Chile and Myrmecopterina (Archihymen) in South Africa (see Enderlein (1914) and Brues (1924)), a distribution which indicates that the family is very old. The wing-venation is much more primitive than that of ants and the antennæ and abdomen are quite different. The female is probably wingless.[2]

[1] If Rohwer's statement in the " Hymenoptera of Connecticut " (1916, p. 620) is correct that one of our North American Sapygas has been bred from the cells of a mud wasp (*Sceliphron cementarium*), the genus would contain at least one entomophagous species.

[2] Bradley (1921) while lecturing to the " Société Scientifique du Chili ", and after speaking of the probable absence of all Myrmosidæ in South America, made the following remarks on Plumarius: " But you have here in their stead, and with similar habits, the most interesting of all Chilean Hymenoptera. Very many years ago your learned fellow-member, the late Dr. R. A. Philippi, described a remarkable hymenopterous insect to which he gave the name *Plumarius*. Because of its peculiar structure all writers down to the present day have been puzzled to know to what family to assign it, and I doubt if any outside of Chile have known more of it than Philippi's figure and description. Even within the past few years Szepligeti has included it in his synopsis of the Braconidæ for the Genera Insectorum. A few years ago a genus was described for an Argentine wasp, under the name of *Konowiella*. Before leaving the United States I became convinced of the fact that *Konowiella*, of which Prof. C. T. Brues had shown me some specimens, was synonymous with *Plumarius*, and that their true position could only be in the family Mutillidæ, Myrmosidæ, or a new closely allied family to be erected for them. I further concluded that they must be nocturnal inhabitants of arid regions, taking in the South American desert fauna a position analogous to that occupied in North America by the Myrmosid genus *Brachycystis* referred to above. On talking to Argentinian entomologists, I could not obtain confirmation of my suspicion that they are

Very little is known concerning the Fedschenkiidæ and Anthoboscidæ, in which the females as well as the males possess wings. Only two species of Fedschenkia are known, both from Turkestan (Ern. André, 1903). The only species of Anthoboscidæ I have seen, our Northern American *Sierolomorpha ambigua*, while very primitive, seems to be too simple to be regarded as near the hypothetical Formicid ancestor. The Mutillidæ, Apterogynidæ, Thynnidæ, Myrmosidæ and Methocidæ all have apterous females and cannot on that account be regarded as ancestral groups. In this I concur with Handlirsch and dissent from Forel and Emery. In his important paper on the Dorylinæ (1896), Emery rejected the Thynnids which Forel had included among the possible ant ancestors but says: "I, therefore, agree in the main with Forel, but would more forcibly emphasize the statement, *that the ancestors of the ants were closely related to the older forms of the Mutillidæ and even belonged morphologically to that family*". The relation to Thynnids he regarded as "merely a collateral one based on their common derivation from Mutillids". Emery was evidently led to take this view on account of the generalized male genitalia of the Mutillids, the apterous condition of the female Doryline ants and the peculiarly ant-like abdominal pedicel of Apterogyna. But the wing-venation of this genus, which Ern. André and Emery include among the Mutillidæ, is peculiarly specialized and very different from that of the Formicidæ. Although Emery was well aware of the obstacle presented by the wingless condition of female Mutillids, he boldly maintained that the females of ants were originally wingless but had later acquired perfectly developed

nocturnal. Accordingly imagine my delight on burning a trap lantern two successive nights at Potrerillos in the Eastern Andes, to obtain nearly one hundred of these insects belonging to several species. Subsequently when Mr. Carlos Silva Figueroa showed me the collection of the Museo Nacional, one of the first things to attract my attention was a specimen of these insects, labelled "*Plumarius*", in the handwriting of Dr. Philippi, proving conclusively the identity of *Konowiella* with that genus. It is my early intention to publish a study of these insects", etc.

wings by inheritance from the males! McClendon and I (1903) criticized this extraordinary opinion many years ago, but in 1920 the Italian myrmecologist states that he is of the same opinion still, though in the next paragraph he shows that he now holds a very different opinion, namely that the ancestral ants really had two kinds of fertile females, one winged and one apterous, an opinion which I am glad to share (*vide infra*).

We are thus reduced to the only two remaining families, the Scoliidæ and Tiphiidæ, which are commonly combined as a single family. They were regarded by Handlirsch in 1893 and 1908 as most closely related to the ants, and in 1924 he says: "I regard it as certain that the ancestral ants had winged males and females and a well-developed sting. At any rate, there was only one segment developed as a petiole. This ancestral group of the ants can no more be derived from the Mutillids than from the Thynninæ, Methocinæ or Apterogyninæ. We must therefore go back to still more primitive forms. The Plumariidæ, however, seem not to represent the point of origin, because their antennæ are much too highly specialized".

It seems to me that the group which deserves the most serious attention in connection with the ancestry of ants is the Tiphiidæ and especially the genus Elis (Myzine) since it resembles the ants in the shape of the eyes, the wing-venation and the generalized tendency to constrictions between all the abdominal segments, so reminiscent of certain Cerapachyinæ (Sphinctomyrmex, Zasphinctus, Euphinctus and Doiylinæ (Leptanilloides).[1]

[1] Apparently none of the taxonomists has noticed that the ants are peculiar in possessing a pair of large glands in the posterior portion of the thorax (see Janet, 1898*b*). They appear to be present in all the species and all the castes and open near the insertions of the hind coxæ. In some ants, e.g., the East Indian species of Crematogaster of the subgenus Physocrema, the chambers and tissues of these glands are so large that the whole metathorax is conspicuously swollen. I have been unable to detect any distinct traces of the structures in question in other Aculeates or in the lower sub-orders of Hymenoptera. Before their taxonomic significance can be appreciated it will be necessary to make a comprehensive, comparative investigation, with surface preparations and sections of the metasternal region in the various Hymenopterous families.

It is interesting to note that this very family Tiphiidæ has affinities in structures and habits with the Bethylidæ, considered in a former lecture. Probably all three convergent groups, the Bethylidæ, Tiphiidæ and Formicidæ, had a common origin among extinct forms, which may be conceived to have resembled the existing Bethylids in habits. They probably hunted insect prey, paralysed and carried it into their burrows and laid their eggs on it, the mother insect remaining with her young till they had matured. From such a sub-social stage we might pass rather easily to the social stage now existing among all ants by assuming that the emerging daughters became attached to their mother and brought their prey into the maternal burrow. Eventually a physiological division of labour gradually developed among the associated individuals through the colony-founding female learning to feed her first brood with her saliva or eggs instead of with prey while her daughters became sterile nurses of her subsequent broods. We may assume also that these ancient Bethyloid or Tiphioid ant-ancestors had both winged and wingless females like Scleroderma (see p. 62). This assumption would furnish the necessary starting point for the three forms of females, fertile and winged, fertile and apterous, and sterile and apterous (workers) which we find among existing ants and thus avoid Emery's inherently improbable derivation of alate from apterous females through inheritance of wings from the males. The present aptery of the females of Mutillidæ, Thynnidæ, etc., might also be traced to the female dimorphism of remote Bethyloid ancestors, but we should have to assume that in them no advance occurred in behaviour pattern even to a sub-social stage like that of Scleroderma. Perhaps the only result of this discussion, which, I fear, has been pushed to the point of ennui, is that the Formicidæ bear much the same phylogenetic relations to the lowermost families of Aculeata, and especially to the Tiphiidæ and Bethylidæ, that the social Vespids bear to the Eumeninæ and the bees to the family Sphecidæ.

Our perspective of the remote origin of the Formicidæ as suggested by their morphology and that of other families of Heterogyna is deepened when we come to consult the fossil record. Here we are on somewhat safer ground than in the paleographic study of the other Aculeate families, because ants are the most abundant insects in the Tertiary formations. We can trace them as far back as the Eocene, but from this period few specimens are known. The number from the Oligocene and Miocene, however, is considerable, and although the described species of Formicidæ are less numerous than the total number of species of the other Hymenopterous families the specimens are so much more abundant that our knowledge of the individual forms is decidedly more satisfactory. Thus up to the present time 11,711 ants, representing only ninety-three species, have been examined from the Baltic amber (Fig. 28). Of these 1461 were studied by Mayr (1868), 690 by Ernest André (1895), and 9,560 by myself (1914). Much uninvestigated material from other formations has been accumulated in museums. Many years ago several American museums loaned me for study some 8,000 ants from the Miocene shales of Florissant, Colorado, but incessant pressure of other work has prevented me from making even a satisfactory generic determination of the specimens. One of my students, Mr. F. M. Carpenter, is now monographing them with my assistance. Nearly eighty years ago Heer (1849) described a number of similarly preserved fossil ants, also of Miocene age, from the shales of Oeningen, in Baden and Radoboj, in Croatia, but his generic allocations are so dubious as to be useless. Fortunately a number of his specimens were later re-examined and reinterpreted by the celebrated myrmecologist, Gustav Mayr (1867). A small but extremely interesting series of species was described in 1891 by Emery from the Sicilian amber of Middle Miocene age (Fig. 27). More recently Cockerell (1915, 1920*a*) and Donisthorpe (1920) have described several species from the Eocene and Middle Oligocene of England.

Plate XIX.

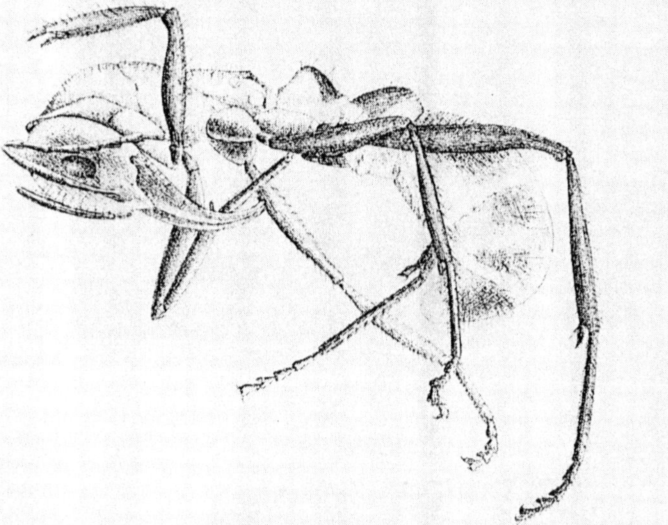

FIG. 28. *Prionomyrmex longiceps* Mayr, a very primitive Ponerine ant allied to the Australian bull-dog ants (Myrmecia), from the Baltic Amber.

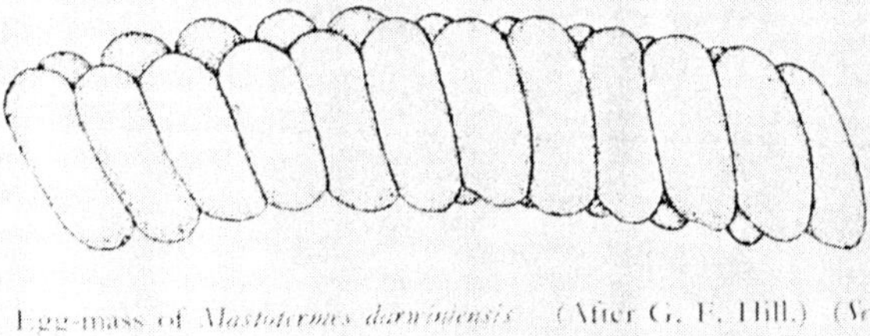

FIG. 29. Egg-mass of *Mastotermes darwiniensis*. (After G. F. Hill.) (*See* p. 138.)

[*face p.* 114

Plate XX.

FIG. 30. Female of European earwig (*Forficula auricularia*) in winter quarters, with eggs. (After B. B. Fulton.) (*See* p. 149.)

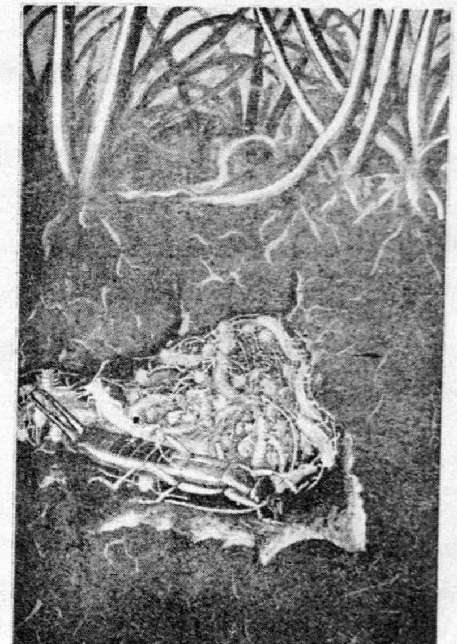

FIG. 31. Section of sod showing female of European earwig (*Forficula auricularia*) in underground nest with young. (After B. B. Fulton.) (*See* p. 149.)

In his monumental work on fossil insects, Handlirsch (1908) listed nearly 300 species of Tertiary ants, but I fail to find among them and the species since described more than 169 with adequate generic determinations, and this is probably a too liberal estimate.

Confining ourselves first to the European Tertiary we find that only two species are known from the Eocene (Bagshot Beds of Bournemouth). They were described by Cockerell (1920) as *Formica heteroptera* and *Œcophylla bartoniana*, both from anterior wings.[1] Although the specimens are undoubtedly typical ant wings, they are inadequate for generic identification. All we can say in regard to them is that true ants, probably belonging to the most highly specialized subfamily Formicinæ, lived during Eocene times in what is now England. For the sake of the utmost brevity I have condensed the available data on the Eocene, Oligocene and Miocene ants of Europe in the table on page 116.

You will notice that the 169 species belong to fifty-four genera, twenty-seven or half of which are extinct' whereas the remaining half are almost equally divided between extant genera now represented by species in Europe and other continents and genera which are at present confined to the Indoaustralian region. To the latter also many of the extinct genera are related. The table shows these facts most clearly for the Lower Oligocene (Baltic amber), which is the oldest formation from which many ants have been obtained. This is largely due, no doubt, to the great number and very beautiful preservation of the specimens. The highest number of Indoaustralian genera is shown in the Sicilian amber (Middle Miocene), in which of the sixteen species representing fourteen genera, eight are Indoaustralian, only three extinct and only three now represented in Europe and other regions. Unfortunately

[1] Cockerell compares the *F. heteroptera* wing with that of *Colobopsis stricta* Jerdon of India, for what reason I cannot imagine, since Colobopsis, like all the other sub-genera of Camponotus, lacks the discoidal cell which is present in the fossil specimen.

ants and other insects are not nearly so abundant in the Sicilian as in the Baltic amber.

Apart from the Miocene of Florissant, to which I have already alluded, the North American Tertiary formations have yielded very few ants. Scudder (1890)

GEOLOGICAL PERIODS.	Number of Species.	Number of Genera.	Number of Extinct Genera.	Number of Palearctic or Cosmopolitan Genera.	Number of Indoaustralian Genera.
Upper Miocene ..	6	3	—	2	1
Middle Miocene ..	16	14	3	3	8
Lower Miocene ..	28	12	2	8	2
Upper Oligocene ..	5	5	1	3	1
Middle Oligocene ..	19	8	3	4	2
Lower Oligocene ..	93	43	19	15	9
Eocene	2	2	—	1	1
Total	169	54	27	15	12

referred four species from the Oligocene of Quesnel, British Columbia to the well-known modern genera Aphaenogaster, Myrmica, Dolichoderus and Formica, and Cockerell (1906*a*) has described a *Ponera hendersoni* from Florissant, but the perusal of the descriptions and study of the figures leaves me very doubtful in regard to the generic identification of all these insects. Six species have been recorded from the Eocene and are of peculiar interest because, with the exception of the two from Bournemouth, England, they are the oldest known ants. Three of them were described by Scudder, namely *Camponotus vetus* and *Liometopum pinque* from the White River Eocene of Colorado and *Lasius terreus* from the Green River Eocene of Wyoming. I regard the generic identifications as very doubtful. The other three species, recently described by Cockerell (1921*a*, 1923*a*, 1923*b*), are *Eoformica eocenica* and *Archimyrmex rostratus* from the Green River formation and *Formica eoptera* from the Upper Eocene (Jackson) of Texas. Eoformica and Archimyrmex are described as extinct genera. The former, which Cockerell believes may possibly be identical with Scudder's *Liometopum pinque* is certainly not related to Formica or Liometopum, Cockerell's restoration of the head as seen from the front cannot be harmonized, at least so far as the eyes and ocelli are concerned, with the profile head on his photograph of the fossil. All that can be said from examination of the figure is that the specimen is a male, possibly of some Formicine. Cockerell believes it to be related to Œcophylla but this is very questionable. He also believes that his *Archimyrmex rostratus* is a Ponerine near Prionomyrmex (Fig. 28) of the Baltic amber and the living Myrmecia of Australia, but to judge from the figure the specimen is more probably a Myrmicine. This is suggested by the shape of the petiole, the blunt or broken (?) spine on the epinotum and the shape of the head, which is unlike that of the existing Ponerinæ. *Eoformica eocenica* is described from a single beautifully preserved anterior wing, which is clearly that of an ant and probably of a Formicine, but a generic identification is impossible.

Although the precise generic position of the few known ants from the Eocene, both European and North American, is so uncertain, they are in no respect more primitive or more generalized than the ants of the amber and these are quite as highly specialized as our living forms.

All eight of the living subfamilies, with the exception of the Dorylinæ and Leptanillinæ are represented in the amber fauna, and so far it has been quite unnecessary to modify their definitions in order to accommodate any of the Tertiary species. Not only were all the castes as sharply differentiated among the amber as among modern ants, but the workers of some genera (Pheidologeton, Pseudolasius, Dimorphomyrmex) may have already become pleo- or polymorphic. The larvæ and pupæ, too, which are occasionally preserved, are precisely like those of living species. And as I wrote in 1914, " there are also unmistakable indications that the habits and instincts of the amber ants were nearly if not quite as advanced as those of existing forms. That many of them had learned to attend plant-lice and had therefore become " trophobiotic " is shown by a block of amber in the Königsberg Collection containing a number of workers of *Iridomyrmex goepperti* together with a lot of their aphid wards. That the amber ants kept myrmecophiles in their nests can scarcely be doubted, since at least three genera of Paussidæ (Cerapterus, Pleuropterus and an undescribed genus) are cited by Klebs (1910) in his list of amber Coleoptera (see also Wasmann, 1919). That these ants also had Acarine parasites is shown by two workers of *Lasius schiefferdeckeri* in the Königsberg Collection, each bearing a mite attached to the base of one of the hind tibiæ (Wheeler, 1914). These specimens also show that the mites had already acquired the peculiar habit of affixing themselves to very definite regions of their host's integument ". The Baltic amber is, indeed, so much like the living fauna that no myrmecologist would be surprised to find any of the extinct genera actually living to-day in the forests of the East Indies or Australia, or to find any of the quaint and archaic genera of these regions in the amber. This has, in fact, happened in two instances. Gesomyrmex was described by Mayr as an extinct genus in 1868, but years later Ernest André (1892) described a living species (*G. chaperi*) from

Borneo, and I have recorded another from Southern China (1921*b*). Ernest André also first described the genus Dimorphomyrmex from Borneo and in 1905 Emery found a species of it in the amber. I have since detected a second species in the same formation (1914) and another living in the Philippines (1916).[1]

Before discussing certain points in connection with the present geographical distribution of ants, attention should be called to some of their general peculiarities which no one seems to have emphasized sufficiently. We have seen that the Formicidæ are one of a series of closely related families, the Heterogyna, many of which are so exquisitely terrestrial in their habits that their females have completely lost their wings. These families as a

[1] It would hardly be necessary to point out that no conclusions based on the absence of certain genera in the amber or other geological formations can be of any use in speculations on the geographical distribution of particular genera at the present time, were it not that zoögeographers are continually overlooking so obvious a truism. Extremely precious as are all *positive* paleontological data, assertions in regard to the *absence* of particular forms in various parts of the world during geological time are utterly worthless. Nowhere is the truth of this statement more clearly illustrated than among fossil insects and especially among the fossil ants. The very nature of the formations in which the fossils occur must have precluded the preservation of large and significant portions of the fauna. Thus the amber has entrapped only small, weak insects and among ants mostly arboreal species or such as occasionally foraged on trees. Hence we find comparatively few fossil Myrmicinæ, Cerapachyinæ and Ponerinæ which are mostly terricolous, and very numerous Dolichoderinæ and Formicinæ, which either live in trees or are fond of visiting aphids, coccids, etc., on their foliage. The complete absence of the two subfamilies Dorylinæ and Leptanillinæ from the amber may be explained in the same way, since most of the modern species of the former and all the modern species of the latter are subterranean, or hypogæic. Of course, it is also conceivable that these subfamilies may not have been represented in the amber forests. The selective action of the environment is shown even better in the case of the Miocene fossils of Florissant, Oeningen and Radoboj. Nearly all of them, unlike the amber inclusions, are males and winged or recently deälated females, and the geological evidence shows that the shales in which they occur were deposited in shallow lakes into which the male and female ants dropped while descending from their marriage flights and into which the workers very rarely happened to be swept by freshets. Failure to see this very obvious explanation led an eminent American geologist, Joseph Leconte (1884), to infer that the social habits, instincts and worker caste had been developed in the Formicidæ since the Miocene epoch, although even a hasty glance at the older Oligocene ants of the Baltic amber would have convinced him of the contrary (Fig. 28).

whole are very dependent on soil (edaphic) and climatic conditions and nearly all of them may be described as decidedly thermophilous. That this is not true of the Hymenoptera in general is shown in the following table from Handlirsch (1913*b*), in which the numbers of species of several of the larger families of the three sub-orders in the temperate, subtropical and tropical zones are recorded :—

	Families.	Temperate.	Subtropical.	Tropical.
Phyto-phaga.	Pamphiliidæ, Cephidæ ..	179	119	1
	Tenthredinidæ	1,964	752	817
Terebrantia.	Ichneumonidæ, Braconidæ, Chalcididæ, etc.	17,011	2,953	6,644
	Cynipidæ	1,181	333	182
	Evaniidæ (includ. Aulacidæ)	168	147	502
	Stephanidæ, Pelecinidæ ..	10	8	106
Aculeata.	Sapygidæ	34	14	5
	Scoliidæ	148	274	592
	Mutillidæ	380	493	1,516
	Formicidæ	458	597	2,888
	Sphecidæ	1,489	1,461	2,243
	Vespidæ	468	591	1,763

It will be observed that the Phytophaga abound in the temperate zone but decrease in numbers towards the tropics, that some families of Terebrantia show the same peculiarity while an increase towards the tropics is shown in others and that the Aculeata, with the single exception of the Sapygidæ, have a similarly pronounced thermophily most conspicuous in the Formicidæ and the allied Scoliidæ and Mutillidæ, which are evidently here taken *sensu latiore*. The Vespidæ show nearly as great a thermophily as the other Vespoids. Had the bees been included in the table we should observe a maximum development in the subtropical zones. We may conclude, therefore, that the Heterogyna as a whole are a strongly thermophilous group of Aculeates.

It might be interesting to construct another table to show the degrees of xerophily in the Aculeata, but it would be more difficult. The great majority of species of many of the families of Heterogyna, notably of the Mutillidæ, Apterogynidæ, Thynnidæ, Plumariidæ, Methocidæ, Myrmosidæ and Fedschenkiidæ, undoubtedly prefer a dry environment. All of these groups therefore abound in and some of them are confined to deserts. The xerophily of the Formicidæ is less obvious because it is overlaid by certain other tendencies. While it is true that the species of ants are very numerous in the humid tropics of both hemispheres, it is also true that individual ants are almost as abundant in deserts, such regions as the sandy pine-barrens of New Jersey and other places which support a more or less xerophytic vegetation. They are, moreover, poorly represented by species and individuals in cool, damp regions like New Zealand, Southern Chile and Patagonia, Northern Europe and the Selkirk Mountains of British America. The fact that the great majority of ants nest in the ground shows that the family was originally terricolous, and this habitat makes it necessary for them to avoid all soils which are not well-drained. Hence their scarcity in wet meadows and swamps and their predilection for hill and mountain slopes, especially those facing east and south in the northern, and north and east in the southern hemisphere. In the rainy portions of the tropics, where we find several large arboricolous genera, they have obviously taken to nesting in or on the trees to avoid the excessive and deleterious dampness of the soil. But even in these regions, as I have observed in Central America and elsewhere in the humid tropics, the terricolous ants greatly prefer slopes or the soil under stones or logs or decomposing wood which absorbs little rainwater.

I have dwelt on these facts because they seem to me to point definitely to a deep-seated xerophily in the ants and to their very probable origin, together with the Scoliids, Mutillids, Thynnids, etc., on the dry interior continental plains and plateaux during the Mesozoic, in the very regions, in fact, which some botanists believe to have been the places of origin of the Angiosperms. It may be noted in this connection that even the amber forests, in which ants must have been extremely abundant, were probably rather xerophytic. The amber itself was produced by a species of pine, and the quantities of stellate hairs scattered through the pieces of fossil resin show that oaks were abundant. In fact, the flora of the Samland must have been somewhat like that of the New Jersey pine-barrens, but with a greater admixture of tropical elements.[1] That the region was mountainous was ingeniously shown by Ulmer (1912) in his monograph on the caddice-flies (Trichoptera) of the Baltic amber. He found that of the thirty recent genera of these insects represented in the fossil resin, the larvæ of 13 per cent. must have lived in strongly agitated water, 8 per cent. in standing water and 4 per cent. in slowly flowing streams, and he concludes that fully thirty-five genera with seventy-three species known only from the amber, passed their larval life in torrents, that seven genera with fourteen species lived in quiet water and that seven genera with seven species were probably indifferent to their aqueous habitat. Of existing continental areas Australia seems to me to present conditions most

[1] For an excellent general account of the amber fauna and flora, see Berry (1927).

like those among which I conceive the Mesozoic ants to have evolved. This continent has a more or less xerophytic vegetation, except in the extreme north where there is a considerable admixture of Papuan faunal and floral immigrants. The great development of the Thynnidæ in Australia is also significant in this connection.

We may conclude, therefore, that the ants are primitively a terricolous and thermophilous family of insects which arose in and still prefer a fairly dry environment. From this condition we can trace three lines of adaptive radiation and specialization. A number of genera in all the sub-families, except the Pseudomyrminæ and Dolichoderinæ, have abandoned life on the surface of the soil and have become subterranean, or hypogæic. Others have increased their native tolerance of drought and have become deserticolous, and it is interesting to note that this category comprises several primitive genera, especially among the Myrmicinæ (Messor, Novomessor, Veromessor, Pogonmoyrmex). A number of genera and subgenera, finally, in regions of great rainfall have become arboricolous. It is noteworthy that this last category is almost exclusively represented by the most specialized genera of the four highest subfamilies, the Myrmicinæ, Pseudomyrminæ, Dolichoderinæ, and Formicinæ, and has no representatives among the Dorylinæ and Leptanillinæ and very few (e.g., Neoponera) among the Ponerinæ. Furthermore, even such exquisitely arboricolous genera as Pseudomyrma and Crematogaster contain some terricolous forms. *Pseudomyrma elegans*, e.g., as was shown by Forel, always nests in the ground, and there are many Crematogasters that nest in rather dry soil in North America, Australia and elsewhere

If my conclusions are correct, Emery (1920) must be mistaken in his interpretation of the sequence of habitats in the Formicine section Eucamponotini, a group embracing the genus Camponotus, with more than 1,000 described species, sub-species and varieties and several allied genera. His views are graphically represented in the following phylogenetic diagram :—

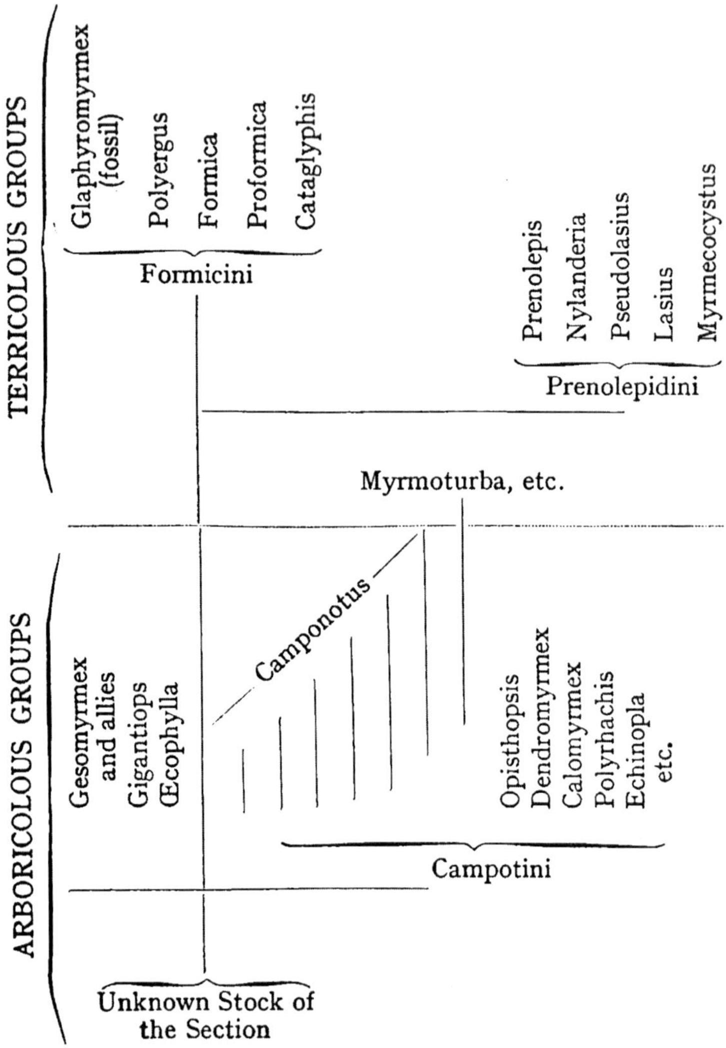

All the arboricolous groups cited by Emery in this diagram occur in Australia, except Gesomyrmex, which is Indomalayan, and Gigantiops, which is neotropical Now Gigantiops, as I have shown (1922*b*), is not arboreal.

In British Guiana I found it nesting in old logs on the ground and Dr. W. M. Mann informs me that he has taken it under stones in Bolivia. Gesomyrmex, too, is probably terricolous. The types of *G. howardi* were sent to me mingled with a lot of well-known soil-inhabiting ants. According to my observations at least some of the species of Opisthopsis and Calomyrmex nest in the earth. Of the various subgenera of Polyrhachis all the Australian species of Campomyrma, Hagiomyrma and Chariomyrma are terricolous and moreover, like Calomyrmex, distinctly xerophilous. In Africa certain members of the subgenus Myrma are terricolous and the same is probably true of some of its East Indian species. Nothing is known of the habits of Echinopla, but it is probably arboricolous. This genus, Œcophylla, Dendromyrmex, and some subgenera of Camponotus and Polyrhachis therefore alone belong wholly below the dotted line in Emery's diagram. And when we confine our attention to Camponotus we find that the terricolous subgenera, Tanæmyrmex (=Myrmoturba), Dinomyrmex, Myrmosericus, etc., are obviously much more primitive than the sub-genera Myrmorhachis, Colobopsis and its allies, etc., which have the head and mandibles of the queen and soldier more or less specialized for living in wood. Here again it will be noticed that Tanæmyrmex, which embraces a considerable number of species, subspecies, and varieties, is highly xerophilous, since most of its forms are confined to somewhat arid or desert regions. In North America Tanæmyrmex occurs only in the drier portions of the South-Western and Western States. The evolution of the Eucamponotini, therefore, in my opinion, has been just the reverse of the one outlined by Emery. I regard as equally fallacious his contention that the ancestral ants which gave rise to Œcophylla, Camponotus and Polyrhachis had the habit of employing their larvæ to spin leaves and detritus together in making the nest. This habit, now so well-known in Œcophylla, occurs only in two among several hundred Neotropical species of Camponotus and in certain subgenera of the paleotropical genus Polyrhachis (Cyrtomyrma, Myrmhopla), and there is nothing to indicate that it may not have been acquired independently in all three genera long after they had become arboreal.

The existence as early as the Lower Oligocene of a large and remarkably diverse series of genera, many of which, representing nearly all the modern subfamilies of Formicidæ, are still extant and even comprise species like *Formica flori* and *Lasius schieffedeckeri*, practically identical with the common *F. fusca* and *L. niger* of Eurasia and North America, indicates that the ant fauna of to-day, even in the tropics, may have declined in the number of genera and species since the beginning of the Tertiary. Certainly there is no evidence of any progressive evolution throughout the Cænozoic Era. We must therefore assume that both the origin and the early and significant development of the family occurred before the Tertiary. But this development owing to the absence of any fossil record, is a matter of mere conjecture. In regard to the period of origin of the Formicidæ, Handlirsch (1908) and Emery (1920) differ. According to the former, " we must adhere to the opinion that the focus of evolution (' Entwicklungsherd ') of the whole family Formicidæ (in the broad sense) is to be sought on the early Tertiary or Upper Cretaceous landmasses of Eurasia and that the forms which there originated reached North America over the Eastern and Western land connections of the northern hemisphere and migrated southward from North America as well as from Europe and Asia ". Emery conceives a more remote Mesozoic origin of the family, believing that the Ponerinæ go back at least as far as the Lower Cretaceous, the Dorylinæ to precretaceous times, and the family perhaps to the end of the Jurassic. In my opinion even this estimate may be too moderate. From what has been said of the thermoxerophily of the Formicidæ it is more probable that they arose in the Triassic, if not in the late Permian, which, according to paleometeorologists (Schuchert, 1914, Huntington (1914), and others), were periods of high temperature and great aridity. The terrestrial habits and the very intimate and plastic relations of the ants to their larvæ and pupæ, as contrasted with the more recently evolved social wasps and bees, seem to point to an origin under rather arid conditions such as are supposed to have existed over the great continental masses during the Triassic and early Jurassic. The arboreal adaptations of certain genera might then be conceived to have arisen during the Cretaceous, a period of continuing high temperature but of greater humidity like that of the tropical rain-forests of to-day.

If we postulate so remote a geological past for the Formicidæ, they would have had plenty of time to acquire not only great generic and specific diversity, but also a cosmopolitan distribution before the beginning of the Tertiary. Many problems suggested by the present geographical distribution of the genera would then assume a different aspect. These problems have given rise to much controversy and cannot here be considered in detail. Emery, apparently influenced by H. von Ihering and Scharff, has made considerable use of hypothetical transatlantic and antarctic intercontinental landbridges to account for the genera now represented only in South America, South Africa, Australia and New Zealand, whereas Handlirsch (1913*a*) and I (1917) are inclined to adopt the older view of a circumpolar and probably Eurasian origin of the Formicidæ and their migration over the well-established Behring Sea land connection and southward in both hemispheres as a satisfactory explanation of the present distribution of such "relict" genera. The latter contention is supported by Handlirsch's elaborate statistics of the present distribution of 16,100 genera and 180,000 species of insects, which show that the evidence for the existence, at least during the Tertiary, of the various landbridges postulated by von Ihering, Scharff, Emery and others, is really extremely meagre. Handlirsch shows, moreover, that in a map of the world including only the more important of the Tertiary landbridges assumed by the analytical school of zoögeographers, the oceans are reduced to a few small lagoons. This would necessitate an entirely different present and past distribution of marine organisms from that actually obtaining, and the tropical faunas of all the continents and islands would be homogeneous and therefore far from exhibiting the enormous divergencies that now prevail.

The study of our Nearctic fauna naturally predisposes American naturalists to defend the hypothesis of a boreal origin of organisms and their southward migration to other parts of the world without the aid of other landbridges than those between Alaska and Eastern Asia and between Eastern North America and Europe. At the present time the most emphatic champion of this view and opponent of the bridge-builders is Dr. W. D. Matthew (1915), a well-known authority on fossil and living Mammalia. But even in Australia, where there has been much advocacy of antarctic landbridges, the tide of opinion seems to be changing, if we may judge from a recent valuable paper by Longman (1923) who finds little support for the contentions of Hedley, Haswell, and others.[1]

In last resort, of course, the geologists of the future will decided whether the landbridges of the earlier geologists and the zoögeographers of the von Ihering school have really existed. If we may judge from such recent very interesting work as that of Kober (1921) on the structure of the lithosphere, there is no geological evidence for the existence either of the bridges constructed

[1] Kolbe (1913) and I have been taken to task by Emery for regarding the insect fauna (including the ants) of Australia as essentially Mesozoic. While it is well known that this fauna comprises also a certain number of Papuan and East Indian elements, there is no proof that they did not enter Australia before Tertiary times. Schuchert (1916) says: "We must, therefore, conclude that Australia has been an island continent at least since late Eocene times, for it is since then that the placental mammals have elsewhere dominated all other land life". And Longmans gives good reasons for assuming that the Marsupials have had a long and peculiarly local pretertiary evolution in Australia.

by von Ihering for the Tertiary or of similar intercontinental landmasses during the Mesozoic. Kober concludes "that without any doubt the six large continental masses, namely North America, Eurasia, South America, Africa, Indoaustralia and Antarctica have existed permanently since the Mesozoic". And the same conclusion may be drawn from his condensed description (p. 283-284) of the continents during the Mesozoic. He does, however, assume the existence of two "hypothetical" land-masses in the Pacific, one with its present centre in Hawaii and one in the Southern Hemisphere, but both are small and very far from constituting land-bridges. Such connections become quite unnecessary on the very different theory of the displacement of the continents during geological time as advocated by F. B. Taylor (1910), Wegener (1924), and Daly (1926). Wegener conceives the continents to have formed a single landmass as late as the Upper Carboniferous, and to have begun to move away from one another by the Eocene. Were it possible to accept this theory, the difficulties we encounter in explaining the present geographical distribution of many organisms, such as the Formicidæ, would vanish.

VI

THE EVOLUTION OF TERMITES

ONE of the most striking peculiarities of the various groups of social wasps, social bees and ants surveyed in the preceding lectures is the constant recurrence in all of them of the same general behaviour pattern or *motif*, with modifications depending on the character of the food and other environmental exigencies. This recurrence, no doubt, is partly due to morphological and ethological similarities among the immediate solitary ancestors of these insects and partly to the more remote common ancestry of all of them. As we have seen, all the social Aculeates are derived from solitary Sphecoids and Vespoids and I have attempted to trace both groups back to hypothetical extinct Bethyloids, the putative primitive stock of the whole Aculeate suborder. The noticeable fact that the social bees, though of Sphecoid ancestry, are nevertheless in many respects more like the social Vespidæ than are the Vespoid Formicidæ, is probably due to the longer and more eventful evolution of the latter.

What may be called the solitary inheritance or endowment of the social Aculeates quite obviously underlies the characters, both structural and behaviouristic, that have been more recently acquired as the result of social life. The fundamental solitary traits most deserving of emphasis are the following :—

(1) The pronounced sexual dimorphism, which, as among the Terebrantia, is even more conspicuous in the behaviour than in the structure of the Aculeates. The complicated instinct patterns are exhibited only by the

female and the male is reduced to a mere fecundating agency. Owing to the perfection of the female spermatheca, which enables her to store and keep alive the sperm of the male for a considerable period of time, she becomes, as it were, a bisexual or hermaphroditic being in complete functional control of reproduction. She alone, therefore, is subjected to the full incidence of environmental stimuli and in her reactions to them is able to advance along the path of progressive specialization or adaptation as far as the plasticity of her constitution will permit. The male, apart from rather slight specific modifications, is condemned to stability of structure and monotony of behaviour. He is, therefore, the conservative sex, the *sexus sequior*.

(2) The storing for the offspring of insect or vegetable food by the solitary Vespoids and Sphecoids. This may be regarded as a prospective or potential social activity since it implies the conditions for social development in a potential, or implicit sense and since the society may become actual, or explicit as soon as the spatial requisite, i.e., the coexistence of the parent with the adult offspring, supervenes.

(3) The nest, ranging from a simple burrow in the soil or pre-existing cavity to cells carefully constructed of earth or vegetable materials.

(4) The development of defence (pugnacious) reactions on the part of the mother insect. These have risen very naturally from the predatory activities of the primitive Bethyloid ancestors and the transformation of the ovipositor into a sting. Even some Terebrants (Ichneumonids) sting when handled.

The solitary Aculeates also use their mandibles for defence, these organs having gradually become more efficient as the result of their employment in seizing and carrying the prey, excavating the burrow, etc. At first they are used in conjunction with, but eventually supersede the sting as the main organs of defence, as Roubaud (1918) has shown.

The more significant modifications and additions that supervene in social behaviour as a result of the actual contact of the mother Aculeate with her offspring, of their association with her and with one another and the consequent physiological division of labour, are the following :

(1) The differentiation of the female into two castes, a fertile (queen) and a sterile (worker) form or phase. The queen therefore departs still further from the stationary, conservative male in increased longevity and stature and in structural modification. Her fecundity is also greatly increased beyond that of the solitary female, while the worker, especially in the ants, tends to depart further from the queen and to become dimorphic in turn as soon as its primitive combined nutritive and protective functions become dissociated (workers proper and soldiers).

(2) The consociation of individuals to form the colony leads to the development of various methods of communication of stimuli and distribution of food or secretions (trophallaxis.)

(3) The collecting, storing and utilization of the food become communistic.

(4) The nest, now the combined and co-ordinated work of the community, becomes much larger, more elaborate and more permanent.

(5) The protective and defensive functions, too, now assumed by the whole of the worker caste or of a peculiarly specialized portion of it (soldier caste), become much more efficient and more formidable.

It should be noticed that with the advent of social life, behaviour necessarily becomes not only more intricate but also accumulative. This is an important matter. Auguste Comte long ago and De Grange (1923) more recently and more clearly have shown that a society is best defined as an accumulating collectivity. Of course, these sociologists, and others who have held similar views, were thinking primarily of human society which is manifestly and essentially characterized by an extraordinary accumulation of objects, thoughts and acts and all that they imply—mores, institutions, languages, inventions, capital, art, science, etc. In his book entitled "The Manhood of Humanity", Count Korzybsky (1921) has also emphasized the accumulative aspects of human society under the term "time binding". He regards it as an exclusively human activity, but as De Grange shows in several passages, it is really a peculiarity of all true societies, whether animal or human. The social Aculeates reveal the accumulations very clearly, though they are, of course, very primitive and rudimental compared with those of man and very largely confined to the sphere of the so-called instincts and the results of their operation. But this must have been true also of the earliest stages of human society. On the material plane savage communities are still far from attaining to a stage as economically accumulative as that of the hive-bee or the higher ants and termites.

During their transition from the solitary to the social stage, there was one peculiarity which the Aculeates seem to have been unable to modify, and that was the status of the male. In all their societies this sex takes no part in the accumulative or social behaviour as such but is left in essentially the same stage as among the solitary ancestors. I say "essentially", because among certain social bees and wasps, he does seem to be slightly socialized, at least to the extent of possessing a feeling of belonging to the hive or nest and of returning to it after foraging or flying about in search of virgin females. But even this spark of sociability is lacking in the males of most Formicidæ although they may be treated with solicitude by the worker personnel. Thus when the nest of *Lobopelta elongata* is disturbed, the males are seized and carried away by the workers. At such times the males assume a rigidly immobile attitude, with their appendages and wings folded in the pupal posture, and are carried under the bodies of the workers in precisely the same manner as the larvæ and cocoons. But, as a rule, in all social Aculeates the males, though often produced in considerable numbers, are merely tolerated for a time as so many parasites on the colony. Being necessary for fecundation, however, they cannot be regarded as good-for-nothings.

We are not justified, of course, in making the anthropomorphic statement that the social Aculeates committed a blunder in not completely socializing the male as well as the female. Even from our human point of view it does not seem probable that such a social employment of the male would have increased the efficiency of the societies of bees, wasps and ants. This is suggested by the termites, which actually accomplished what the Aculeates could not or at any rate did not accomplish, since the various morphological specializations, and activities (division of labour) in termite colonies have been taken over equally by the individuals of both sexes. To this extent they seem to us to be more perfectly social insects than the social Aculeates, and yet I doubt whether any one of us, if given the choice in some future metempsychosis of becoming either an ant or a termite, would choose the latter.

It must be admitted that on superficial examination the termities exhibit striking resemblances to the social Aculeates and especially to the ants. Like the Formicidæ, they are all social and there is a very pronounced polymorphism among the individuals of the colony (Fig. 35), some being wingless and behaving like the workers and soldiers of ants. There is typically a single mother queen and although she is accompanied by a king, both lose their wings before establishing the colony and after a dispersion flight which resembles the nuptial flight of the ants. Even the foraging, seed-storing and fungus-growing habits of certain ants are paralleled by the higher termites (Fig. 36). But though very interesting and suggestive these and several other less conspicuous resemblances are merely convergent, or parallel developments. When we come to study the termites more closely we find a number of fundamental idiosyncrasies, one of the most significant of which is the above-mentioned bisexual composition of the colony as a whole.

The termites constitute an independent order, the Isoptera of Brullé. Owing to the fact that they are nearly all tropical insects and do not make attractive cabinet specimens, they were much neglected by entomologists till the beginning of the present century. Since that time, however, they have attracted a number of careful investigators so that we are now in possession of a mass of literature on their anatomy, behaviour and taxonomy. More general works are also beginning to appear, such as the treatises of Holmgren (1909-1913), Escherich (1909, 1911), Snyder and Banks (1920) and Hegh (1922), which contain such an amount of information that I refrain from attempting to compress a synopsis of even the most important data into this lecture. Hegh's very recent comprehensive and profusely illustrated volume, "Les Termites," Part I., furnishes just the reading I should desire for the student who is interested in this course. As in previous lectures, I shall confine myself to a discussion of certain evolutionary problems.

Down to the present time somewhat more than 1200 species of Termites have been described mostly from the warmest portions of the tropics and more than one-third of them from the Ethiopian Region, i.e., from

Africa South of the Sahara. In North America we have some thirty-four species, only one or two of which extend to the Northern boundary of the United States and into British America. Europe seems to have only two species, which are confined to Central and Southern France and the Mediterranean littoral.[1] At present the delimitation of the main groups and their names are somewhat unsettled. Holmgren recognized three families, which he called the Protermitidæ, Mesotermitidæ and Metatermitidæ.[2] They embrace respectively the most primitive, moderately specialized and most highly specialized forms, but, unfortunately, his family names are invalid and have had to be abandoned because there are no generic names corresponding to them.[3]

His Protermitidæ are now separated into two families, the Mastotermitidæ and Calotermitidæ and Banks and Snyder have united his Meso- and Metatermitidæ in one family, the Termitidæ. If the separate rank of Holmgren's two families is to be maintained, the Mesotermitidæ would take the name "Rhinotermitidæ", while the term "Termitidæ" would be reserved for the Metatermitidæ. To these Sjöstedt (1925) has recently added the Hodotermitidæ, formerly included among the Calotermitidæ. The Mastotermitidæ and Calotermitidæ, being the most primitive termites, will require special consideration in connection with the phylogeny of the Isoptera as a whole. In the preceding lecture I failed to state that there is among the Formicidæ an

[1] Dr. Alfred Emerson has kindly ascertained for me the total number of termites described to date from the various regions of the globe. There are 123 genera and sub-genera comprising 1237 species. Excluding those with overlapping distribution and unknown habitat, the species are distributed as follows: Ethiopian 452, Malagasy 38, Indomalayan 272, Papuan 30, Australian 111, Palearctic 22, Nearctic 34, Neotropical 282.

[2] This was Holmgren's classification as published in 1910 in the "Zoologischer Anzeiger", where the Mastotermes group is treated as a sub-family of Protermitidæ. In his monograph (1911) he accepted Silvestri's view (1909) of the family rank of the Mastotermitidæ.

[3] There is a genus Protermes, but it belongs to Holmgren's Metatermitidæ.

obvious and significant increase in the size and polymorphism of the colony and the complexity of behaviour as we pass from the primitive genera, especially among the Ponerinæ, to the more specialized genera of this and the other sub-families. The same is in general true of the termites, the colonies of the Mastotermitidæ and Calotermitidæ being usually smaller and less definitely polymorphic than those of Termes and Eutermes, the highest genera of the Termitidæ.

The Mastotermitidæ are represented by a single living species, *Mastotermes darwiniensis*, described by Froggatt (1896, 1905) from Northern Australia. The species was based on the sexual forms which are three-quarters of an inch long, but Froggatt also described the soldier (1897) under the name *Termes errabundus*. The colonies are very populous and live in timber, without making a true nest. They are often found in the woodwork of houses and in such situations, like other termites, do considerable damage. Desneux (1904) and later Silvestri (1909) and Holmgren (1911) studied Mastotermes very carefully and found it to be a veritable synthetic type of the greatest interest. Although it is certainly a true termite it has nevertheless so many Blattoid characters that there can be no doubt about its phylogenetic descent. Holmgren showed that it must be derived from the Protoblattoidea, a group which, so far as we can learn from the fossil record, became extinct in the Permian and gave rise to two branches, one terminating in the Blattoidea, or cockroaches, the other in the Isoptera. The male and female Mastotermes have a well-developed anal lobe to the hind-wing, a character which has been lost in all other termites, the wing-venation is like that of Protoblattoids, the tarsi are distinctly five-jointed, the tibiæ are spined, there is no frontal gland and the female possesses three pairs of distinct gonapophyses, in the form of an ovipositor (Crampton 1920). According to Silvestri, there is a true soldier but no true worker caste. He describes a worker-like form but regards it as a nymphal stage.

Although Holmgren and Hill (1925) regard the latter form as a true worker, the conditions in the Calotermitidæ would seem to support Silvestri's interpretation. Holmgren and Jucci (1924) have shown that the hypodermis of Mastotermes has the same peculiar structure as in Blattoids, and Jucci has found that the fat-body has bacteriocytes containing symbiotic bacteria like those described by Blochmann, myself and others in the roaches. Cleveland (1923*b*) and Jucci have also shown that the hind gut of Mastotermes contains infusoria like those occurring in all the other lower groups of termites. Hill (1925) has recently described the eggs of *M. darwiniensis*. Unlike those of all other termites they are "extruded in masses, each comprising sixteen to twenty-four eggs, cemented together laterally in two parallel rows" (Fig. 29), a condition strangely reminiscent of the arrangement of the eggs in the oöthecæ of roaches.

The Calotermitidæ, which are somewhat more specialized than the Mastotermitidæ comprise a number of genera, several of which form small colonies in rather sound wood and make no true nests. Miss Thompson (1919) has shown that their eggs are considerably larger than those of the higher termites Termitidæ). She gives the following measurements for the ova of four of the genera: Termopsis 1.3-1.7 mm.; Calotermes 1.2-1.4 m.m.; Cryptotermes 1.2-1.3 mm.; Neotermes 1.6 mm. I shall consider very briefly only the three most primitive genera: Archotermopsis, Termopsis and Calotermes.

Archotermopsis wroughtoni, which is the sole living representative of its genus and inhabits the North-Western Himalayas at altitudes varying from 4,000 to 9,000 feet, has been conscientiously studied by Imms (1919). The fully developed colonies comprise besides the royal couple, winged males and females, soldiers, worker-like forms and various nymphal stages, but no true workers. Eyes are present in all the forms, though small in the soldier. The eggs are few in number and of large size

(.95-.98 mm. long, .45-.5 mm. broad). "The soldiers rank among the largest known, and are chiefly remarkable in that they exhibit well-defined external secondary characters, a feature which has been lost in almost all other Termites. The reproductive system exhibits no indications of degeneration, or arrest of development and is in a condition similar to that found in the winged sexual forms, prior to swarming". The tarsi of Archotermopsis are five-jointed as in Mastotermes, Termopsis and Hodotermes but in no other termites and the styles and cerci are very long, the latter being six to seven jointed. Imms calls attention to a number of Blattoid traits in *A. wroughtoni*, e.g., the structure of the digestive and reproductive organs of both sexes, the presence of cæca in the midgut, not recorded for any other termites, eight Malpighian vessels (the number in young roaches), the structure of the salivary glands, salivary receptacles and the hypopharynx, the resemblance of the vesiculæ seminales of the male to the mushroom gland of roaches, the spines on the tibiæ, etc. Cutler (1921) has called attention to the fact that the intestinal Protozoa of Archotermopsis are of a more primitive type than those inhabiting the intestines of higher termites.[1]

The genus Termopsis is represented by three species which are confined to the Pacific and North Western States of North America. Two of the species, *T. angusticollis* and *nevadensis* (Fig. 54), have been carefully

[1] "In all the protozoa examined from the gut of *Archotermopsis wroughtoni* the nuclear division is very different from that found in related species. Further, in *Ditrichomonas termitis* the nuclear division and the locomotor complex is of a more primitive nature than that described for other Trichomonads: a statement probably true for *Ioenopsis polytricha* and *Pseudotrichonympha pristina.* It appears that the protozoa to which *Archotermopsis wroughtoni* is host are in general more primitive than those inhabiting other species of termites. Imms describes *A. wr.* as 'one of the most primitive of living termites'. The association, therefore, of primitive parasites or 'guests', whichever the case may be, with a primitive host is extremely interesting, and is suggestive that the two groups of organisms have remained associated together for a long period, neither having developed into more complex species, as has occurred with other termites and their associated protozoa".

studied by Heath (1902, 1907, 1927) who found that their colonies may sometimes grow to a considerable size.[1] There is no true worker caste and the soldiers may at least occasionally become fertile and lay eggs which give rise to viable offspring, though their later development has not been followed. They often possess small wing-vestiges like those of the fertile *Archotermopsis*, figured by Imms and the soldiers of *Calotermes occidentis* (*vide infra*). Miss Thompson (1922), who also studied Termopsis, believed, on what now seems to be insufficient evidence, that the soldiers are always infertile.

Calotermes has been investigated by Grassi and Sandias (1893), Heath (1902), Jucci (1924) and others. The genus, of which more than fifty species have been described, is cosmopolitan and is represented even in New Zealand and the Galapagos Islands. The colonies are usually small and comprise the same castes as in Termopsis and Archotermopsis. That the soldiers may be fertile, at least occasionally, is indicated by Grassi's observations and by Snyder's statement (1924) that in the most primitive and largest species, *Calotermes occidentis* of Central America, the soldiers uniformly have vestiges of wings. It will be noticed, therefore, that the development of the soldier caste has preceded the incidence of sterility. This was clearly seen by Imms in Archotermopsis and the fertile Termopsis

[1] Heath's former estimates of colony growth in Termopsis were too great. In his latest paper (1927) he says: " Royal pairs have been found of such size and in burrows so spacious (that) at least one year's residence is indicated, and yet they were without offspring. On the other hand, other colonies, taken in the same general period (the fall of the year) and in favourable situations, comprised as many as twenty inhabitants. By the close of the second year, this number is doubled in the case of *T. nevadensis*, averages upward of one hundred the third season, and increases to more than 400 by the end of the fourth year. There are also six records of colonies ranging in number of individuals from 786 to 1,120; these may be in their fifth year. Four other colonies vary in number from 2,213 to 2,881, and may be in their sixth year. A few other communities, close to the 3,500 mark in number of inhabitants, have also been examined when the extent of the burrows strongly suggests an occupancy of an even longer period than six years, although definite proof is lacking. All of these communities were headed by the original king and queen ".

soldiers found by Heath (Fig. 54*b*) and the normally sub-apterous soldiers of *Calotermes occidentis* point to the same conclusion. Probably a study of fresh material of *Mastotermes darwiniensis* will show that also in this most primitive of termites the soldiers are always or at least occasionally fertile. In the African Hodotermes and the Japanese Hodotermopsis, which were formerly also included among Calotermitids, a worker as well as a soldier caste has been developed and we therefore have the typical number of castes of the Termitidæ, or higher termites. In the latter, however, the soldiers and workers are always sterile and in Termes even complemental royal forms seem not to be developed.

When we turn to the paleography of the termites we find that they are recorded from nearly all the Tertiary formations in which ants have been preserved but that they are still unknown from the Mesozoic. Handlirsch (1908) cites quite a number of fossil species. More recently von Rosen (1913*a*) and Snyder (1925) have published revisions of the known forms with very illuminating results. Nearly all the species from the Eocene, Oligocene and Miocene of Europe and North America prove to belong to the primitive genera we have been considering, namely Mastotermes, Archotermopsis, Termopsis and Calotermes and to closely allied extinct genera. Thus the only termite recorded from the European Eocene is *Mastotermes bournemouthensis*, from Bournemouth, England. Two other species, *anglicus* and *batheri*, are described from the Middle Oligocene of the Isle of Wight and one, *M. croaticus*, is recorded from the Miocene of Radoboj. Von Rosen also interprets Scudder's *Eutermes fossarum*, from the Miocene of Colorado, as being in all probability a Mastotermes. Some time ago, Prof. E. W. Berry showed me a drawing of a beautifully preserved Mastotermes wing which he had found in the Lower Eocene of Tennessee, and which has since been described by Collins (1925) as *M. wheeleri*. All this is very interesting because it shows that Mastotermes, now represented by a single species confined to Northern Australia, was during early and middle Tertiary times a flourishing, cosmopolitan genus. To a closely allied genus, Miotermes, von Rosen assigns four species from the European Miocene (Radoboj, Oeningen and Württemberg) and one from Florissant, Colorado (*M. coloradensis* Scudder). In the Baltic amber there is a species of Archotermopsis (*A. tornquisti*), a genus which, as we have seen, is now restricted to the Himalayas, and three extinct genera allied to Archotermopsis and Termopsis, namely Xestotermopsis, Proëlectrotermes and Electrotermes. Furthermore, the four species of Parotermes from Florissant, three described by Scudder (1883, 1890) and one by Cockerell (1913), are obviously related to Hodotermes, now confined to Africa, Central and Southern Asia, and also belonging to the Calotermitidæ. Cockerell (1916) has described a *Termopsis swinhoi* from the Burmese amber (Miocene). This genus is now restricted to Western North America. Von Rosen has recognized a true Calotermes (*C. œningensis*) in the Miocene of Oeningen. These data indicate very clearly that the primitive termites were represented by more genera and species in Tertiary times than they are at present. Of the highest termites there are only a few species in the Miocene: *Odontotermes pristinus* from Radoboj, *Eutermes fraasi* from Württemberg and two species of Termes, *obscurus* and *croaticus*, from Radoboj, and the generic references of some of them are doubtful. Of the lower Termitidæ the modern genus Leucotermes is represented by one species (*L. hartingi*) in the Miocene of Europe and one in Colorado (*L. meadi*). The Baltic amber contains three species of Reticulitermes, a genus to which the most abundant of living European and North American termites belong (*R. lucifugus* and *flavipes*). The very few remains of the family Termitidæ in the early and middle Tertiary of Europe and North America may indicate either that the family was just developing or that the climate of those regions was too cool for them. Even to-day the

Termitidæ flourish only in the warmer portions of the Ethiopian, Oriental and Neotropical regions.

The paleographic data thus confirm the general phylogenetic conclusions derived from the morphological and taxonomic study of the living termites and suggest that they must have originated at least as far back as the Mesozoic. There is nothing to prove that they may not have begun their evolution in the Permian, when the last Protoblattoids became extinct. Handlirsch (1903) derives the termites from Blattoid stock in the Cretaceous, but this is probably too recent. Warren (1919) has pointed to certain peculiarities in their present geographical distribution, which show a greater survival of the most primitive genera in the Northern and Southern sub-tropical zones and the greatest representation of higher Termitid genera in the tropics. After selecting as the most primitive genera Mastotermes, Termopsis, Archotermopsis, Calotermes, Leucotermes and Coptotermes, he says: "If the species contained in these genera in any particular region are added together, and compared with the total number of species of that region, then the percentage will indicate the general *facies* of the termite fauna". His results are given in the table on page 144.

I also quote his general conclusions in regard to the present geographical distribution of the Isoptera, because they agree in the main with those suggested for the Formicidæ in the preceding lecture:

"(1) There is an ancient or primitive aspect to the Northern and Southern portions of the termite area, taken as a whole. Thus North America, North India, Japan and China, along the Northern boundary, and Australia in the South, possess such a fauna.

(2) Africa possesses the highest termite fauna, since it contains a greater number of species, and has a larger percentage of species belonging to the more specialized genera than any

Region.	Percentage of primitive species.	Region.	Percentage of primitive species.	Region.	Percentage of primitive species.	Region.	Percentage of primitive species.
N. America ..	67	S. Europe and N. Africa	18	N. India ..	22	Japan and China	63
S. and Central America ..	18	Ethiopia ..	9	Indian Peninsula and Ceylon ..	13	Malay ..	9
		Ethiopia including Madagascar ..	7			Australia and Tasmania ..	30

other region of the world. For the comparison drawn it may be objected that the comparative areas of the different regions should be taken into account. If such is done the broad results would not be essentially changed, since it may be noticed that in the relatively small area of the Malay Region some 185 species have been recorded, while from the Indian Peninsula and Ceylon only seventy-seven and from the whole of Australia only fifty-three.

(3) Next in order comes the fauna of the Indo-Malay region, which includes the Indian Peninsula, Assam, Malay Peninsula and the Malay Archipelago. These regions possess a greater proportion of species of less specialized genera than is found in Africa.

(4) South America, including Central America and West Indies, possesses a fauna which is somewhat more primitive than that of the Indo-Malay region, and the number of species recorded is considerably smaller, being 188 to 262.

(5) The Termites of North America are few in number and primitive in character.

(6) The Termites of Japan and Southern China are few, but there is a high percentage of unspecialized species.

(7) In Australia the number of species in the termite fauna is somewhat low, but there is a high percentage of unspecialized and archaic forms.

The above facts suggest the following hypothesis: Since in the Tertiary Epoch the genera *Termes* and *Eutermes* were fully established, we must suppose that in Pretertiary times, when the contour of the great landmasses was not the same as at the present day, a fauna of unspecialized termite genera extended over considerable portions of Australia, Africa, America, Europe, and Asia. Failing any other direct communication between the Old and New World, we must suppose that the fauna was continued to the North, even as far as Behring Straits. To explain the distribution of termites it is necessary, as in the case of mammals, at least to assume a land-connection in these regions. As the climate became colder, and the land-masses gradually approached their present contour, evolution of new species occurred, especially along the central warmer portions of the world (South America, Africa, India, and Malay Region), while North (North America, Europe and Middle Asia) and South (Australia) the remnant of the old termite fauna persisted and shrank towards the equator.

It is thus possible to explain the present distribution of termites without the assumption of a land-connection by an Antarctic continent.

According to this view, the genus *Porotermes*, which occurs in both Chile and in Australia, is to be regarded as a portion of the old cosmopolitan fauna which has persisted along the Southern border of the termite area.

The relatively primitive character of the Madagascar termite fauna may be due simply to the fact that a forest country is favourable to *Calotermes*, but the condition is paralleled by the persistence in Madagascar of the ancient lemur fauna, which was largely replaced on the main-land."

The termites certainly appear to be very archaic insects—far more archaic even than the ants. This is not surprising when we consider that they are descended from such primitive forms as the Protoblattoids. But even the living orders to which they are most closely related have a similarly archaic aspect, or habitus and constitute the surviving twigs, so to speak, of an ancient branch of the original Hexapod stock (Palaeodictyoptera). It may be of interest to glance at these allied orders, because some of them have only recently come to light and because most of them suggest that even the most primitive insects in past ages may have manifested social proclivities. That societies as complicated and as highly integrated as those of the termites could arise among such archaic forms confirms the conclusion, not so readily inferred from the richly endowed Aculeates, that simple physiological and instinctive activities may be quite sufficient to produce and maintain very elaborate social organizations.

The five orders of extant insects which seem to be most closely related to the Isoptera are the Blattoidea, Embiidina, Dermaptera, Zoraptera and Grylloblattoidea, or Notoptera. They may be briefly discussed *seriatim.*

(1) The Blattoidea, being derived from the Protoblattoidea, have had the same ancestry as the termites and though they have acquired the peculiar habit of enclosing their eggs in an oötheca nevertheless sometimes show a vague approach to the social habits of the termites. One of the best examples is *Cryptocercus punctulatus*, our sole North American representative of the sub-family Panesthinæ. It is known to occur in the Atlantic States from the Catskill Mountains to Georgia and Westward to Kentucky and also in the Pacific States. Its distribution, therefore, is discontinuous. One other species is known from Japan. Concerning *C. punctulatus* Hebard (1917) says: "This remarkable insect is distinctive among the North American roaches in the peculiar pronotum, which is thickened and somewhat hooded cephalad, in the absence of tegmina and wings and in the production of the sixth dorsal and ventral abdominal segments which completely conceal the genitalia in both sexes". Hence the sexes cannot be distinguished externally, except when the internal surface of the sixth ventral sclerite happens to be visible. The insect always occurs in similar situations, living in colonies consisting of the parents and offspring, eating their way like termites through the dead sap-wood of logs and stumps in places where the wood is soft,

soggy and decayed. Undoubtedly Cryptocercus is a very ancient relict, like several other North American and East Asiatic animals, such as Limulus, the giant salamander of Japan, our North American Menobranchus, Necturus, etc. In New Zealand I have noticed another roach, *Polyzosteria novæ-seelandiæ*, with very similar habits, though sub-apterous and belonging to a different sub-family, the Blattinæ. Snyder (1924) calls attention to the fact that in the roaches of the genera Salganea and Panesthia "a break often occurs following in general the anal sulcus at the base of the wing, as at the humeral suture of termites, where the wing breaks off after the colonising flight; in the primitive termites this suture is also often poorly defined". He also says that "it is interesting to note that roaches and termites mate in a similar manner", but this statement, as I shall show in the sequel, is somewhat misleading.

(2) The Embiidina, a small order of some eleven genera and fifty-six described species, are confined to tropical and sub-tropical regions and live in small colonies that spin silken, anastomosing galleries either in cavities in the soil under stones or on the bark of living trees. The silk is spun with the fore feet, the swollen metatarsi of which contain peculiar sericiparous glands. Several observers have published notes on the habits of these insects, which feed on dead vegetable substances such as grass, lichens and fungi. Imms (1913) describes the social behaviour of a Himalayan species, *Embia major* (Fig. 33), which is nearly an inch long, as follows: "Maternal care on behalf of the ova and young larvæ is strongly exhibited by the females, in very much the same manner as has been long known to occur among the Dermaptera from the observations of Frisch, De Geer, Xambeu, Green, and others. The female *Embia major* shows very marked solicitude for the welfare of her offspring after her first few eggs have been deposited. She takes up her position in close proximity to the ova and usually concealing them, so far as possible, by means of her body. If alarmed and

driven away, she returns sooner or later to take up the same attitude. When the young larvæ are hatched, they remain around the parent female, who conceals them, so far as she is able, by means of her body, very much after the same manner as a hen guarding her brood of chickens". . . . "As the larvæ approach their second stage of growth, they exhibit a tendency to wander away from the female and construct small tunnels for themselves. They are markedly social, the whole of a brood living together within a complex silken mesh-work of tubes".

(3) The Dermaptera, or earwings, are so well-known that I need not dwell on their behaviour. A couple of recently published figures (Figs. 30 and 31) are here reproduced to show the female of the common European *Forficula auricularia* caring for her eggs and young. This species has lately been introduced into the United States and is proving to be a nuisance in some localities. Jones (1917) and Fulton (1924*b*) have published good accounts of its habits which are very similar to those described by Bennett (1904) for *Anisolabis maritima*, which has also been introduced into North America.

(4) The order Zoraptera remained unknown till 1913, when it was established by Silvestri on a single species of the genus Zorotypus. Several additional forms have since been described from various tropical countries—Africa, Ceylon, Java, Costa Rica, and Florida. Professor A. Emerson has taken a species in British Guiana and I have found one in Panama. Nothing is known concerning the habits of these insects, which bear a greater resemblance to the termites than any of the groups above mentioned, apart from the facts that they live in small colonies under bark or in dead wood and sometimes in the vicinity of termitaria, that these colonies comprise winged and wingless females and wingless males, and that the former may drop their wings like termites. The winged adults have well-developed eyes but the apterous adults are blind. The venation of the wings is very unlike that of termites, and the tarsi are two-jointed.

(5) In 1914, Professor E. M. Walker described under the name *Grylloblatta campodeiformis* (Fig. 32) a slender, thysanuriform, wingless insect which he found running about like a centipede under the stones on a talus slope at an altitude of 6,000 feet at Banff, Alberta. Since then a related species, *G. barberi* (Caudell 1923, 1924), has been taken in the mountains of Plumas County, California and more recently two species *nipponensis* (Caudell and King 1924) and *ishiana* (Silvestri 1927) have been discovered in Japan. These insects are undoubtedly very ancient relicts like Cryptocercus and Termopsis, which occur in the same regions. According to Walker, Grylloblatta combines characters of the Blattoidea and Dermaptera. Crampton (1915) regards it as a link betewen the Dermaptera and Isoptera, but Imms (1927) and Silvestri (1927) would place it near the Blattoidea and Mantoidea. Walker made it the type of a new family of Orthoptera, but Brues and Melander (1915) erected an order, Grylloblattoidea, for its accommodation. Others have adopted as ordinal names "Notoptera" or "Grylloblattaria". Comstock (1924) leaves the taxonomic status of the insect uncertain; Handlirsch (1924) assigns it superfamily rank among the Orthoptera and between the Locustariæ and Tridactylidæ. An interesting paper on the behaviour of Grylloblatta has been published by Miss Ford (1926).

You will have observed that, with the exception of the Grylloblattoidea, whose breeding habits are unknown, all the orders allied to the termites exhibit vague but unmistakable social proclivities. We are justified, therefore, I believe, in inferring that they and the Isoptera, though primitive and archaic insects, nevertheless indicate a possible prevalence of this mode of life among the ancient Orthopteroid ancestors. And since, as I have shown in my second lecture, there is now good reason to assume that the Hymenoptera are not descended from Orthopteroids, as Handlirsch (1904, 1908) supposed, but from very different, Mecopteroid ancestors, we must disavow any genetic relationships

FIG. 32. Adult male *Grylloblatta campodeiformis* cleaning antenna. (After Norma Ford.)

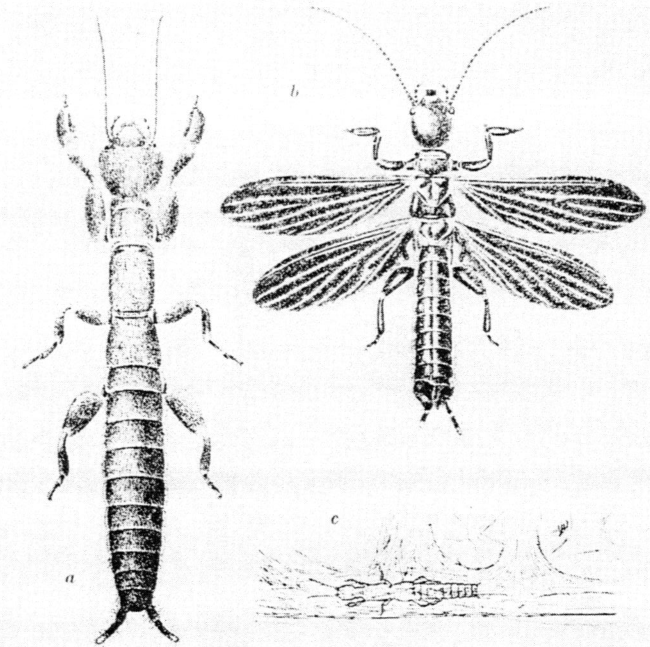

FIG. 33. a. *Embia major* of India, female. b. Male. (After A. D. Imms.) c. *Anisembia texana* of Texas, female in web. (After A. L. Melander).

[face p. 150

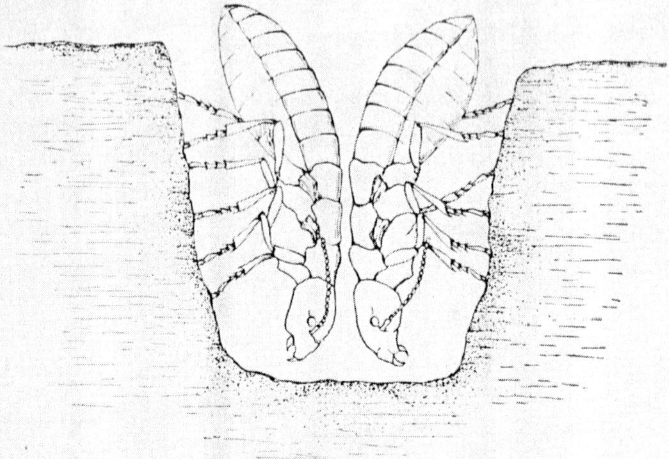

FIG. 34. Young king and queen of *Hodotermes turkestanicus* beginning to dig their burrow in the soil after the nuptial flight. (After G. Jacobson.)

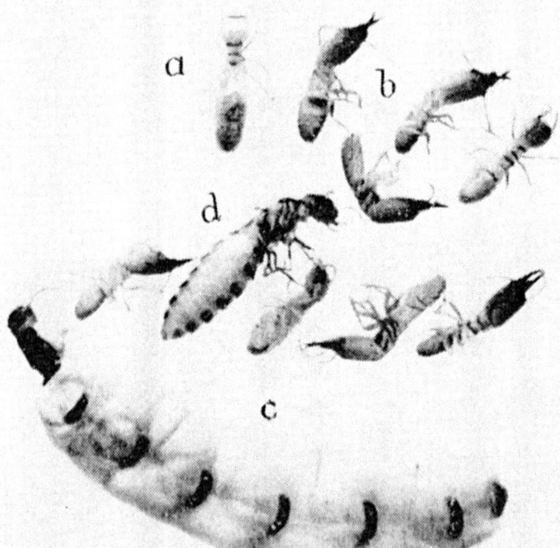

FIG. 55. *Leucotermes tenuis* of the Bahamas. X 2. a. Worker. b. Soldier. c. Physogastric queen. d. King.

[*face p.* 151

between the Aculeate and termite societies. The fundamental differences between the two groups are clearer in their methods of colony formation. In the Aculeata the colony is typically established by the fecundated female, either alone after the marriage flight or accompanied by a company, or swarm of workers. In the termites there is also a flight of winged males and females but each colony is established by a couple (royal pair) after deälation (Fig. 34), a process which, among ants, occurs only in the female. Fuller (1915), in his important paper on South African termities, has shown that the males and females of each species exhibit a rather elaborate courtship after the dispersion flight. Of his several accounts I select for illustration only those of *Termes natalensis* and *T. latericius*. The males and females of *T. natalensis* "flew high with well-sustained flight, and many passed overhead. The females, as appears usual, alighted first; selecting free and high-standing grass spears. Upon gaining a foothold (fifteen to twenty inches from the ground) they at once reversed their position and stood head downwards. Then all four wings were half opened and the apex of the abdomen became visibly swollen. Some males were flying low over the herbage when this act took place, and in a short space of time the females were discovered by them. The male alights as a rule, directly on the dorsum of the female, flying to her with much precision. He rapidly alligns himself, with wings closely folded and head upward, and combs across and across from one cercus to the other the dilated abdomen of the female. During this the female sits quite still, but presently she deälates and gives evidence of restlessness; when her movements sufficiently stimulate the male, he deälates and crawls from her back. The female then moves forward down the stem, the male close behind with mouth organs always closely brushing the anal plates of the female. There is no doubt in connection with this mating that, whilst the male is attracted to the female by the sense of smell, as is the case with *T. vulgaris*, the directness with which he alights indicates further that, within a circumscribed radius, the female is visible to him ; the play of the rays of the setting sun upon her half-spread wings rendering her quite a conspicuous object ".

Concerning *T. latericius* Fuller writes as follows: " The mating of this species was observed in Pretoria at dusk on December 4th, 16th, and 18th, 1914. The females were seen to alight first, taking up an inverted position on pendulous parts of grasses and herbage. Immediately on assuming this attitude they begin to agitate their wings violently, and keep them in motion until a male becomes associated with them. How long a female may go on agitating the wings without a male finding her cannot be said, but one watched for twenty minutes never stopped, and after dark was still unmated. The males seem to have great difficulty in finding the females ; they fly low and incessantly over the tops of the grasses, and again and again approach quite near to a female, and even circle within a few inches around, only to fly off again far afield. When mating was accomplished during twilight it was noticed that the male flew in narrowing circles around the female and alighted near to her. Directly the male touches the female the wings of both drop with inexplicable suddenness, those of the female seeming to fall whilst still being agitated. After this the female leads the way to the earth, the male following close behind. Here they soon burrow into the soil, and there form a cell. Two pairs which burrowed into observation cells on December 16th had both produced fifteen to twenty eggs eight days later ".

Thus the royal couple does not really mate till the initial nest-cell has been excavated by their co-operation. Owing to the fact that the penis of the male is vestigial or absent, copulation takes place by simple apposition of the genital orifices. The act, first observed by Grassi (1893-94), was seen to be frequently repeated during the life of the royal couple by Heath (1902) and Feytaud (1912). In the course of time the queen of the higher termites is known to become enormously physogastric (Fig. 35). She is constantly accompanied by the king, who exhibits only a feeble enlargement of the abdomen. Fuller gives the following description of his majesty's behaviour (*Termes natalensis*) :—" This creature displays an extraordinary devotion to his mate. He seldom wanders more than an inch away from her ; constantly approaching her head and paying apparent court ; then inspecting her flanks and eggs with pre-occupied attention ; behaving, in short, as most male termites seem in the habit of doing, but never revealing the purpose of his existence. The male is always somewhat inflated with body-fat ; the conjunctiva of the sides of the abdomen being distended, but not that of the dorsal and ventral sclerites ".

The courtship behaviour described above may, I believe, be regarded as a further indication of the Orthopteroid affinities of the Isoptera. It is well-known that elaborate types of courtship are not exhibited by the Hymenoptera, though they are frequent in Diptera, Lepidoptera and Orthoptera. I have recently (1925) published an account of some singular cases among the flies. Fabre's fascinating descriptions of the courtship of Mantids and Locustids are well known, and Hancock (1905), von Engelhardt (1914) and Fulton (1915) have described it in the Gryllids of the genus Oecanthus. In the latter insects the female is enticed to the male by a peculiar gland on the back of the thorax and feeds on the secretion before or after mating. Recently Wille (1920) has described in detail the somewhat similar courtship and copulation of our common roach, *Phyllodromia germanica*. In this case the female feeds on the secretion of the male, but the glands open on the dorsal surface of the abdomen. In the termites the rôles are reversed, the male being attracted by the glandular secretion emanating from the tip of the female's abdomen and copulation is a much less elaborate process than in the Blattids. Snyder's remark above quoted therefore requires qualification.

That Fuller's observations on the courtship and founding of the colony among South African termites are also of more general behaviouristic significance is evident from the following passage: " It seems almost certain that this phase in the history of a termite is a succession of events which follow upon one another in regular order, and, if the chain is broken at any point, it remains so. This may be illustrated by several specific cases. The night-flying *Eutermes trinervius*, if attracted by illuminating a white sheet, alights, and both sexes deälate after meeting. If some are captured at once before meeting and placed in a tube, they do not deälate. Some, so captured, have been kept over four days alive in confinement, and whilst a few deälated, or, to be more correct, lost their wings, no attempt at pairing was made. The same holds good of the dusk-flying *Termes incertus*, but to a modified extent only. Further, with these two species and three others (undetermined) no attempt is made at burrowing if the sexes are kept separate. In the case of two males confined together it was noticed that one would frequently follow the other as if it were a female. The stimulation induced them to burrow together, but they soon desisted from doing so, although stimulated thereto several times by one another; ultimately they died on the surface of the soil. Again, a number of pairs of *T. natalensis* were captured one evening and placed in a small box; in the morning it was found that the couples had lost all regard for one another; they were allowed to wander over a table, but none again mated voluntarily. When, however, artificial burrows were made, and the pairs placed in them side by side, the environment reacted upon them at once and they burrowed to the bottom of the jar. In 1914, similar results attended the mating and re-mating of this species ".

In the remainder of my account of the archaic aspects of the termites I shall have to be brief. Next to the sexual peculiarities, already discussed, perhaps the most conspicuously primitive traits are exhibited by their ontogeny and their feeding habits, which have had far-reaching effects on all their social activities. Owing to their rigid retention of the gradual, ametabolic type of ontogeny, so characteristic of the whole Orthopteroid series, the termites have developed less intimate relations between the parents or other adult members of the colony and the young than we find among the social Aculeata. In these insects the larvæ and pupæ are quite helpless so that the former have to be carefully fed and both, in the ants, have to be transported to favourable and away from unfavourable conditions of temperature, moisture, etc. In termites such care is restricted to the eggs and very young nymphs. Moreover, in the gradual development of the nymphs with their series of active instars separated by moults there is the possibility, non-existent among Aculeates, of permanent arrest of growth at a number of different and very definite periods of development. This arrest may affect both the soma and the gonads as in the soldiers and workers of the higher termites or it may extend to the soma only. In such cases "neotenia" may result and we have the complementary, or substitute males and females, which in the colonies of certain termites may, wholly or in part, replace the royal couple when it dies of old age or from other causes. In the Aculeata, of course, no such developmental resources are available, although atypical adult forms may be produced (ergatomorphic females, pseudogynes, pterergates, etc.). The gradual, ametabolic development of termites, together with the equal representation of both sexes in the personnel of the colony, certainly presents the basis for a more elaborate polymorphism than is found among the female colonies of the social Aculeata.

The termites, probably because their immediate ancestors never passed through a predatory, carnivorous stage, failed to acquire many of the extraordinary instinct endowments which we observe in the social Aculeates. Unlike these aggressive and high-strung insects, they have lived and for the most part still live within their food, like the wood-boring beetles. Even the ability to forage for vegetable food and to store it is confined to certain Hodotermitidæ and Termitidæ, such as the species of Hodotermes which collect grass in the daylight, the highly specialized species of the genus Eutermes, e.g., the Ceylonese *E. monoceros*, which, according to Bugnion (1909, 1914) forages for lichens, various African and Australian species of the same genus, which collect dead grass, and *Termes latericius*, which harvests seeds (Fuller). But even the species of Termes, the only forms that have learned to grow fungi on masses of their own excrement as food for their young and the royal couple, still feed largely on humus, dead grass and wood, and are apparently very far from having attained to the skill of the Attine ants in cultivating, propagating and utilizing the fungi (see Morstatt, 1922).

The ultra-conservative habit of devouring wood and living in its cavities (Fig. 37) is the key to an understanding of most of the social peculiarities of the termites. It has long been suspected that they are unable to digest the cellulose unaided and several investigators have surmised that the ciliated protozoa which occur in great numbers in the intestines of the lignivorous species, actually ingest the cellulose and render it assimilable by their hosts.[1]

[1] The following statement in regard to the occurrence of the protozoa in termites is made by Cleveland (1923*b*). " Careful examination of the intestinal contents of five workers of each species of termites in the U.S. National Museum revealed that wherever protozoa were present wood was also present and, *mutatis mutandis*, protozoa were present only when wood was present; thus confirming Imms' postulatum. Four families of termites are known. Among the eighteen genera and sixty-four species examined from the family Termitidæ, protozoa and wood were present in only one of the twenty-one species of the genus *Nasutitermes* and two of the eight species of the genus *Mirotermes*; but protozoa and wood were present in eighteen genera and seventy-six species, or all that were examined, in the other three families (Mastotermitidæ, Kalotermitidæ and Rhinotermitidæ). Thus the correlation between a strict wood-feeding habit and the presence of intestinal protozoa is perfect and positive ". This distribution agrees with that given by Grassi in 1917 and quoted by Jucci (1924, p. 209). In another paper (1923*b*) Cleveland gives an extensive taxonomic list of the termites in which protozoa have and have not been found.

FIG. 36. Mushroom gardens of *Acanthotermes militaris* from a nest at Malela, Belgian Congo. The white dots are the food bodies. (Photograph by H. O. Lang.)

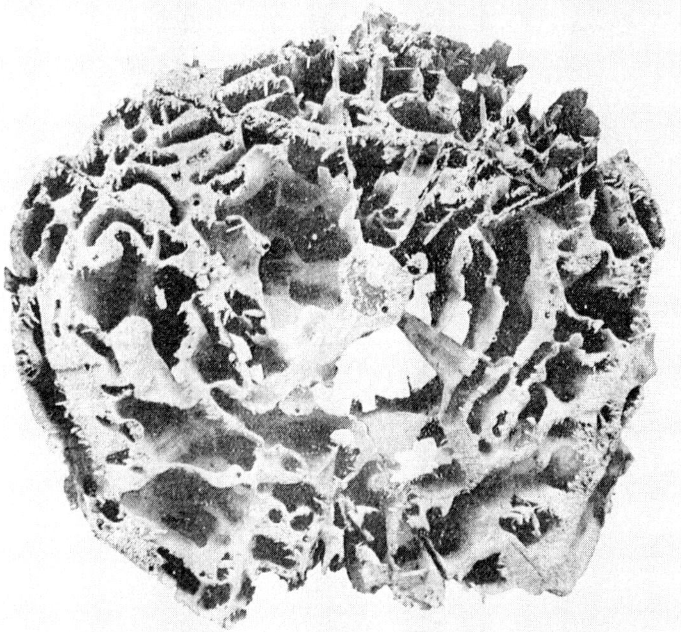

FIG. 37. Damage by *Calotermes hubbardi* to rafters in an "adobe" building in Arizona. (After Thos. E. Snyder.)

[*face p.* 156

Plate XXIV.

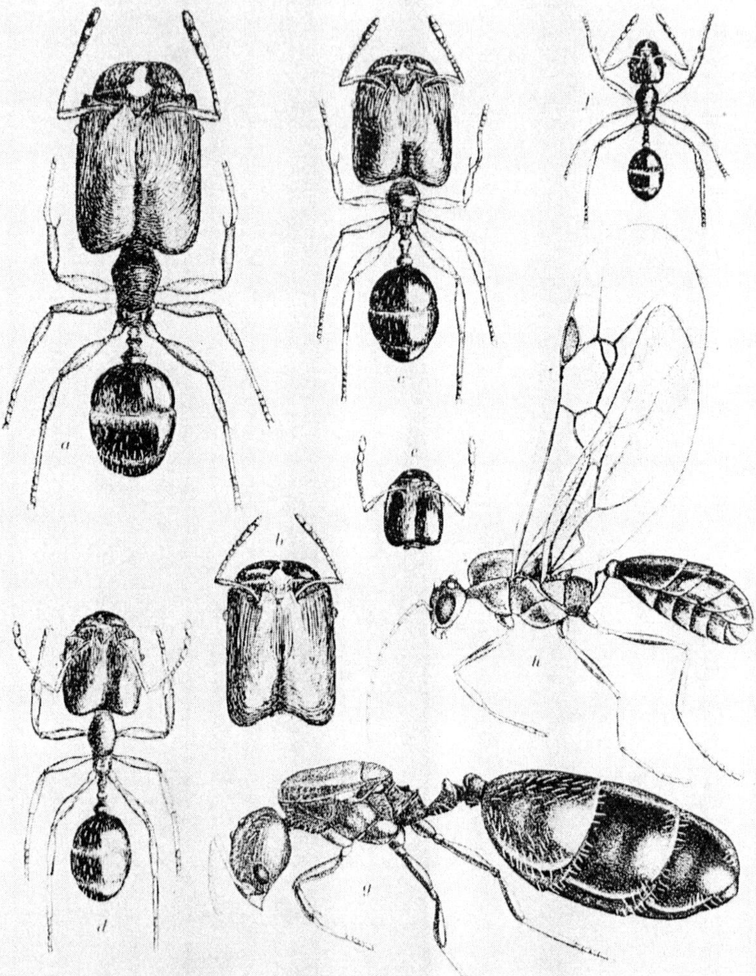

Fig. 38. A small Myrmicine harvesting ant of Texas, *Pheidole instabilis*, with polymorphic worker caste. a. Soldier. f. Worker. b. to e. Forms intermediate between the soldier and worker (lacking in most other species of the huge genus Pheidole). g. Queen (dealated). h. Male. The Figures are all drawn to the same scale. (*See* p. 172.)

[*face p.* 157

A considerable body of literature has been published on the protozoa, of which many species have been described since they were first discovered by Lespès in 1856. More recently Buscalioni and Comes (1910), Imms (1919), Jucci (1924), and Cleveland (1923*a*, 1923*b*, 1924, 1925*a*-1925*e*) have devoted considerable attention to the interesting problems suggested by these unicellular organisms and their possible rôle as symbionts. Grassi in 1911, by heating termites to about 35° C., succeeded in killing off the protozoa in their hind-guts without injuring the termites themselves. He claimed that they lived several months and inferred that they could digest wood without the aid of the protozoa. Cleveland, however, who has carried the heating experiments further with *Reticulitermes flavipes*, Termopsis, and other North American species, succeeded in completely "defaunating" the termites, i.e., in killing all the intestinal protozoa by keeping the insects at a temperature of 36° C. for twenty-four hours. Such individuals, when fed with a normal wood diet, die in the course of ten to twenty days, but when given digested wood (humus) or fungus-digested cellulose live indefinitely. They also continue to live when reinfected with protozoa. Cleveland also found that while the other intestinal organisms of termites, the bacteria and fungi, are unable to digest cellulose, the protozoa of the genera Trichonympha and Pyrsonympha, so common in *Reticulitermes flavipes*, possess this ability. It is absent, however, in some of the other intestinal protozoa. He believes that the protozoa derive their proteins from their hosts, but Jucci (1924) finds that the protozoa may be digested, so that the hosts probably derive much of their proteins from the bodies of the symbionts. This indirect and dependent method of obtaining their carbohydrates and proteins seems to have been abandoned by the higher Old World Termitidæ which have come to live on humus (Cubitermes, Anoplotermes, etc.) and fungi grown on their excrement (Termes) (Fig. 36).

Xylophagy has had at least two other far-reaching effects on the social organization of the termites. In the first place, the bulky and at best rather slowly digestible food seems to be responsible for their habit of feeding one another with regurgitated, partially digested "stomodeal" and voided, more completely digested, "proctodeal" food, a form of trophallaxis, which certainly favours, if indeed it be not necessary to the thorough infection of the various individuals of the colony with the symbiotic protozoa. Brunelli (1904) calls this feeding of the termites on the digested and partly digested contents of the intestines of other individuals not inaptly "social rumination". In the second place, the habit of tunnelling through the wood and devouring it has naturally led to the development of a timid cryptobiosis and all that it implies, the degeneration or loss of the eyes, especially in the soldier and worker castes, the loss of pigment and the thinning of the integument, except in the winged forms and the head and thorax of the soldiers, and the very general photophobic and hygrophilous proclivities of all the members of the colony. The defencelessness thus acquired has in turn led to the development of a diversity of protective structures and behaviour like the mandibles, hard cranium and cephalic glands of the soldiers and in the higher Termitidæ the agglutination of earth, fæces, etc., to form a solid nest, or termitarium. Such protective adaptations have proved to be necessary on account of the numerous enemies, particularly the ants, many of which live largely or entirely on termites. This is true especially of the predatory Ponerinæ, e.g., *Megaponera fœtens*, *Paltothyreus tarsatus* and *Euponera senaarensis* in Africa, *Neoponera commutata*, *Pachycondyla crassinoda* and *Paraponera clavata* in Central and South America and a number of species of Leptogenys in the tropics of both hemispheres. In a paper describing the various methods of defense among termites, Bugnion (1923) says: "In tropical countries the termites have no enemies more redoubtable and none they actually fear more than the ants. Between these little creatures there is an incessant struggle, an implacable warfare which has been waged without mercy for thousands of years. Hence the varied and ingenious methods of defence which the termites have developed and which are nearly all designed to protect them from the attacks of their hereditary foes, the ants." At the same time it should be noted that some tropical ants will not devour termites. This seems to be true of our formidable Neotropical army ants (Dorylinæ) of the genus Eciton. The African Dorylines, Anomma, Typhlopone, Rhogmus, etc., however, seem to be very fond of them. There is a host of species (Carebara, Pædalgus, Aëromyrma, etc.) which live as thief-ants in the walls of the termitaria and prey on their inhabitants, and an even greater host which drives the termites partially or completely out of their nests and takes possession of their galleries. In Australia this is the regular habit of *Iridomyrmex sanguineus* and of many ants of the genera Opisthopsis, Calomyrmex, Polyrhachis, Camponotus, etc. Another series, too numerous to cite, behaves in like manner towards the African Termitidæ. One who has witnessed this incessant and implacable warfare between the ants and termites of the tropics can only wonder whether eventually the former may not be destined to exterminate the latter.

Reverting, in conclusion, to De Grange's definition of a society as an accumulative collectivity, we see that the most conspicuous accumulative activities, at least in the higher termites, are architectural and the result of their having early resorted to a food-supply which was abundant and readily obtainable but not easily assimilable. The peculiar food also furnished very simple conditions in which to establish their colonies, originally in all probability like those of various subsocial beetles (Passalus, Phrenapates, Parandra, Coccidotrophus, Micromalthus, etc.).[1] All of these insects live in cavities of plant tissues.

[1] On re-reading an interesting paper by Lameere (1909) since this paragraph was written, I find the following similar but more detailed account of the hypothetical origin of the termite colony : " Let us endeavour to picture in imagination the roaches adapting themselves to a woody diet and tunnelling the tree trunks which become at the

The later more elaborate developments or accumulations (polymorphism, trophallaxis, mycetophagy, etc.) were probably acquired very naturally as the result of more complicated interactions among the individual termites and between them and their chosen ligneous environment.

---

same time their natural habitat. What energy they must expend and what precautions to insure the welfare of their progeny! The oötheca may disappear; it has become useless since the eggs are as well protected by being simply laid in the galleries hollowed in the wood and now constituting the nest. This is a saving that can be easily realized. But the young on hatching are unable to obtain sufficient nourishment by attacking so refractory a substance as wood; they are therefore fed by the mother and the great advantage of the father's co-operation in this activity is obvious. We therefore have a situation very much like that of the Lucanid beetles of the genus Passalus, which burrow in all directions in rotten tree-trunks and both the male and female of which masticate the wood as food for the larvæ". We should now have to substitute "Protoblattoids" for "roaches" in this passage.

VII

POLYMORPHISM

POLYMORPHISM was frequently mentioned in the preceding lectures but I refrained from discussing it because it keeps recurring in very nearly all the social insects and could not be viewed in proper perspective till their idiosyncracies had been surveyed. It is twenty years since I first discussed this perplexing subject (1907*a*). In the meantime it has been constantly in my thoughts and I have, of course, consulted all the pertinent works to which I could gain access. At no period have the students of the social insects been so numerous, evinced a keener interest in or been better equipped to deal with the subject. And yet we have made little progress towards the solution of the many problems suggested by the phenomena of polymorphism. It is, in fact, so intricate that although I have set apart two lectures for its discussion I shall be able to present only a small number of its many aspects and difficulties. Perhaps it will conduce to clearness, if I confine this lecture to an account of some of the leading facts and more obvious inferences and in the next delve into the speculations which they have suggested and will no doubt long continue to suggest.

It is difficult to define polymorphism, because the phenomena to which zoölogists apply the term in its broadest sense grade insensibly into those of variation, mutation and metamorphosis, of alternation of generations (heterogony), pædogenesis, seasonal and other types of heteromorphosis and heterochromatism. If we restrict the term to the cases in which two or more different kinds of individuals of the same species coëxist, it includes, of course, sexual dimorphism and the differences between young and adults, but as employed by entomologists, it is properly applied to the coëxistence of two or more adult forms of the same sex, the "pœcilandry" and "pœcilogyny" of de Peyerimhoff (1897). There are also polymorphic larval forms, cases of what Giard (1891, 1894, 1905) called "pœcilogony." I shall have occasion to refer to these and to polymorphism in various insects but my main concern is with the social species. The latter are unfortunately so peculiarly constituted that their polymorphism is often regarded as necessarily bound up with certain other phenomena, such as those of sex-determination, parthenogenesis, neotenia, sterility and arrest of development, some of which may be briefly considered for the purpose of eliminating them from the general discussion. The problem of the sex of the offspring of parthenogenetic female insects has been much befogged by the eternal discussion, especially in Germany, of the honey-bee, an exceptional creature which it is very difficult to dislodge from zoölogical text-books and lecture rooms as the paragon of social insects. It is, of course, well known that the offspring of unfecundated Hymenoptera are usually males, but the exceptions are becoming so numerous that the so-called Dzierzon "law" should be regarded merely as a common occurrence to be noted in the particular species in which it happens to be observed. It is not a law, but at most a rule. That it has no necessary connection with the polymorphism of the social insects is sufficiently proved by the termites, in which both sexes develop from fertilized eggs and are equally polymorphic. Winkler (1920, p. 76-108) has recently taken the pains to record the sexes produced by parthenogenetic females among insects and especially among the Hymenoptera. Among the Phytophaga he cites twenty-four species as definitely known to be thelytocous, two which are ampherotocous and more than forty which are arrhenotocous. But many of the cases recorded as thelytocous are not pure since a few males may be produced among the females. He cites among the Terebrantia, excluding the Cynipidæ, in which ampherotocy alternates with bisexual reproduction and including the Bethylidæ, fourteen species, representing thirteen genera and several families, as certainly thelytocous and several which are incompletely so, since the broods comprise also a few males. There are, moreover, a few ampherotocous forms, though undoubtedly arrhenotocy is the most prevalent type of parthenogenesis. A few peculiar cases may be mentioned, which have come to light in America since Winkler's work was published. After breeding many thousand individuals of *Habrobracon brevicornis*, Whiting (1921) has demonstrated that males may occasionally develop from fertilized eggs. He informs me that he has also secured a few females from unfertilized eggs, and Leiby (1922) asserts that the fertilized eggs of *Copisosoma gelechiæ*, a polyembryonic Chalcidid, produce either male or female offspring, though unfertilized eggs produce only males. Mr. S. M. Dohañian who has been experimenting for many years with various parasitic Hymenoptera at the Federal Gypsy Moth Laboratory at North Melrose, near Boston, sends me the following interesting notes :—" In our studies of the life history and habits of the hyperparasites of one of our important gypsy moth parasites, *Apanteles melanoscelus*, Mr. C. F. W. Muesebeck and myself have bred at least three secondaries which have produced exclusively females in parthenogenesis. The following are excerpts from our prepared manuscript, to be submitted very shortly for publication, and treating of species entirely thelytocous in parthenogenesis : Females of *Hemiteles tenellus* Say are invariably secured in parthenogenetic reproduction. In the laboratory we have obtained several pure lines of females through twelve generations, over a period of three years. The male, if this sex exists, is unknown. In the course of the rearing of many thousands of parasites from field-collected cocoons of *Apanteles melanoscelus*, nothing has been obtained that could be the male of this species, although females have always been secured in large numbers. It is interesting to note at this point that males of the European *Hemiteles areator*

(Panz.), which is not clearly distinguishable from *tenellus* morphologically, and of which *tenellus* has sometimes been regarded as a subspecies or variety, occur abundantly in museum collections. Having the opportunity, during the summer of 1924, to determine the result of parthenogenetic reproduction with *H. areator*, we found that unfertilized females produce males. It is evident from the disparity in the biology of the two forms that they are quite distinct. Parthenogenesis, or reproduction without fertilization, has been so often noted among the parasitic Hymenoptera, that it may be regarded as of general occurrence in these groups. All the hyperparasitic species attacking *Apanteles melanoscelus* were found to be capable of reproducing without previous fertilization, as would be expected. But of particular interest in this connection was the observation that different species of a given genus may differ with regard to the sex of the progeny resulting from parthenogenetic reproduction. One of the two common species of Hemiteles reared from Apanteles was found to be arrhenotocous, while the other is always thelytocous. The two species of Pleurotropis and those of Anastatus obtained from the Apanteles cocoons differ in the same way, one species in each genus always producing males, the other females. Males of the thelytocous species have not been encountered or at least have not been recognized."[1] Picard (1922) has also studied two of the European species of Hemiteles, *fulvipes* and *longicauda*. The former is a general parasite, attacking other parasites, especially *Apanteles spurius*, various Microgasters, some Tenthredinidæ (*Blennocampa pusilla*) and apparently even the eggs of spiders. The virgin females produce only male offspring. *H. longicauda*, which is a hyperparasite of *Apanteles glomeratus*, however, is regularly, if not exclusively thelytocous.

Finally, in the Aculeata, there are several cases in which females are said to have developed from unfertilized

[1] The observations of Muesebeck and Dohanian have been recently published (1927).

eggs. Without stopping to cite the works of other antidzierzonists, attention may be called to those of Onions (1912, 1914) and Jack (1917), according to whom the workers of the South African race of *Apis mellifica* (var. *kaffra* or *intermissa*), though possessing a spermatheca, produce parthenogenetically not only males and and workers but even queens! In solitary and other social bees, according to the recent ingenious investigations of Descy (1924) on *Osmia*, of Stöckhert (1923) on Halictus and many observers of Bombus, parthenogenetic females are always arrhenotocous. The same results are recorded for unfecundated ant queens and workers by Forel (1874), Lubbock (1888), Miss Fielde (1901), Janet (1909), Tanquary (1913), and myself, but there are exceptions. Tanner (1892) claimed that in his artificial nests workers of the large leaf-cutting ant of Trinidad, *Atta cephalotes*, produced males, workers and queens, and Reichenbach (1902), Mrs. A. B. Comstock (Wheeler, 1903) and Crawley (1912) claim to have reared workers from the unfecundated eggs of *Lasius niger* workers. The instances adduced show clearly that the phenomena of parthenogenesis have no necessary connection with polymorphism and that whatever may be the ultimate solution of the problem in common races of the honey-bee, whether unfecundated queens and workers produce males only or occasionally also workers, or whether fertilized eggs may sometimes produce males, can contribute nothing of theoretical importance to the problem of sex determination that has not been contributed already by other Hymenoptera in which all the possibilities mentioned are realized under more favourable conditions of experimental control.

We can hardly speak of polymorphism in the social Aculeates till a worker caste makes its appearance as a distinct morphological expression of the behaviouristic and physiological division of labour among the originally monomorphic female members of the colony. Among the bees we noticed the presence of such a caste in the Bombinæ, Meliponinæ, Apinæ and Halictinæ. In the social Halicti it is represented by the sterile summer generation of females, which are sometimes sufficiently different from their mother to have been described as distinct species (Stöckhert, 1923). The social species of Allodape are still in too primitive a stage to have developed a morphologically recognizable worker caste, although it may be said to be represented in an early, purely behaviouristic, phase by the daughters that associate themselves with their nest-founding mother. The Vespidæ exhibit a very gradual evolution of the worker in the five social subfamilies. In the Stenogastrinæ and some Polybiinæ no worker has been detected; in other Polybiinæ dissection alone reveals whether a female is functioning as a worker, i.e., behaving as a sterile individual, or nurse. In the Ropalidiinæ and Polistinæ the workers differ little externally from the queens but in the Vespinæ the difference becomes well established in both stature and colouration. In ants and termites the worker caste is wingless, usually sharply differentiated and frequently secondarily dimorphic or pleomorphic as the result of a more intensive physiological division of labour. In these insects also the worker and soldier castes evidently first appeared as merely functional, i.e., physiological and behaviouristic castes before they assumed the morphological characters which now enable us to recognize them so readily.[1] If my contention, reiterated in the preceding lectures, is correct, that the ants, termites and various groups of social bees and wasps represent phylogenetically independent societal developments, we must also assume that the worker castes which appear in no less than ten of them, are likewise independent and have derived their similar and convergent peculiarities from the similar trophic and other

[1] As a physiological caste among ants the plerergates, or "honey-pots" of various species of Myrmecocystus (Fig. 60), Prenolepis, Plagiolepis, Camponotus, Melophorus, Pheidole, etc., in arid regions may be cited. These are workers or soldiers which are at first indistinguishable from other members of their respective castes but eventually become sluggish and acquire a great distention of the crop and abdomen as the result of storing liquid food. The gynæcoid workers, which develop their ovaries and assume the rôle of queens in queenless colonies, may also be regarded as constituting a physiological, emergency caste.

societal conditions to which they have been constantly subjected during their ontogenetic and racial history.

On closer examination the ten independent societies may be divided into four groups, according to the idiosyncracies of their worker castes : first, the Halicti, humblebees and four subfamilies of wasps, second, the Meliponine and Apine bees, third, the termites and fourth, the ants. In the Vespidæ, Bombinæ and Halictinæ the worker form has developed concomitantly with an increase in its sterility while the queen retains unimpaired the morphological and physiological status of the primitive solitary Vespid or Apid female. Here the conditions appear to be quite perspicuous, and Marchal's (1896, 1897) " nutricial castration " (from *nutrix*, a nurse) would go far towards accounting for the phenomena. It will be recalled that he gave this name to functional atrophy or arrest of development of the ovaries brought about by the nursing, or nutricial proclivities of the worker. He showed that by eliminating the queen from the Vespa colony as many as a third of the workers may become fertile. A similar result is produced by a suppression or merely by a temporary suspension of the oviposition of the queen. This can be due, as Marchal maintains, only to an abolition of the nursing function and the appropriation by the workers of the food which under normal conditions they would feed to the larvæ. Nutricial castration, of course, leads to " alimentary " castration, i.e., the inhibition of the development of the reproductive organs in the larvæ from which the workers develop. Thus we have a circular process the effects of which are aggravated by the tendency of the adult workers to distribute the limited amount of available food among many more larvæ than can be adequately fed (see Roubaud, 1916). The feeble polymorphism of the wasps and in all probability also that of the humblebees, are due to well-known physiological and instinctive processes peculiar to the social medium.

While the worker caste in wasps and bees seems to acquire its sterility gradually and before or during its morphological differentiation, the corresponding caste of termites, or at any rate the soldiers, seem to have developed morphologically before their fecundity was impaired or lost. In very primitive species like Archotermopsis (Imms, 1913) and evidently to some extent also in Termopsis (Heath, 1927), the soldier caste is fertile (Fig. 54) and there are fertile neotenic males and females but no workers. We may say that in Archotermopsis each sex is trimorphic and represented by three fertile forms, completely winged individuals, partially winged neotenics and sup-apterous or apterous soldiers. With increasing social development a worker caste makes its appearance as an arrested early nymphal stage and like the soldier is normally sterile.

At first sight the conditions in the Formicidæ would seem to be the same as those of wasps, except for a more profound somatic modification and a more pronounced sterilization of the worker caste. A comparative study of primitive ants, however, especially of the Ponerinæ, Cerapachyinæ, Dorylinæ and Pseudomyrminæ, suggests that this resemblance may be superficial. The ants, as we have seen, are much older than the social wasps and seem to have arisen from a section of the solitary Vespoidea much nearer the archaic ancestors of the Aculeata, the Bethyloidea, several genera of which still retain two forms of fertile females, winged and apterous. Here, too, therefore, even in the solitary ancestry polymorphism had preceded sterility, and it is not improbable that the early Formicidæ may have retained the female dimorphism after they became social and have utilized it in the production of winged queens and wingless workers. The winged female simply took on the function of disseminating the species and starting the colony, while her apterous but still fertile daughters remained with her and specialized in rearing her and their own subsequent broods and in foraging for food. The winged queen acquired the ability to drop her wings and the voluminous vibratory muscles could then be broken down and used in nourishing her ovarian eggs and in lengthening her life till the wingless offspring emerged. Janet (1907) and C. Pérez (1912, 1920) have, in fact, shown that deälation is actually followed by these processes in ant queens after their nuptial flight. The next step was the reduction of the apterous fertile females to the rank of workers. This hypothesis might account for the conditions seen in certain subfamilies and genera of ants, e.g., the retention of the wingless as the only queen in Dorylinæ, many Ponerinæ and Cerapachyinæ, the presence of both winged and wingless queens in the same colony (Ponera) or in different localities (Harpagoxenus), and the not infrequent occurrence of ergatomorphic queens in several genera (Leptogenys, Onychomyrmex, Diacamma, etc.).

In this connection I may mention the opinions of Mrázek (1916), who endeavours to show that polymorphism in the social Hymenoptera is due to a masked or telescoped alternation of generations like that of the Cynipidæ. At first sight there would seem to be little to recommend this hypothesis, which was long ago suggested by Steenstrup and von Siebold, but it is supported at least by Stöckhert's observations on the social Halicti (1923). Obviously, *Halictus malachurus* and the allied species may be most readily derived from species, of which there are several in the genus, with two annual generations through the survival and increasing longevity of the hibernating autumn females and the progressive sterilization of the summer generation till it was reduced to the status of a worker caste. The ant-colony, too, if we accept the interpretation which I have given above, might be conceived to have arisen by a telescoping of two generations, originally with fertile winged and wingless queens, respectively. But the social wasps, e.g., the Polybiinæ, seem to me to present an insuperable obstacle to Mrázek's hypothesis, because they are so closely related to and so naturally derived from the solitary Eumeninæ, in which there is no hint of the occurrence of an alternation of generations. Nor do Mrázek's attempts to find indications of such a condition in the honeybee seem to me to be either clear or felicitous, but the honeybee seems destined to drive biologists to distraction. There is nothing, so far as I can see, in the constitution of termite societies to suggest an alternation of generations.

Instead of resorting to a contracted or telescoped heterogony I should prefer to regard polymorphism, at least in the ants and termites, as phylogenetically prior to the conditions in which one or more of the forms became sterile. This view implies that the phenomenon is essentially the same as the pœcilogyny and pœcilandry of many non-social insects. Although many of the latter cases were succinctly reviewed by de Peyerimhoff as long ago as 1897, it may be of interest to glance at some of the cases that have been described within recent years. The most familiar are the male Lucanid beetles which are often singularly like the pleomorphic workers and soldiers of certain ants and termites in the structure of the head and mandibles. Among the army ants of the genus Eciton *s. str.*, e.g., we might recognize extreme telodont and priodont forms with intermediates as in many Odontolabine Lucanidæ, as described by Griffini (1905), Champy (1924) and others. This resemblance which has not escaped Viehmeyer (1923), might be traced in detail, did my time permit.[1] A case which recalls the variation in the eyes in pleomorphic series of worker ants or among the castes of termites is that of the small Pselaphid beetles of the genus Bythynus. In *B. algericus*, according to de Peyerinhoff (1910), the female has minute eyes, whereas the males are dimorphic, one of the forms having rather small, the other large eyes. Polymorphism of the wings, resembling that of the neotenics among termites, is common among Hemiptera of both sexes and has been carefully studied in a number of aquatic species by Poisson (1924). Another example, in certain respects more like that of ants than any of the preceding, though it also occurs only in the male, is furnished by the peculiar fig-insects (Chalcididæ) of the genus Philotrypesis, as described by Grandi (1921, 1923, 1925). The heads of the

[1] After measuring series of soldiers and workers of *Anomma nigricans* and *Camponotus gigas*, Julian Huxley (1927) concludes that this resemblance to the pœcilandry of Lucanids and other animals (chelæ of crabs

various males of *Ph. erythræa* and *unispinosa* var. *ornata*, as figured, might readily be mistaken for a series of heads of pleomorphic Camponotus workers. Grandi adopts Berlese's terms for the extreme and intermediate forms, the "epimegetic" males being the largest, the "eumegetic" the transitional and the "hypomegetic" the smallest in the series. But the pœcilandry of Philotrypesis may become even more complicated than the pœcilogyny of ants, e.g., in *Ph. minuta*, the males of which form two series, an "acanthocephalous" and a "mutic," "each of which comprises eumegetic and hypomegetic and furthermore eumegetic macrognath (heterodont) and eumegetic brachygnath (heterodont) forms." The resemblance of the species of Philotrypesis to ants is enhanced, moreover, by their extreme subaptery.

That polymorphism is due to other causes than those determining sexual dimorphism is, I believe, clearly indicated by the foregoing examples and the occurrence, already mentioned, of an identical polymorphism of both sexes in termites. This conclusion becomes a certainty when we consider the cases of polymorphism ("pœcilogony") in the larvæ of certain insects which as adults are homomorphic. Twenty-two years ago Giard (1905) reviewed all the known cases in a comprehensive paper. I shall therefore cite only two, one that has been more recently discovered and one that has been more carefully investigated. Silvestri (1906) found that there are two very different types of larvæ in the polyembryonic Chalcidid *Litomastix truncatellus*, one possessing reproductive organs and developing into the adult insect, the other ("asexual larva") without reproductive organs and dying without further development. The same types of larvæ were later found by Patterson (1918, 1921) in another polyembryonic Chalcidid, *Paracopidosomopsis floridana*. Among the cases of pœcilogony, Giard briefly referred to the pædogenetic Cecidomyids and Chironomids as "determined in part by the primary factors of temperature and food." The more recent studies of Springer (1915) and Harris (1923, 1924) have demonstrated

antlers of stags, horns of Titanotheres, etc.) can be expressed as a mathematical equation, namely $y = bx^k$, in which $y =$ the organ measured, $x =$ the rest of the body and b and k are constants.

that the larvæ of the Cecidomyids of the genera Miastor and Oligarces are distinctly trimorphic. Springer calls them "typical pædogenetic," "vagrant" and "pupa larvæ" respectively, and believes that their peculiarities may be due to differences in the external conditions. Harris designates the same three types as "typical white pædogenetic," "yellow pædogenetic" and "pupa larvæ." The last develop into adult flies without pædogenetic reproduction. He believes that the various larval types and their sex are determined by genetic and not by environmental factors. In regard to the sexes he says: "Males as well as females arise from mass cultures of larvæ that have been produced by pædogenesis in Oligarces sp. But males and females are not produced by the same individual under normal conditions. In the descendants of members of a colony taken from nature there were found to be two types of pædogenetic larvæ in respect to the sex and pupæ and adults which they produced, (1) male producing and (2) female producing. These two types of pædogenetic larvæ are not morphologically distinguishable." This would seem to mean that there are really six kinds of larvæ, three for each sex.

Polymorphism has acquired such a degree of complexity in the ants (Fig. 38 and 39), and especially in the termites that it will now be advisable to concentrate our attention on these groups. The male ant is an extraordinarily conservative and stable organism, although it is known to be dimorphic, i.e., to have a winged and an apterous phase in a few species (e.g., *Ponera eduardi*) and exclusively apterous forms in a few genera (Anergates, Formicoxenus, Symmyrmica, Cardiocondyla). So far as size, or stature is concerned, it may be used in each species, as Emery has shown, as a rough primitive standard with which to compare the much more variable queen and worker. The extraordinary range of variation in the female sex of the Formicidæ is indicated by the fact that no less than twenty-one different forms have been recognized. Of these eight are clearly pathological and produced either by the presence of parasites in the

Plate XXVI.

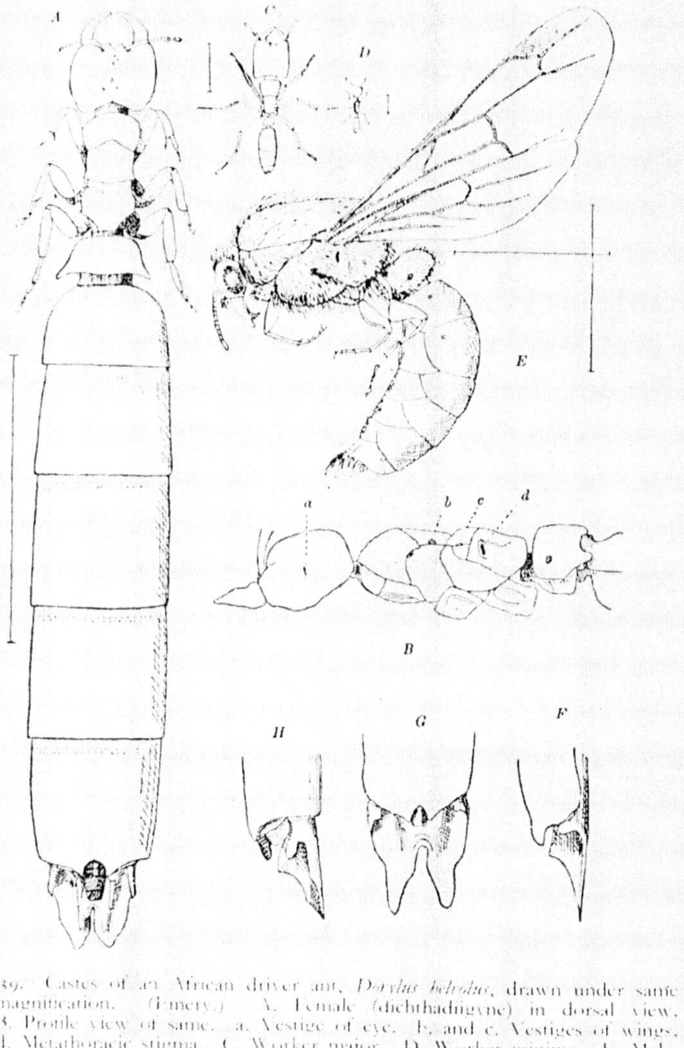

FIG. 39. Castes of an African driver ant, *Dorylus helvolus*, drawn under same magnification. (Emery.) A. Female (dichthadiigyne) in dorsal view. B. Profile view of same. a. Vestige of eye. b. and c. Vestiges of wings. d. Metathoracic stigma. C. Worker major. D. Worker minima. E. Male. F. Tip of gaster of female in profile. G. Tip of gaster of female *D. furcatus* in dorsal view. H. Same in profile. (After C. Emery.)

[*face p.* 172

Plate XXVI.

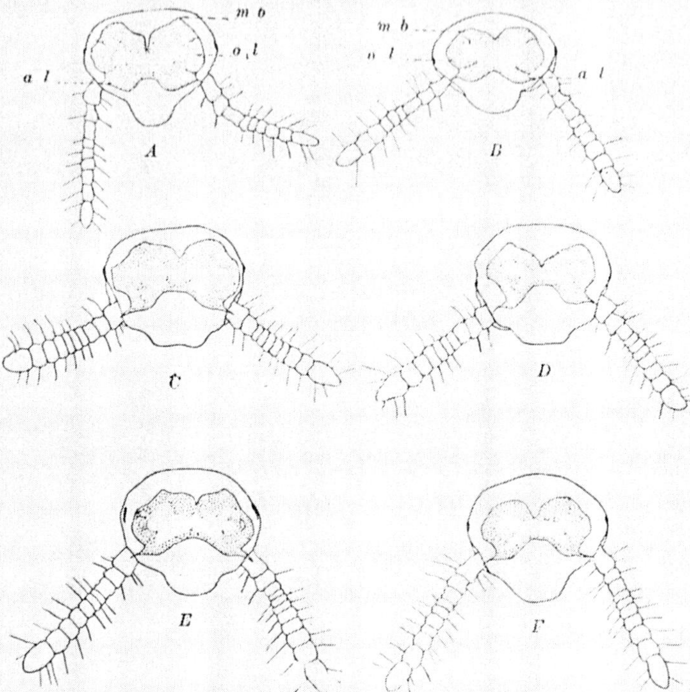

FIG. 40. Sections of heads of recently hatched termites to show relative size of brain and eyes. *A*, *C* and *E* just hatched reproductive forms. *B*, *D*, *F*, just hatched worker-soldier forms. *mb*. Mushroom body of brain. *ol*. Optic lobes. *al*. Antennary lobes. (After Caroline B. Thompson.) (*See* p. 182.)

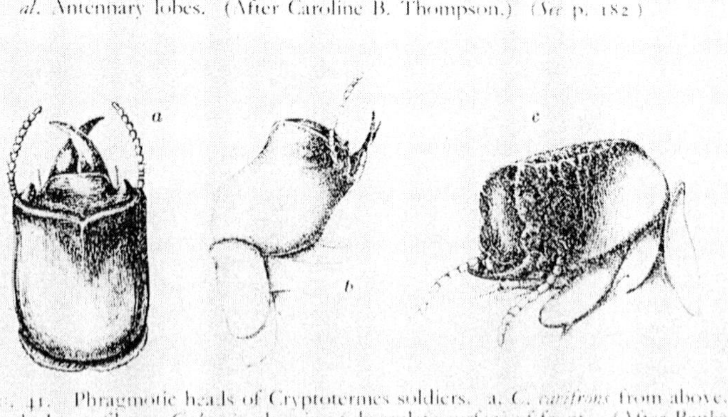

FIG. 41. Phragmotic heads of Cryptotermes soldiers. a. *C. cavifrons* from above. b. In profile. c. *C. brevis*, showing tuberculate surface of front. (After Banks and Snyder.) (*See* p. 188.)

[*face p.* 173

colonies or abnormalities of embryogeny or larval growth (chromosomal (?) or nutritional), four are merely excess or defect developments in stature (nanism and giantism) and the remaining nine are either normal forms of very general occurrence or specialized forms peculiar to certain genera. If we classify the twenty-one forms as typical, atypical and pathological we find that there are really only three typical forms or phases of the female, the queen, soldier and worker, and that each of them may be regarded as the centre of development of a number of atypical and pathological types, some of which are considered in the next lecture.

A comparative study of hundreds of species and all the known genera enables us to draw up the following list of hypothetical evolutionary stages in the pœcilogyny of the Formicidæ :—[1]

(1) The male, queen and worker are of the same or very nearly the same size, and the worker differs from the queen only in the absence of the wings, somewhat smaller eyes and ocelli, simpler thorax and more or less abortive reproductive organs. From this stage, which is represented by the most primitive subfamilies of ants, the Ponerinæ, Cerapachyinæ and Pseudomyrminæ, all the following stages may be derived.

(2) The queen has disappeared and the worker is fertile and capable of mating with the male. This stage occurs in a number of Ponerinæ genera (Diacamma, Rhytidoponera, etc.) and probably in Leptomyrmex among the Dolichoderinæ and Ocymyrmex among the Myrmicinæ.

(3) The queen increases in stature, while the worker retains more nearly the stature of the male. (Lasius, Brachymyrmex, Crematogaster, Iridomyrmex, etc.)

(4) The queen and worker are sharply differentiated, the former increases in size and the latter, becoming very unstable or variable, is represented by a series grading from large macrocephalic to small, microcephalic individuals (Figs. 38 and 39.) In taxonomic works the

[1] The stages and their arrangement differ somewhat from those published by Emery in 1894 and by myself in 1907.

members of this series are described as "workers major" and "minor," or "maxima," "media," and "minima" (Camponotus, Atta, Pheidologeton, Dorylinæ, etc.).

(5) Of the series in stage (4) the maxima and media workers alone survive and the former is converted into a soldier (*Machæromyrma bombycina*).

(6) The mediæ, or desmergates connecting the maximæ and minimæ of stage (4) are no longer developed, so that only the extremes persist. The maxima is then known as a soldier (miles, dinergate), the minima as the worker (proper). The female is therefore now distinctly trimorphic (Pheidole, Ceratopheidole, Acanthomyrmex, Ischnomyrmex, Oligomyrmex, etc.)

(7) The queen disappears and the soldier, becoming fertile, assumes her rôle, while the minima alone represents the worker caste. This stage is problematical but seems to exist in some species of Pheidole and Oligomyrmex (*Ph. lamia* and *O. panamensis*).

(8) Only the queen and the soldier forms persist, the workers being suppressed. This seems to be the condition in the peculiar slave-making ants of the genera Polyergus and Strongylognathus.

(9) All the worker forms in the series of stage (4) except the minima, disappear so that we have enormous queens and very minute workers of a single type (Carebara (Fig. 72), Pædalgus, many Solenopsis, etc.).

(10) The entire worker caste is lost, so that the species returns to a simple sexual dimorphism, i.e., is represented by single male and female forms as in most insects and other Metazoa. The female of this stage, which occurs only in certain permanently parasitic ants (Anergates (Fig. 78), Wheeleriella (Fig. 77), Epœcus, Pseudoatta, etc.) is scarcely larger than the male.

We must not interpret these stages as succeeding one another in a definite linear series but as radiating from the first stage which is basic. Stages (2) to (10) have evidently arisen independently and repeatedly in unrelated genera. It can be shown, furthermore, that most of the stages are responses to peculiar environmental and ethological requirements, in last analysis mainly of a trophic nature. This is perhaps most obvious in stages (9) and (10), of which we may take Carebara and Anergates as examples. The species of Carebara (Fig. 72) are " thief-ants " which live in the walls of the termitaria of African and South American Termitidæ. The nest is connected with the chambers of the termites by very tenuous galleries, too small to permit the termites to pass but permitting the diminutive Carebara workers to enter and devour the termite brood. So abundant is the food supply that these workers are able to rear gigantic queens, but are themselves, on account of their cleptobiotic propensities, condemned to perpetual dwarfhood. In the monotypic *Anergates atratulus* (Fig. 78), which is a permanent parasite in the colonies of the common European *Tetramorium cæspitum*, there is no need of a worker caste, because the broods of both species are brought up by the host workers. The Anergates workers have therefore long since disappeared and the queens have been reduced in size. Owing to this secondary return to the sexual dimorphism of the solitary Aculeates, the number of females that can be reared is greatly increased, an obvious advantage in a parasite, and their decreased stature enables them to be more readily adopted by the Tetramorium colonies. The latter actually seem to prefer the minute parasites to their own huge queens, just as some ladies, for æsthetic or economic reasons, prefer the tiny Pekinese dogs to huge mastiffs as pets. These few facts and many others that might be adduced, show that the number and character of the female castes can be regulated by the ant colony, just as the number and character of its cells are regulated in the body of a Metazoan. This regulation, which is seen both in the development and abolition of castes, has an ontogenetic as well as a phylogenetic aspect and is in some respects a very plastic, in others a rather rigidly determinate process, depending apparently on the phylogenetic age of the castes in the particular species under observation.

That the various female castes are adaptive can

scarcely be doubted since they have arisen in the colony as an expression of a physiological and behaviouristic division of labour. Considerable confusion seems to have been introduced by the recent work on the morphology of the brain of the ants and other social Aculeates and a failure to take the facts of behaviour adequately into consideration. It would seem that the fertile female, or queen of the non-parasitic social Aculeate is to be selected as a fundamental type of which the various worker and soldier forms are to be regarded as partial or more or less modified expressions or differentiations. The instincts of the queen embrace, so to speak, all the essential instincts of the sterile or substerile castes. This has never been disputed in Vespa and Bombus, and there is no doubt that the queens of these insects have a larger and more perfectly organized brain than the cospecific workers (von Alten, 1910). The contention of most authors that the differentiation of the corpora pendunculata of Dujardin, or " mushroom bodies," as they are often called, which are situated in the anterior portion of the protocerebrum, is directly proportional to the behaviouristic capacities of the insect, holds good of the queen, worker and male brains of Vespa. In the honey-bee, however, as Jonescu (1909) has shown, the brain of the queen is inferior to that of the worker, but in that insect the queen is really a parasite on her own colony and, unlike Vespa, quite unable to bring up her offspring without worker aid. In the ants most queens exhibit a behaviour and brain capacity like that of Vespa, but some queens, mostly parasites in the colonies of other species, are deficient in behaviouristic endowment and possess smaller and less highly differentiated brains than the workers of their own species. The latter type, according to my observations, is exceptional, the former the more prevalent, and this is the basis of Miss Thompson's and my disagreement with Forel (1874), Pietschker (1910), Ziegler (1912, 1920) and Brun (1923*a*). Forel first advanced the generalization (1874) that the brain of the worker ant is superior to that of the queen in his studies on *Lasius fuliginosus*, now known to be a temporary social hyperparasite of *L. umbratus*, which in turn is a parasite of *L. niger*. And although I called attention to this fact, Brun has recently reached the same conclusion as Forel, mainly from a study of *L. umbratus!* Miss Thompson's researches on three species and genera of non-parasitic ants, *Camponotus pennsylvanicus*, *Formica schaufussi* and *Lasius americanus*, support the opinions which I advanced in 1910, although she finds that the brain of the *F. schaufussi* queen is degenerate compared with that of the worker. She summarizes her results as follows :—" The queen brain seems to me to represent the generalized type from which the worker caste has departed, and, while some queens are notably degenerate in brain structure, others have remained in a far more generalized condition, for example, the queen of *Lasius niger*. If we select as a standard the degree of development (1) the optic apparatus, including the eyes and optic lobes, (2) the mushroom bodies ; and compare the castes of the three genera under consideration with the queen of *Lasius niger*, we shall obtain evidence of divergence from this type in the two opposite directions of increase, and of reduction of parts, or degeneration. From this comparative study two facts are worthy of note : (1) the Lasius queen has a more highly developed, generalized type of brain than either of the queens of Camponotus or Formica. (2) the queen brain in its most highly developed typical condition, as in *Lasius niger*, is superior to and more highly developed than that of the worker. Therefore, the conclusion is justified that the queen brain is the primitive type from which by degeneration and specialization of structure the worker brain has been derived."

The discrepancy between the results of Miss Thompson and myself on the one hand and those of Forel, Pietschker, Ziegler and Brun on the other are difficult to explain. Two of the species investigated by Pietschker and Miss Thompson are really identical (*Camponotus pennsylvanicus* and *ligniperda* being subspecies of *C. herculeanus* and *Lasius americanus* a variety of *L. niger*), and all of us used material in the same stages, i.e., late pupæ before pigmentation. I mention this matter because Holmgren (1909) has shown that the brain of the old physogastric queen of *Eutermes chaquimayensis* shrinks to two-thirds of its size in the virgin queen, showing that profound involutionary changes occur during the imaginal instar. Not improbably similar changes may occur in the brains of deälated queen ants at the time of the dissolution of their thoracic musculature. Hence the comparison of the brains of such old queens with those of their workers might seem to support the views of Forel, Pietschker and Brun. It may also be advisable in future work on the ant brain to pay more attention to the behaviour of the queens. It is singular that although *F. schaufussi* is one of the most abundant of our North American ants, I have never found its queen in the act of founding a colony independently. If, as seems possible, the young queens establish their colonies by emigration with companies of workers, we should have a condition like that of the honeybee (or rather like that of the Meliponinæ) and we should be able to account for the smaller and more degenerate brain in the queen.[1]

[1] I have been criticized by Forel (1921-23) and Brun (1923*a*) for my skepticism in regard to the close relationship between the psychic behaviour and development of the pedunculate bodies in ants (1910*a*). More than twenty-five years ago, one of my students, the late Dr. C. H. Turner undertook a comparative study of the corpora pedunculata in the Annelids and the lower Arthropods and in 1899 published a preliminary but no final paper on the subject. Since this paper seems never to have been consulted by investigators of the insect brain, it is only fair to glance at its contents. Before its publication, Hamaker (1898) had shown that there are corpora pedunculata in the brain of the Polychæte Annelid *Nereis virens* and Dietl (1876), Berger (1878) and Viallanes had noticed their occurrence in Decapod Crustacea. Turner also found them in Annelids (Nereis, Polynoë, Lepidonotus) and Decapods (Cambarus). His figures show that they are much more highly developed in Polynoë and Lepidonotus than in Nereis or in many of the lower insects, e.g., Tomocerus and Lepisma, according to Kuehnle (1913), Boettger (1910) and Bretschneider (1915). But the most astonishing condition is seen in Limulus, in which, as Patten (1894) and Turner have shown, the corpora pedunculata are so large and complicated that they resemble the cerebral hemispheres of a higher vertebrate! They have, in fact, advanced much further in relative volume and morphological differentiation beyond the ant corpora pedunculata than have the latter beyond those of the lower Orthopteroids. I ask again, therefore: what is such a stupid and archaic

The polymorphism of the termites, to which we may now turn, is more complicated than that of the ants, both because it affects the two sexes equally and because it is more elaborate in each of them. We may distinguish the following eight castes, with sixteen different kinds of individuals :—

(1) First form males and females (true kings and queens)
(2) Second form males and females (neotenic kings and queens)
(3) Third form males and females (ergatoid kings and queens)
(4) Large male and female workers
(5) Small male and female workers
(6) Large male and female soldiers
(7) Medium male and female soldiers
(8) Small male and female soldiers

Probably no single colony of termites ever produces all of these castes but five or even six of them are frequently encountered. Moreover, the soldiers may be of two very different types, most species having large-headed "mandibulate" forms (Figs. 38 and 54), while some of the genera of Termitidæ have a very different kind, the "nasuti" (Figs. 52 and 53), with small head and mouth-parts and a long snout, with the frontal gland opening at its tip. The matter is complicated by the fact that the workers, soldiers, and neotenic individuals, especially in the lower termites, may be differentiated at different instars, so that they vary in the number of antennal joints and in other minor characters. It has been usually

creature as Limulus doing with the most highly developed corpora pedunculata in the whole Arthropod phylum? Forel says he knows nothing about Limulus and seems to imply that he cares less, so convinced is he of the precise function of the structures in question, but if they are really an infallible index to the "plastic-psychical and individual mnemic capacities of insects", as Brun contends, we must assume either that Limulus, somewhere in the depths of the sea and quite unknown to us, exhibits an astonishing wealth of "plastic mentality", or that its extraordinary corpora pedunculata are a wonderful depository of "mnemic engrams", perhaps painfully acquired by the enterprising Palæozoic Protolimulus and transmitted as useless heirlooms to its modern moronic descendants!

supposed that as we advance from the lower to the higher termites polymorphism tends to increase, but Holmgren believes that the reverse development really occurs. He gives the following series of the more common stages, which afford an interesting comparison with those above enumerated for the ants :—

(1) Soldiers (mandibulate) highly polymorphic, of a single type ; neotenics and ergatoids frequent ; definitive workers lacking (Calotermes, Termopsis).

(2) Soldiers (mandibulate) polymorphic, of a single type ; neotenics and ergatoids frequent ; definitive workers present, of a single form (Leucotermes)

(3) Soldiers polymorphic, of two types (mandibulate and furcate nasuti) ; definitive workers probably present, of a single form (Rhinotermes).

(4) Soldiers (mandibulate) dimorphic, of two types, large and small ; neotenics and ergatoids unknown ; definitive workers dimorphic, of two sizes (Acanthotermes, Termes).

(5) Soldiers (mandibulate) monomorphic, of one, larger type ; neotenics and ergatoids unknown ; definitive workers dimorphic, of two sizes (Syntermes, Odontotermes).

(6) Soldiers (mandibulate) monomorphic, of one, smaller type ; neotenics and ergatoids unknown ; definitive workers dimorphic, of two sizes (Odontotermes).

(7) Soldiers (mandibulate) and workers monomorphic ; neotenics and ergatoids of occasional occurrence (the great majority of higher termites)

(8) Soldiers lacking secondarily ; workers monomorphic (Anoplotermes)

(9) Soldiers (nasuti) secondarily dimorphic, of a single type ; workers monomorphic ; neotenics and ergatoids unknown. (*Eutermes bivalens*, *aquilinus*, group of *E. trinervius*).

(10) Soldiers (nasuti) secondarily trimorphic ; workers monomorphic (*E. diversimiles*, *castaneiceps*).

(11) Soldiers (nasuti) secondarily dimorphic ; workers monomorphic (*E. heteropterus*, *velox* and others).

(12) Soldiers (nasuti) secondarily monomorphic; workers monomorphic (*E. cyphergaster, tenuirostris, hospitalis*, etc.).

To this list Holmgren adds the following commentary: " Evolution seems here to proceed from a very comprehensive polymorphism towards simpler conditions. The simplification culminates in Anoplotermes, in which soldiers no longer occur and the workers are monomorphic. In the simple, trimorphic stage (7), however, a secondary caste differentiation, stage (10), supervenes, which is followed by another increase in the number of castes. This more incisive, secondary polymorphism (pentamorphism) is again in turn reduced to a simple trimorphism (stage (12)). The development from stages (1) to (7) proceeds from more labile (fluctuating) to more stable conditions of caste, because the soldier class in the first stages of this series may arise by breeding from a greater number of different larval forms. The germ plasm of groups (1) and (2) is, so to speak, in a " trial and error " phase (" Versuchsstadium "). Their mobility is due, perhaps, to the soldiers and " workers " being in great part fertile and this may have brought about a complex germ plasm mixture. This is only wild speculation, but it agrees with the conditions of species formation which I have already discussed. We learn, however, that such lability may also arise without fertile soldiers and workers from the species of Eutermes of stages (9) and (10), where a secondary " lability " sets in. When among the ants a polymorphic worker class becomes dimorphic the condition is often said to be due to the dying out of the intermediate forms. It does not seem to me to be necessary to assume such an extinction, since the course of events may very well be due to an omission of differentiation. When, e.g., in Rhinotermes a sharply defined soldier dimorphism arises, it is probably not produced by a suppression of intermediate soldier forms but to an independent differentiation of those already existing." The question here touched upon by Holmgren would seem to be rather academic, since a phylogenetic loss can only express itself in the ontogeny as a failure to develop or to differentiate.

Having treated of the Aculeate brain it will be of interest to glance at that of the termites, especially as some significant observations on caste differentiation have been based on that organ. It has been studied by Holmgren (1909), Kuehnle (1913), Miss Thompson (1913, 1916, 1917, 1919) and Jucci (1923). Miss Thompson, after a comparative study of the various castes of *Reticulitermes flavipes*, sums up her observations on the corpora pedunculata as follows: " The mushroom bodies differ very little in size, by actual measurement and in the estimated number of cells. They are largest in the worker, smallest in the soldier and are intermediate in size in the sexual forms, although the mushroom bodies of the true adult are nearly as large as those of the worker. A comparison of the Reticulitermes " mushroom body " with those of ants and bees shows, as one would naturally expect, that the termite mushroom body is much more simple and primitive. This primitive condition is apparent in the small and uniform size of all the nerve cells especially in the cell group I; in the presence of three zones of cells instead of the four found in ants; in the incomplete differentiation of the two lobes whose cells are not separated by a deep furrow, as in ants, and whose two cups or calyces are completely fused by intervening masses of glia cells; in the shallowness of the cups; finally in the smaller size of the entire mushroom bodies and their slight differentiation in the different castes." This is in harmony with the archaic Orthopteroid affinities of the termites, although according to Kuehnle, the termite brain has considerably larger corpora pedunculata than that of the Dermaptera.

In her later papers (1916, 1917, 1919) Miss Thompson describes the brain and reproductive organs of the sexual forms as being perceptibly larger than those of the worker even on hatching from the eggs (Fig. 40). Her interesting results are summarized in a paper by her and Synder (1919): " Thompson (1917) finds that in *Reticulitermes flavipes* the newly hatched nymphs are externally all alike, but are differentiated by internal structural characters into two clearly defined types, the reproductive and the worker soldier types, which give rise respectively, to the three adult reproductive castes and two adult sterile castes. The two types of newly hatched nymphs are distinguishable by four structural characters: the bulk of the brain, the relative size of the brain and head, the compound eyes, and the sex organs. When the young reproductive nymphs have attained a length of 1.3-1.4 mm. other structural differences are observable that further differentiate them into two kinds of individuals which later develop into two of the three adult reproductive castes; namely, adults of the first form with long wings, and adults of the second form, with short, scale-like wing pads. The ontogeny of the third reproductive caste, without wing stubs or wing pads, is yet to be worked out in *R. flavipes*. At a later period in the ontogeny (body length 3.75 mm.) the worker soldier nymphs differentiate into two kinds of nymphs which develop into the two sterile adult castes, the soldiers and workers." Miss Thompson and Snyder (1920) also found that the adult sexual individuals of the first form (royalty) really have a larger brain than the other castes, that the brain of the second and third forms is larger than that of the worker and the brain of the soldier smallest.

Of course, this discovery of the intrinsic differentiation of fertile and sterile castes in the egg is of unusual interest. Bugnion (1912, 1913, 1914) described and figured the soldier of *Eutermes lacustris* as hatching from the egg as a nasute, although Knower (1894) had claimed that the nasute of *E. pilifrons* was not differentiated at the time of hatching. On reinvestigating the matter, Miss Thompson (1919) concluded that Bugnion must have mistaken the projecting labrum of the hatching *E. lacustris* for the cephalic horn of the nasute! Her observations on just-hatched nymphs of *E. pilifrons, morio* and *sanchezi* confirm those of Knower, but in her paper of 1919 she claimed to have detected in thirteen species belonging to nine genera of termites the same differentiation which she had previously found among hatching nymphs, into two forms, corresponding to the two series of adult fertile and adult sterile castes. More recently her results have been seriously questioned by Jucci (1923) and Heath (1927). The former's elaborate study of the neotenic sexual forms of *Reticulitermes lucifugus* led him to support Grassi's contention that the hatching termites are all alike and acquire their caste distinctions as the result of differential feeding. Jucci's statements are very positive, not to say dogmatic, and the portion of his work published to date, though containing measurements of the brains of the adult winged and neotenic castes does not deal with the soldiers or their earliest nymphal stages and is therefore disappointing so far as the problem under dicussion is concerned. Heath, after a careful study, extending over many years, of the Californian *Termopsis nevadensis* and *angusticollis*, is also very positive that Miss Thompson was mistaken in believing that there are two structurally recognizable types among larval termites at the time of hatching. In fact, " measurements of scores of recently hatched young disclose no differences other than those of ordinary variation, either in width of head, number of antennal segments, or size of brain and gonad. The first visible signs of caste differentiation appear at a relatively late stage." Termopsis possesses no worker caste and " only soldiers are produced during the first three or four years. When a normal colony comprises approximately twenty inhabitants, the first adult soldier is probably in the fifth instar ; the second, somewhat larger, certainly is in the sixth. Later, others are produced in the seventh, and after the reproductive caste becomes differentiated, with about 450 inhabitants, they are in the eighth. Finally, in very old communities, some at least, are in the ninth. The winged insect invariably is in the eighth. No sign of wings exists until the reproductive caste appears, after which every member of the sixth, and probably the fifth, instar possesses small, yet distinct wing rudiments. These persist and enlarge in the later stages of the reproductive caste, and, of smaller size, they frequently occur in the soldier nymph, and to a less degree in the adult. The third form adult is probably a sexually mature soldier nymph; the second form adult seemingly has the same origin and, though possessing wing buds, it does not belong to a special caste." Heath has found four fecund Termopsis soldiers, which differ from those of the typical form in the smaller and shorter head (Fig. 54). Of one of these he says: "When this fourth specimen was discovered it was associated with a very small substitute male (as subsequent dissection proved), and about 150 individuals of the first four instars together with twenty-six fully developed soldiers. The colony was located in a small branch that had fallen from some tree, and evidently was a portion only of a larger community. This colony, from which the adult soldiers had been removed, was placed in a shallow crystallizing dish, and, nearly five weeks later, the fertile soldier and substitute male were seen to copulate. Eggs appeared two weeks later, and, due to close observation on the part of several of my students, the egg-laying activities of the 'soldier' were actually witnessed—a phenomenon I had observed some years previously. The young hatching from these eggs were carefully examined and, so far as could be determined, they resembled in all essential respects the first instar of normal parentage."

The observations of Imms on Archotermopsis and Heath on Termopsis indicate very clearly that in termites each sex was originally represented by two fertile forms, a winged small-headed form and an apterous or subapterous large-headed soldier, which was differentiated from the small-headed type and later became sterile. This view, which is opposed to that of earlier writers, who regarded the soldier as a specialized worker, is shared by Emerson (1926), who states his opinion as follows: "I hold the opinion that the soldier caste has developed pylogenetically from the first reproductive adult and not from the worker as suggested by Thompson (1917). The evidence for my opinion is found in the following facts: 1. Primitive soldiers are sometimes fertile or approach fertility (Heath, 1903; Imms, 1919; Thompson, 1922). 2. Soldiers are often found with wing buds which may be pigmented in the Kalotermitidæ. None with wing buds have ever been found in the higher termites, however. If the soldiers evolved from the third form, how could they possess wing buds? 3. The primitive soldiers always possess small compound eyes and may even possess traces of the lateral ocelli in some cases. 4. Soldiers are found among all the primitive termites, even where the worker is lacking. The only genus which lacks soldiers (Anoplotermes) has obviously lost them secondarily as all its close relatives possess soldiers." Emerson seems to have been led to take this view by his rather startling discovery in British Guiana of the direct transformation by a single moult of what were obviously functional workers of *Nasutitermes* (*Constrictotermes*) *cavifrons* into nasute soldiers. (Figs. 52 and 53.) In this case, of course, the worker must be interpreted as a differentiated stage in the life-history of the soldier. Unlike Miss Thompson and Snyder (1919) and Imms (1919), who regarded the nasute as a separate caste, Emerson believes it to be a specialized form of the mandibulate soldier. The process of specialization is actually shown in the genus Rhinotermes. "The facts indicate that the large mandibulate soldier of Rhinotermes gave rise to a minor soldier which gradually developed an elongated labrum for dispersal of the secretion of the frontal gland, and lost the large mandibles which gradually degenerated into functionless minute points. The evolution finally resulted in the establishment of two types of soldiers within the same colony, one specialized for biting, the other specialized for the emission of a defensive volatile liquid."

In conclusion I wish to add some remarks on the different types of soldiers which are of unusual interest both among ants and termites. Although they have arisen independently in both groups they exhibit certain marked similarities and in both constitute at least primarily a defensive or protective caste. While this caste is universally present in termites, except in the genus Anoplotermes, where it has been lost secondarily, it has been definitively developed in comparatively few genera of ants. And whereas in termites as shown by Mastotermes, Archotermopsis, Termopsis, Calotermes, and perhaps Hodotermes, it has preceded the appearance of the worker in phylogeny, in ants it has as certainly differentiated out of the more primitive worker caste. In the latter there is only one general type of soldier, the mandibulate, which is also very generally developed among the termites, but in some of the higher Termitidæ (Eutermes), as we have seen, this is replaced by a second type, the nasutus. There is, moreover, a difference between the ants and termites in that the soldier of the former differs much less from the cospecific queen and worker in the size and conformation of the head and mandibles than do the soldiers of the latter from the cospecific worker and sexual forms.

Now the methods of defence may be very different in different species and accordingly we find in both ants and termites soldiers adapted to particular offices, or "professions." In one form, occurring in both groups, the soldier may use its head for merely occluding, or blocking the entrance gallery of the nest in order to keep out intruders. The soldier may, however, and this is the most frequent condition, have an enormous cranium with powerful jaws for aggressive attack on invaders. Owing to the excessive development of the mandibles and their muscles, which in turn determine the shape, greater size and hardness of the cranium, these soldiers may also be utilized by certain species of ants as crushers of seeds or the harder parts of the insect food, which without this assistance would be inaccessible to the feeble worker caste (Pheidole, Pheidologeton, etc.). The soldiers of some ants (Hingston, 1922) and termites seem also to act as directors of the foraging commissariat. Many mandibulate termite soldiers and all the nasuti have a large frontal gland which produces a sticky liquid, or latex, that can be projected against and disable their enemies, especially marauding ants. This gland in certain species, e.g., in *Rhinotermes taurus*, according to Holmgren (1909) and *Coptotermes ceylonicus*, according to Bugnion and Popoff (1910) and Bugnion (1914, 1923), is so enormously developed that it extends back into the abdomen and occupies a considerable portion of its cavity. Other singular conformations of the head in soldier termites, very unlike those of the adult and workers, are seen also in some mandibulate types, like Capritermes, the jaws of which are long, narrow, twisted and very asymmetrical and enable their possessors to leap into the air (Bugnion, 1923).

Very singular is the type of ant and termite soldier which has the head modified to fit the entrance of the nest and render the ingress of enemies difficult or impossible. The only cases known to me among the termites occur in the genus Cryptotermes (Fig. 41), though the soldiers of the allied Calotermes use the head in the same manner. As usual, the modification occurs neither in the nymphs of Cryptotermes, which function as workers, nor in the sexual castes, all of which have heads of the ordinary simple conformation. In ants with soldiers of this type, the head of the queen is very similarly but less perfectly adapted. The most familiar examples among the latter insects are the Camponoti of the subgenus Colobopsis, several of which occur in Eurasia and North America. Their head is short, cylindrical and anteriorly sharply truncated, with the truncated surface, including the mandibles, circular, indurated and more coarsely sculptured than the remainder of the body, so that it fits very snugly into the circular entrance of the nest, which is excavated in hard wood, ligneous galls or the stems of rushes. When a Colobopsis worker wishes to leave the nest she strokes the soldier's abdomen with her antennæ and the animated door moves back and as soon as she has passed out returns at once to its former position. On returning she knocks with her antennæ on the exposed truncated surface of the janitor's head and a similar response permits her to enter. I find this same type of head in single exotic species of three other unrelated genera, namely Pheidole (*Ph. colobopsis* of Brazil), *Crematogaster* (*Colobocrema*) *cylindriceps* of the Philippines and *Epopostruma* (*Colobostruma*) *leæ* of Australia, all of which probably have much the same habits. There are also several lignicolous subgenera of Camponotus (Paracolobopsis, Pseudocolobopsis, Manniella, Neomyrmamblys) which exhibit a similar structure of the head in the queens and major workers. In many species of Cryptocerus, which live in hard wood, the heads of the soldiers are broad and shield-shaped, and are also used for closing the nest entrances.

Very similar adaptions for occluding burrows, etc., are found not only in a number of other Arthropods but even in animals belonging to other phyla. In some cases the head, in others the posterior end of the body is thus adaptively modified, and in both instances the truncation, its circular outline and the induration of its integument are often strangely similar. Thus in larval tiger-beetles (Cicindelidœ) and the burrowing bees of the genus Halictus, the head is round and plug-shaped, whereas in other forms, like the bark-beetles (Ipidæ, Platypodidæ) and some carpenter bees (Xylocopa), which live in burrows in hard wood, and in the caterpillars of a North American moth (*Cicinnus melsheimeri*, according to Harris (1862)) and an allied South American moth (*Perophora sanguinolenta*, according to Sharp (1899)), which inhabit tubular cases made of leaves, the posterior end of the body is more or less truncated, hardened and spinulate. Similar modifications of the head are also seen in some tubicolous Annelids (Maldanidæ, Amphictenidæ), and Barbour (1926) and Dunn (1926) have described certain toads (*Bufo empusus* of Cuba) and tree-frogs (*Hyla lichenata* of Jamaica) which occlude their burrows with their hard, peculiarly modified heads. Still other cases, in which the posterior end of the body is similarly adapted are the small armadillos of the genus Chlamydophorus, certain Uropeltid snakes, according to Barbour, and the Mexican spider *Chorizops loricatus* (Fig. 42), which instead of making a trap-door like the allied species, closes its burrow with the sharply truncated and corrugated tip of its abdomen. Hentz (1875) long ago called attention to a similar habit in a related spider from the Southern States (*Cyclosoma truncatum*).

Since there is no general term to cover all these peculiar, sporadic and convergent modifications of the ends of the body for closing tubular cavities, I have suggested the word " phragmosis," from φραγμός, a fence or barricade. From the evolutionary and behaviouristic points of view the phenomenon, as one of the most striking and effective methods of protection and defence, would seem to deserve more careful investigation than it has received. The phragmotic insect, instead of secreting or constructing a stopper, like the operculum or epiphragm of snails and the earthen or silken barricades or doors erected at the entrances of their burrows by ants, wasps and trapdoor spiders, actually employs for the purpose a specialized portion of its own body, thus affording a proof that no hard and fast line can be drawn between behaviouristic activities on the one hand and physiological and morphogenic processes on the other. The phylogenetic development of phragmosis is obscure. The ants and termites, at least, seem to show that it cannot have arisen as a sudden, saltatory variation, or mutation, but must have developed gradually, since we have among many lignicolous Camponoti and species of Calotermes various approximations to the perfected conditions seen in Colobopsis and Cryptotermes. I have dwelt on the sporadic cases of phragmosis, because they show very clearly the adaptive character of at least one of the soldier types among ants and termites and therefore make it seem probable that this caste has not arisen as a sport, or mutation. The significance of this statement will be shown in the next lecture.

VIII

POLYMORPHISM (*continued*)

WHEN we turn from the facts of polymorphism as ascertained by observation and experiment to their interpretation we at once find ourselves involved in the physiological problems of ontogeny and growth, of phylogeny, or evolution and of behaviour, or the activities of the organism—in fact in all the really fundamental biological problems. For there is no *a priori* reason to suppose that the development of the castes among social insects requires any new or unusual explanation since intelligent behaviour, which plays a rôle so considerable and so unique in the interpretation of human societies and gives such scope for diversity of opinion among sociologists and philosophers, is so feeble even in the social Aculeata as to be quite negligible in comparison with their physiological and instinctive activities. In my paper on polymorphism, published in 1907, I endeavoured to treat the subject from a number of points of view, but here it seems best to reduce the problem to the alternative between predetermination and epigenesis. This problem, as is well known, has long divided biologists into two camps comparable to those among philosophers, who are either Platonists or Aristotelians, idealists or realists, nativists or empiricists, instinctivists or environmentalists, etc. The morphologist, primarily an observer, naturally inclines more to the predeterministic, the physiologist, primarily an experimentalist, to the epigenetic view. Each tends to a one-sided interpretation of the facts, and both, we are now convinced, are right—there is predeterminism and there is epigenesis in organic development, but there is still abundant opportunity for divergent opinions on the scope of the two factors in any particular case. In fact, present day geneticists and physiologists are nearly as widely separated as were the old preformationists, or evolutionists and the epigenesists.

The difficulty of deciding as to how far differentiation of the two or three female castes of the social Aculeates and the five or more castes of termites may be intrinsic or predetermined in the egg or as to how far it may depend on extrinsic conditions, arises from the fact that the castes develop only in a very special and, to use Child's term (1924), "standardized" environment, which cannot readily be brought under experimental control. They are, indeed, produced only in living societies, whose activities are so multiform, so interdependent and so exquisitely equilibrated that the various factors can be isolated only in thought. Even the maintenance of colonies for considerable periods of time under artificial conditions is far from easy, and the small size of the insects presents almost insuperable obstacles to a physiological inquiry into many of their activities with our existing methods. All their subtle, intimate and significant behaviour, moreover, normally takes place in complete darkness, and even the letting in of the amount of light requisite for observation is a very disturbing factor. Nevertheless, in our search for the causes of caste differentiation, we may exclude from the outset such extrinsic factors as temperature, moisture, and darkness, not because they are inoperative, but because they act uniformly on all the castes in their various ontogenetic stages. It is obvious, moreover, that the efficient extrinsic factors must emanate from the adult worker caste, because in all social insects, after the colony is established, it is not only far and away the most numerous in individuals but is exclusively concerned with the acquisition and distribution of the food and the education of all the castes, its own, of course, included. Hence the differentiating factors if not blastogenic, or in the eggs, must be trophogenic, or due to differences in the nutricial behaviour of the workers which determine the social medium. But even with these restrictions the problem of polymorphism still retains an exasperating complexity, since there are at least six possible interpretations, which may be formulated as follows :

(1) The eggs are all alike and the sexual, worker and soldier castes are differentiated or produced by the social medium by means of larval feeding (Grassi, Marchal, and all melittologists).

(2) The various castes are predetermined in different eggs but the predetermination is so feeble, or so incompletely established, that the feeding of the larvæ may reverse or abolish it, with results essentially like those of (1) (Bouvier).

(3) The castes are predetermined as such in the eggs, i.e., each caste develops from a different kind of egg and the trophic influence of the social medium has no morphogenetic but at most a quantitative (megetic) effect in determining stature (Bugnion, Thompson, Imms, etc.).

(4) The castes are predetermined to the extent that the eggs are all alike, but each has either

(a) a set of ids, determinants or genes which may be activated by a particular kind of food and thus produce a specific caste (Weismann), or

(b) differential susceptibilities, or potentialities, which may react specifically to a particular kind of food. (Emery, O. Hertwig.)

(5) The castes may be blastogenic (3, 4) in some groups of social insects, and trophogenic (1, 2) in others.

(6) Some of the castes in the same species may be blastogenic, while others are trophogenic.

You will observe that the distinctions between these possible interpretations are rather subtle, especially in regard to the nature of the blastogenic determinations. Since they must be biochemical and physiological and are still unknown, I shall not waste your time in discussing the essentially morphological hypotheses that have been advanced by geneticists to account for them, hypotheses which are, in my opinion, essentially revampings of Weismannian fallacies and at best photographs of the problem in pseudo-mathematical language. All we really know is that the egg of any species of organism has a specific colloidal constitution and gives rise to an organism determined in part by that constitution and in part by the action of the medium in which it always lives and develops. And although eggs of the same parent undoubtedly differ more or less in their constitutions, it is equally certain that the precise nature of the differences has not as yet been revealed to our too statically and too morphologically minded geneticists. Let us therefore turn to the extrinsic factors or methods of larval feeding, which though extremely diverse, nevertheless admit of a certain amount of direct observation and experiment.

Obviously larval feeding can be distinguished as either quantitative or qualitative, and it has been usually assumed that the former effects only growth in size, or stature whereas the latter may determine morphological differences. If we take nutrition in a broad physiological sense, we may distinguish foods that are utilized merely in the metabolism of growth, maintenance and replacement and those which have a stimulative or excitatory effect on the metabolic processes. The latter, even in extremely minute quantities, like the vitamines and hormones, may be decidedly morphogenic. On the other hand, qualitatively different foods may have no appreciable effect on form differentiation, and in plants pronounced morphological differences may result from quantitative differences in food as in the various forms of leaves on shoots of different vigour. Although it is difficult to find similarly unequivocal examples in animals, they probably occur. It should be noted, furthermore, that it is not the food administered that is effective but that which is actually assimilated and, as we all know from experience, this depends on the physiological state and appetite of the organism.[1]

That differences in the quantity of food consumed by the larva affect the size of the imago is known to all entomologists. Weismann (1892) reared blowfly larvæ on different amounts of food and obtained very small, but fertile individuals. Whether they actually laid as many eggs as unstarved individuals seems to be doubtful.[2]

Similar experiments have been performed by Herms (1907) and Whiting (1914) on flies and by Nelson and Sturtevant (1924) on honeybees. Popovici-Baznosanu (1910*a*, 1910*b*) succeeded in rearing bees of the genus Osmia about one-sixth normal size by leaving them only small portions of the bee-bread which their mothers had stored with the eggs. These examples illustrate a very general ability of starving insect larvæ to produce dwarfed but otherwise normal adults. A much more remarkable resistance to starvation was shown in Wodsedalek's experiments (1917) with *Trogoderma tarsale*, a common pest in museums. He succeeded in keeping the larvæ of this beetle alive for five years and nearly two months without food. During that time the full-grown individuals (8 mm. long) gradually became smaller and smaller till they reached practically their size at hatching (about 1 mm.) when they had only about one-six-hundreth of their original size. When fed these reduced larvæ again

[1] " Those who have reared animals will have observed that the young hatched from the same batch of eggs do not grow uniformly though they may be living in the same medium, breathe the same air and receive the same food abundantly. This has impressed me especially in breeding axolotls; certain larvæ grow conspicuously and soon become vigorous specimens, while others remain small and weak and others pine away and die. These differences are due in part no doubt to defective aquarium conditions which affect different individuals differently. But the same facts are observable in all other cultures, whether of fish, amphibians or insects. *The most successful individuals are the most voracious, those that eat most or have the best digestions and therefore the best appetite* ". (Emery, 1896.)

[2] Ezikov (1926) claims to have demonstrated that in blowfly larvæ " insufficient food brings about a decrease in the number of ovarian tubules because the parts of the ovary are quantitatively reduced in the pupa ". And Reynolds and Macomber (1921*a*, 1921*b*, 1923) find that starvation has a tendency to produce sterility even in rats and man.

began to grow. "A number of the larvæ which were half grown when placed under starvation for the first time, have through alternating periods of 'fasting and feasting' attained that size three times and are now on the way to their fourth 'childhood,' and even some of the large specimens have started dwindling down to their third 'childhood' after having twice attained the practically maximum larval size." Although there is nothing like so great a resistance to starvation in the larvæ of the social Aculeata, there are cases of an extraordinary ability to pupate precociously on a very small amount of food. Thus in certain ants of the genera Dorylus and Carebara (Fig. 72) the minute worker pupæ are several thousand times smaller than those of the queens. By whatever means these differences may be determined, it is certain that the mere quantity of food administered to the larva must be an essential factor in their attainment.

It is well known that the social Vespidæ feed their larvæ with pellets consisting of malaxated insects moistened with saliva or other oral secretions. There is some doubt as to whether honey is fed to the larvæ even by the wasps which habitually store it in their combs. But the malaxated soft parts of insects must be of very unequal food value. The differences between the workers and queens are so small and intermediate forms are so numerous that quantitative feeding would probably account for their production. The normal sterility of the workers seems to be due to acceleration of the development of the somatic and retardation of the reproductive tissues, a condition the reverse of neotenia, probably due to precocious pupation and therefore of the nature of prothetely. The ovaries of the workers are inhibited but not aborted and may produce viable ova if the adult insect be able to obtain sufficient food and does not undergo nutricial castration as the result of over-indulgence in nursing. The queen wasps undoubtedly arise from the larvæ which are most abundantly fed. In the social Vespidæ, therefore, there would seem to be no reason to appeal to qualitative differences in larval feeding to account for the feeble dimorphism of the female sex.

Caste development in the humblebees, notwithstanding the great difference in the quality of the food, is so similar to that of the Vespidæ that we should naturally expect the causes to be similar and to be merely quantitative. In the honeybees and stingless bees, however, the matter is more complicated. Most authors seem to assume that the somatic differences between the queen and worker honeybee are directly determined by qualitative differences in larval feeding, but comparison with the Meliponinæ suggests that the differences in food primarily affect the fertility of the two castes and that the somatic differences depend on or are correlated with the development of the ovarian tissues. Silvestri (1902) and H. von Ihering (1903) showed that there are two different conditions in the Melponinæ. In Melipona the cells in which the males, queens and workers are reared are all alike and of the same size. They are provisioned with the same kind of food, pollen and honey, and an egg is laid in each. They are then sealed up, and although the larvæ are not fed from day to day as in the honeybee, but like those of the solitary bees subsist on stored provisions, their uniform treatment nevertheless results in the production of two sharply differentiated female castes. But on emerging the queen Melipona has very small ovaries with immature eggs. In the species of the allied genus Trigona, however, which differ from the Meliponæ in constructing large queen cells and in storing them with a greater quantity of honey and pollen than the worker cells, the queen emerges with her ovaries full of ripe eggs. These facts would seem to indicate that the large size of the queen cell and its greater store of provisions are primarily adaptations for accelerating the development of her ovaries. In the honeybee we may, perhaps, adopt a similar explanation for the feeding of the queen larva with a peculiar substance, the "royal jelly." It has long been known that the larvæ of queen and worker honeybees are fed on different substances after the first two or three days of their postembryonic lives. After that time the worker larva is given pollen and honey, whereas the queen larva receives a peculiar pap, the " royal jelly " which was by Schiemenz (1883) regarded as a salivary secretion of the worker nurses but by von Planta (1888) and Schönfeld (1897) as partially digested food regurgitated from the clylific stomach. More recently Metzer (1910) and Zander (1911) have shown that the structure of the proventriculus is such as to preclude regurgitation of the stomach contents, so that the royal jelly must be regarded as a salivary secretion (see also von Buttel Reepen, 1915, Nelson and Sturtevant, 1924, and Lineburg, 1924). The queen emerges in about sixteen days from the time the egg is laid, whereas the worker, though a smaller insect and possessing imperfect ovaries, requires four or five days longer to complete her development. That this interpretation of the queen's development is correct is also indicated by the fact that she may lay within ten days of the date of her emergence. It would seem therefore that the qualitative feeding of the queen larva is not primarily concerned with morphogenesis but with the growth of her gonads. She differs from the worker in certain degenerative characters, such as the shorter head and tongue, shorter wings, simplified hind legs, absence of certain salivary glands, etc. These deficiencies may perhaps be the results of ovarian hypertrophy, according to Emery's law which will be discussed later.[1]

[1] Jucci (1924, p. 221 *nota*), in an interesting discussion of the effect of saliva on caste differentiation in the social insects, states that Aeppler (1922) found cystine in unusually large amounts among the protein components of the royal jelly, and that " in bees, according to Bishop, this salivary nutriment may eliminate the factors that inhibit the development of the reproductive organs, factors which may consist in an excess of certain proteins." Pezard (1919), who seems to be of the same opinion , says à propos of the alimentary castration of hens fed exclusively on flesh : " The exclusive flesh diet produces in the organism a slow intoxication to which the reproductive glands are particularly sensitive and in consequence of which they may gradually atrophy and not develop ". Jucci cites also several other cases from which one might infer that the ovaries of worker honey-bee larvæ may be inhibited by a diet too rich in proteins (pollen), whereas in the queen larva the cystine in the royal jelly may counteract the inhibitory action of such proteins and permit the ovaries to develop. He adopts

Experimental bee-keepers have repeatedly demonstrated the fact that an egg or young larva transferred from a worker to a queen cell where it is fed on royal jelly, or saliva, becomes a queen. The experiments have been more recently repeated by Klein (1904) and von Buttel-Reepen (1911). The latter cites Klein's results as follows: " If a larva half to one or one-and-a-half days old be taken from a worker cell and placed in a queen cell, a *perfect* queen will develop from it, since up to that time the worker larva has been given the same kind of food. But if the same experiment be performed with larvæ two-and-a-half to three-and-a-half days old, there is a certain small difference, the resulting queens being slightly smaller and the legs, which are always paler in the queen than in the worker, being spotted with black, especially the hind tibiæ, where there is in the latter caste a concavity, the corbula. . . . But in larvæ that have remained in the worker cell about four-and-a-half days and are then transferred to queen cells, there is a decided difference (i.e., they are more like workers) owing to the effects of the larval food which for the queen has a different composition after the second day." Klein also transferred a one-half to one-and-a-half day old larva from a worker to a queen cell, left it in the latter two days and then transferred it back to the original cell where it was fed as a worker for one-and-a-half days more. The emerging adult had very pronounced worker characters, although the hind legs had shorter hairs and were shaped somewhat more like those of the queen. There can be little doubt that by varying these feeding experiments it would be possible to produce artificially a finely graded series of annectant forms between the

---

a similar interpretation for the development of the termite neotenics. It would not account for the conditions in Trigona unless we suppose that a considerable amount of saliva is mixed with the pollen and honey stored in the large queen cells. According to Ezikov (1927), G. A. Koschewnikow has shown that the pupa of the worker honeybee " suffers a regressive metamorphosis of its ovaries, owing to the fact that most of the tubules, which are present at this time, degenerate ".

typical worker and queen, comparable to the series which is normally present in the colonies of wasps and humblebees.

In the ants it has been impossible hitherto to demonstrate a precise dependence of caste differentiation on food, because these insects not only themselves eat such a variety of substances but also feed them to all but their youngest larvæ. The substances range all the way from portions of the most diverse insects to the softer parts of seeds, the nectar of plants, the saccharine excreta of aphids, coccids, etc. and the hyphæ of fungi. Some species are exquisitely entomophagous, others largely aphidocolous or coccidophilous, others exclusively mycetophagous and a few, like *Solenopsis geminata*, might be called omnivorous. And yet notwithstanding all this variety of nutriment and apart from certain exceptions due to parasitism, the whole family exhibits a uniform female dimorphism or pleomorphism. The following examples will make this clear :—

The larvæ of the tropical subfamily Pseudomyrminæ (Fig. 43) as I have shown (1918, 1920, 1922, Wheeler and Bailey, 1920), are peculiar in possessing just back of the mouth on the ventral surface of the first abdominal segment a pocket, the trophothylax (*t*), in which the worker nurses place the food consisting of a pellet (*pe*) moulded in their own infrabuccal pocket. This pellet is really refuse and is made of triturated fragments of insects, from which the juices have been extracted by the worker, plus the fungus spores, pollen grains and other particles of detritus combed from her body by means of the strigils on the fore feet. The ants of other subfamilies cast the pellet away as worthless, but the Pseudomyrminæ are so parsimonious that they feed it to their young. It is precisely as if we had attained to such a degree of household thrift that we fed our children exclusively on the contents of the garbage-pail, dust-bin and bath-tub. Although such a method of feeding must preclude any qualitative regulation of the larval diet, there are distinctly marked queen and worker castes, though of nearly the

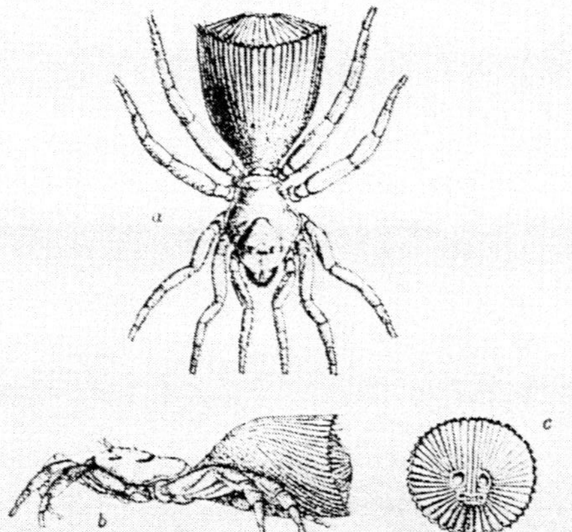

FIG. 42. *Chorizops loricatus*, a Mexican spider with the posterior end of the body modified for closing its burrow. a. Dorsal view. b. Lateral view. c. Posterior end of abdomen. (After A. Petrunkevitch.) (See p. 190.)

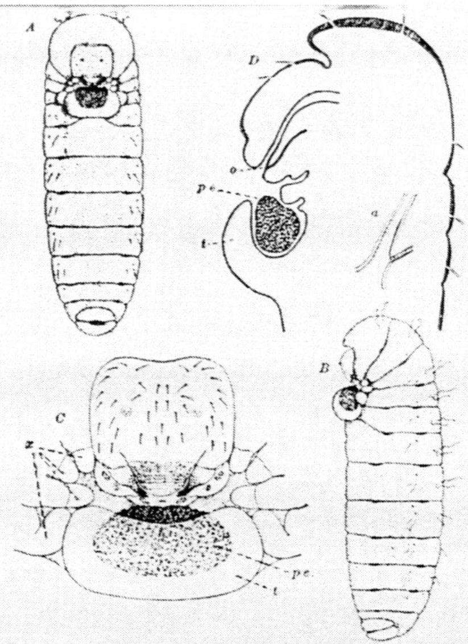

FIG. 43. Larva of *Pseudomyrma gracilis*. A. Ventral. B. Lateral view. C. Head and adjacent portions of same enlarged. D. Sagittal section through anterior portion of larva. o. Oral orifice. x. Exudatoria. t. trophothylax, or pocket, which holds the pellet, pe, deposited by the worker nurses and which is eaten by the larva. Note the hooked dorsal hairs of the larva, which serve to suspend it from the walls of the nest. a. Mouth cavity, more enlarged to show the fine spinules (also seen in C.), which serve to triturate the pellet and probably also as a stridulatory organ.

face p. 200

FIG. 44. Vertical section through the center of a nest of the Argentinian leaf cutter, *Atta vollenweideri*, showing the chambers containing the fungus-gardens. (Photograph by Dr. Carlos Bruch.)

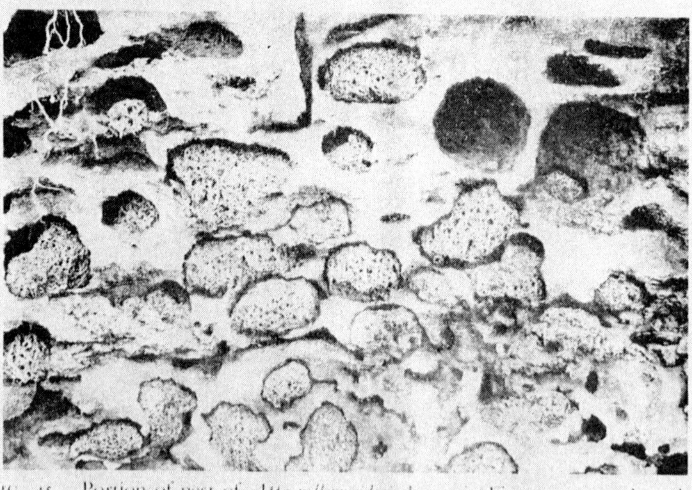

FIG. 45. Portion of nest of *Atta vollenweideri* shown in Fig. 44 more enlarged to show the sponge-like fungus-gardens *in situ* in the chambers. About one-eighth natural size. (Photograph by Dr. Carlos Bruch.)

[*face p.* 201

same stature, in all Pseudomyrminæ.[1] The youngest larvæ, till their jaws are strong enough to triturate the pellets, are fed on saliva, and perhaps the same substance may also be given to some of the older larvæ and determine their development as queens, but of this there is no evidence.

As a second example we may select the Attine ants of the American tropics, a tribe of Myrmicinæ which uniformly feed, both as larvæ and adults, on the mycelium of particular fungi cultivated on a substratum of triturated leaf-tissue or the excrement of caterpillars or beetles. In Atta, the most highly specialized genus, the colonies are very populous, the queens are very large and the worker caste exhibits a long series of forms grading down from huge-headed soldiers to minute forms which never leave the nest but devote themselves to caring for the fungus-gardens (Figs. 44 and 45). In Acromyrmex, the queens are smaller and the pleomorphism of the workers less pronounced, and in the lowest genera, Trachymyrmex (Fig. 46), Cyphomyrmex, Sericomyrmex, Apterostigma, etc., the colonies are very small, the workers monomorphic and often nearly as large as the queens. In the latter genera, as I have recently observed in colonies of a Panamanian Sericomyrmex, the queen mother scatters her eggs among the fungus and the hatching larvæ feed on the mycelium within reach of their mouth parts, which, as in other Attini (Wheeler and Bailey, 1920), are beautifully adapted for puncturing the hyphæ and imbibing their protoplasmic contents. The workers occasionally transport the larvæ to fresh pastures but are never seen to feed them directly with cropped fungus. Atta, however, according to Tanner (1892), employs the latter method of nursing the young. But whichever method is employed, the feeding of the larvæ among the Attini is quantitative, although the possibility of an administration of saliva to certain larvæ and their conversion into queens as a result of such treatment, cannot be altogether excluded.

[1] Except *Viticicola tessmanni*, which for some unknown reason has intermediates between the queen and worker (Wheeler 1922).

The feeding of ant-larvæ with regurgitated liquids has been observed by a number of myrmecologists (Miss Fielde, 1901, Newell, 1909, Newell and Barber, 1913, Emery, 1918, myself and others) and in some species (Dolichoderinæ, Formicinæ) this may, perhaps, be the only method employed. In many Ponerinæ and some Myrmicinæ (Aphænogaster) and Formicinæ (Lasius) the larvæ may be given crude pieces of insects and those of Messor and Pogonomyrmex fragments of seeds. But the very youngest larvæ are always fed on regurgitated food and Emery's observations indicate that at least the older queen and soldier larvæ are similarly treated. The composition of the regurgitated liquid, however, is not so easily ascertained. It may consist of saliva or liquid food from the crop or a mixture of both. There is one instance in which we know that the larvæ are fed throughout their larval life with pure saliva and that is the first brood of the colony-founding queen. Although she may take no food for eight or nine months after leaving the maternal nest for her nuptial flight, she is able nevertheless to bring up a brood with her saliva alone. But this brood comprises only a few individuals, which are, moreover, always small and always workers! Here the salivary diet certainly does not produce soldiers or queens and the diminutive size of the brood and of its members shows that the quantity of the food may be more important than its quality.[1] That the conditions are not due to some constitutional incapacity of the queen is proved by the fact that in the parasitic ants whose queens lay their first eggs in the colonies of other species, the stature of the emerging adults is not diminished, since they are adequately nourished during their larval stages by the host workers.

The intimate dependence of the differentiation of the queen and worker castes on food is beautifully shown in the growth of all Aculeate colonies that are established by single queens. The first brood always consists of small workers, those of succeeding broods gradually increase in size, and only after the largest workers have appeared are

[1] These statements now need considerable qualification, owing to the researches of Meyer (1927), who finds that the recently fecundated queens of several ants (notably *Messor structor*) while establishing their colonies not only devour many of their own eggs but also feed them to their first brood of larvæ. More observations are needed, however,

the queens produced. This is most clearly shown in ants with polymorphic workers, like Atta, Pheidole, Pheidologeton, Camponotus (Fig. 47), etc. Pricer (1908) has investigated the growth of *Camponotus pennsylvanicus* colonies with the following results: The workers of the first brood are of the very smallest type (minima workers) and the largest, or maxima workers are not produced till about the third year, when the colony comprises about 900 to 1,200 individuals. The queens and males do not appear till the third to sixth year and not till the worker population reaches 2,000. A very similar slow succession of small workers, large workers, soldiers, neotenics and royal forms is also observed in the growing colonies of termites. The phenomenon is so constant and so obviously dependent on the gradual improvement of the trophic status of the ant and termite colony as to furnish one of the strongest arguments in favour of the trophogenic hypothesis of caste differentiation.

Apart from observation of the feeding of ant larvæ and experiments in rearing them under controlled conditions, there is also a third method of attacking the problem under consideration, namely by observing the effects of certain parasites of the individual larvæ and the disturbances introduced into the trophic behaviour of the colony as a whole through the presence of certain myrmecophiles. Four examples of such "experiments of Nature" may be briefly described :—

In my second lecture I dwelt on the Eucharine Chalcidids, whose planidium larvæ when introduced into ant nests, settle on the larvæ or semipupæ and suck out their juices. In the case of *Orasema viridis* (Fig. 6), which I observed in Texas (1907*a*) infesting colonies of *Pheidole instabilis*, the planidia attach themselves to the neck of the larva or semipupa and after withdrawing sufficient nutriment, like so many leeches or vampires, and attaining their own semipupal stage, drop off and complete their development as independent organisms. The affected ant larva or semipupa is not killed immediately but may actually pupate though it is unable

before this practice, which had previously been observed by J. Huber (1905), in the large fungus-growing *Atta sexdens* of Brazil, can be regarded as general among the formicidæ.

to become an adult. Owing to the withdrawal of so much of its substance, the pupa is peculiar in exhibiting certain characters of inferiority such as the smaller size of the head, thorax and eyes as compared with the pupa which it would have formed had the larva remained unparasitized. It therefore belongs to a new nonviable or " phthisergate " caste, the result of mere mechanical withdrawal of the body juices and of the ability of the larval organism to form a symmetrical but microcephalic whole. Were the adult capable of emerging and taking part in the activities of the colony, von Buttel-Reepen and many geneticists of the Neodarwinian school would probably say that it derived its singular characters from special determinants or genes in the egg from which it developed, but such an inference would be quite erroneous and would take no account of the extraordinary regulatory powers of the ant larva. I may say that similarly inferior pupal forms (phthisodinergates and phthisogynes) seem to be produced by the Orasemas when they parasitize larvæ that would normally develop into soldiers or queens.

Many years ago (1901*b*, 1917*a*, 1910*b*) I described another case of parasitism of a Texan ant, *Pheidole commutata*. The larvæ of this insect are sometimes infested with Nematode worms of the genus Mermis and develop aberrant forms, which I have called " mermithergates " (Fig. 48 B and C). They are considerably larger than the normal workers (A), which they nevertheless resemble in structure and the small size of the head, although they possess small ocelli. In thoracic structure they approach the soldier while the gaster is enormously distended with the Mermis and retains scarcely any vestiges of fat, reproductive organs and other viscera. The behaviour of such parasitized individuals is also peculiar, since they never excavate the soil, nor care for the brood like normal workers but keep running about in a chronic state of hunger, begging their normal nest-mates for food. Similar mermithergates have been described by Emery (1890, 1904) in quite a series of ants, including *Pheidole absurda* and various Ponerinæ of the genera

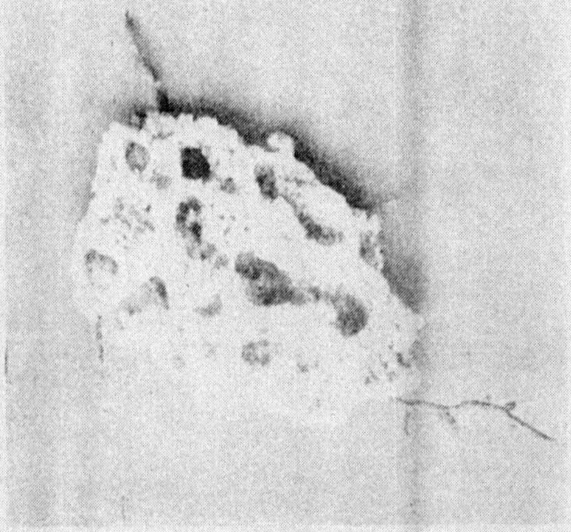

FIG. 46. Pendent Fungus-garden of a small species of *Trachymyrmex* from Panama, removed from the fungus chamber, but still attached to a rootlet. Natural size. (Photograph by D. Fairchild.) (*See* p. 201.)

FIG. 47. Queen of a carpenter ant, *Camponotus pennsylvanicus*, with incipient colony in abandoned cocoon of *Rhagium lineatum* under pine bark. Slightly enlarged.

[*face p.* 204

Plate XXX.

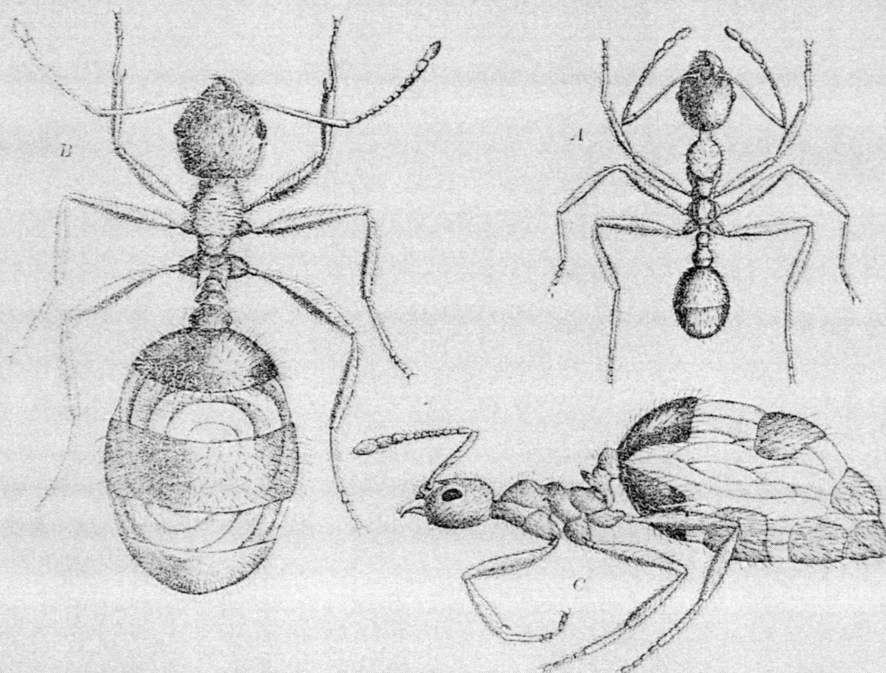

FIG. 48. A. Normal worker of *Pheidole commutata*. B. and C. Mermithergate of same in dorsal and lateral view.

face p. 205

Odontomachus, Neoponera, Ectatomma, Pachycondyla and Paraponera.

Since all of the anomalies cited were workers of Neotropical species it was supposed that some exclusively American Mermis confined its attacks to worker larvæ, but I have recently described (1922*a*) a mermithergate of *Camponotus* (*Dinomyrmex*) *cassius* from the Congo, and Mrázek (1908) has shown that the virgin queens of the common European *Lasius alienus* may be infested with Mermis and that such individuals, which he calls "mermithogynes," develop abnormally small wings. These mermithogynes had been seen in 1898 by Crawley (1910*a*) and described as brachypterous to distinguish them from the ordinary macropterous females. I have also seen several mermithogynes of our North American *Lasius neoniger* (1910*b*). The parasites were 53-55 mm. long and filled the whole abdomen of the ants, so that in the living specimens there could have been little left of the reproductive organs and fat body. There was nothing unusual about these queens, except the somewhat smaller head and thorax and small size of their wings, which measured only 6 to 6.5 mm. in length, whereas those of normal *L. neoniger* measure 10 to 11 mm. In 1920 the British myrmecologist Donisthorpe (1921) found numerous mermithogynes in a colony of the typical *L. niger* and in the same year Crawley and Baylis (1921) published a detailed account of a number of mermithogynes of *L. flavus* and *alienus*, with a description of the Mermis (*M. myrmecophila* Baylis). Emery (1924) described two mermithogynes of *Lasius umbratus* var. *meridionales* from Italy, and more recently Vandel (1927) has described mermithergates in *Pheidole pallidula* from Southern France. These had been previously seen, but interpreted as a new species of parasitic ant ("*Ph. symbiotica*") by Wasmann (1909*d*, 1910*c*).

The brachyptery of the mermithogynes is very probably due to the withdrawal of thoracic materials by the Mermis during larval life, but the various mermithergates are not so easily interpreted. The Ponerinæ described by Emery seem to be in part abortive ergatoid females and in part modified workers, but owing to the similarity between the normal queens and workers in that subfamily, we need much more material for an adequate understanding of the phenomena. In the two species of Pheidole (*absurda* and *commutata*) in which there are normally three sharply defined female castes, queens, huge-headed soldiers and diminutive workers, the mermithergates exhibit a singular mixture of their various characters. We may, perhaps, conjecture that Mermis has a twofold effect on its host: first, in increasing the appetite of the larva and thereby stimulating the trophic attentions of the worker nurses, and second, in modifying the proportions of the different parts of the body, possibly through some secretion of the parasite acting like a hormone on the growing tissues of the larva. But whatever interpretation we adopt, the blending of characters of all three castes in the adult mermithergate seems clearly to point to potentialities for all of them in the young larva.[1]

The third case of interest in this connection is that of the beetles of the Staphylinid tribe Lomechusini, which have been studied by Wasmann for more than thirty years and concerning which he has written nearly as many papers. Since the latest of these (1920) contains a full bibliography I need not cite the literature and may here confine my remarks to the brief summary of the observations published in my "Social Life among the Insects." The Lomechusini comprise only three genera:—Lomechusa (Fig. 50) and Atemeles, peculiar to Europe and Northern Asia, and Xenodusa (Fig. 49), known only from the United States and Mexico. The species of Atemeles and Xenodusa have two hosts, those of the former living during the summer and breeding in Formica colonies but hibernating in colonies of Myrmica, the latter also breeding with Formica but hibernating with ants of the genus Camponotus. Lomechusa, on the other hand, has only one host, *Formica sanguinea*, with which it lives throughout the year. The adult beetles of all three genera look much alike. They are of an oily red colour,

[1] I have reconsidered this whole subject of Mermis parasitism in ants, in connection with several new cases, in a paper to be published during 1928.

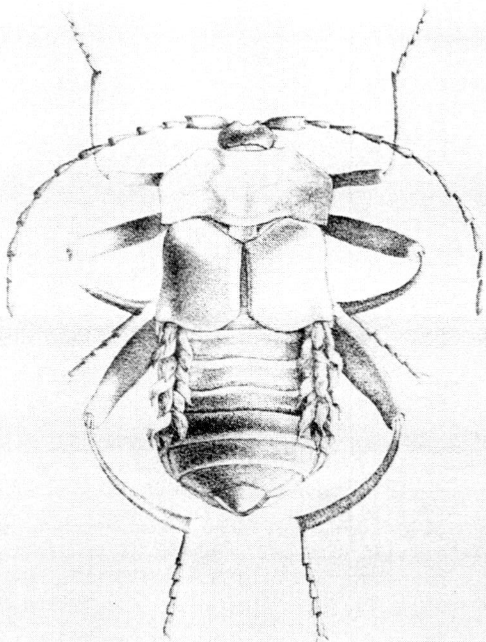

FIG. 49. *Xenodusa cava*, a North American beetle which breeds in the nests of Formica during the summer and passes the winter in the nests of Camponotus. Note the tufts of trichomes along the sides of the abdomen.

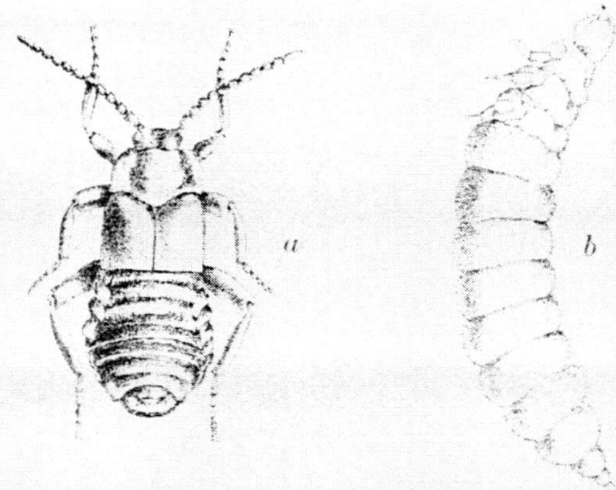

FIG. 50. a. A European guest-beetle (*Lomechusa strumosa*) and b. Its larva, which live with colonies of the blood-red slave-maker (*Formica sanguinea*).

[*face p.* 206

Plate XXXII.

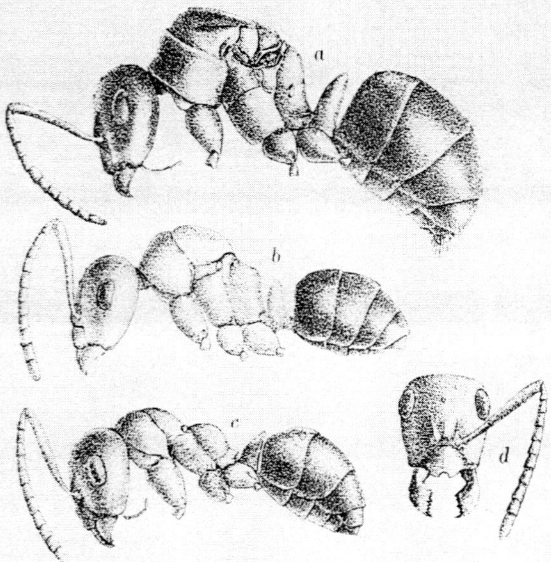

FIG. 51. Blood-red slave-maker (*Formica sanguinea*) in profile. a. Queen, with wings and legs removed. b. Pseudogyne. c. Worker. d. Head of same from above, showing the characteristic notch in the clypeus.

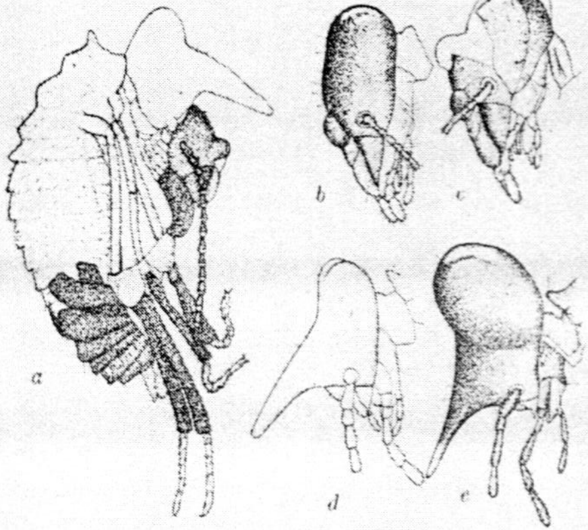

FIG. 52. a. Nasute soldier nymph of *Nasutitermes (Constrictotermes) cavifrons* Holmg emerging from a pigmented, worker-like skin. b. Lateral view of head of worker-like individual. c. Head of moulting individual showing the "nose" pushing out through the split Y-suture. d. Head of nasute soldier after moulting. e. Head of same mature and pigmented. (After Prof. A. E. Emerson.) (*See* p. 186.)

[*face p.* 207

have long, mobile antennæ, short elytra and a voluminous abdomen which can be curled up over the thorax and is provided on each side above with a segmental series of beautiful golden trichomes. Atemeles and Xenodusa beg their food from the ants by stroking their cheeks with the forefeet (Fig. 67). The larvæ of Xenodusa are active, have long legs and employ the same method as the beetles in persuading the ants to regurgitate. They also devour the defenceless Formica larvæ. The adult Lomechusa is more passive in its behaviour and uses its antennæ in soliciting food. Its larvæ (Fig. 49*b*) have very short legs and being unable to run about lie quietly among the ant brood. They devour the brood and are also fed by regurgitation. In all probability they secrete fatty exudates which are greatly appreciated by the ants. At any rate, the ants seem to prefer the predatory larvæ to their own, or perhaps regard them as unusually promising ant larvæ. In consequence of this infatuation the Lomechusa larvæ often destroy the greater part of the brood, so that in *sanguinea* colonies heavily infested with the parasites the queen larvæ develop abnormally. Either they are neglected or the ants actually endeavour to convert them into workers, because they feel that this caste is inadequately represented in the colony. But whatever be the treatment of such queen larvæ, they develop into pathological adults, known as " pseudogynes" (Fig. 51*b*), abortive creatures, resembling workers in size and in the shape of the head and gaster, but with a more voluminous and convex thorax, approaching that of the queen. They are paler than the normal workers and very lazy, cowardly and incompetent. Usually they constitute 5 to 7 per cent., less frequently 20 per cent. or more of the personnel of an infested *sanguinea* colony. Their appearance in a nest indicates that the colony is in a diseased condition and on the road to extinction. Similar psuedogynes are also produced in the Formica colonies infested with Atemeles and Xenodusa, but not in the Myrmica and Camponotus colonies in which the beetles hibernate, because they do not breed among their winter hosts and cannot therefore interfere with the normal development of their brood.

It will be seen that unlike Orasema and Mermis, Lomechusa is not a parasite on or in the individual larva of its host, but a predatory symphile, or social parasite, which acts primarily on the threptic, or brood-rearing behaviour of the worker personnel, causing it either to neglect the queen larvæ or actively to alter the normal method of feeding them. Wasmann has shown that these effects may persist in *sanguinea* colonies long after the disappearance of the parasites as a proclivity of the workers to keep on rearing pseudogynes till the colony succumbs as the result of a veritable social malady. Viehmeyer (1904) has demonstrated experimentally that the production of pseudogynes is really due to a parasitically induced modification of the worker behaviour and not to some pathological condition in the eggs laid by the mother queen. In one of his *sanguinea* colonies the production of pseudogynes continued for three years without the presence of Lomechusa. During the fourth year he separated the mother queen from her workers and succeeded in bringing about her adoption by a body of workers from a colony that had never been infested by the parasites and had therefore never acquired the habit of rearing the anomalous forms. The fresh worker personnel forthwith reared the larvæ that hatched from the queen's eggs as normal individuals.

When ants themselves become parasites in the colonies of other ants, the behaviour of both species is necessarily modified, because the social environment of each is altered. Leaving a fuller consideration of this matter for a future lecture, I will give only one example which shows that the modifications of the threptic behaviour of the parasites may be reflected in the structure of their queen and worker personnel. Years ago (1901*c*, 1903*c*, 1907*c*) I found in the hills of New England and later in the mountains of Colorado a small ant, *Leptothorax emersoni*, which is always associated with a larger Myrmicine, *Myrmica canadensis*. The Leptothorax inhabit small chambers near the surface of the Myrmica nest and connect them by means of tenuous galleries with the chambers of their neighbours. The Leptothorax workers spend much of their time in the Myrmica nest where they keep mounting the backs of the workers and assiduously licking their bodies and especially their heads and mouthparts. The Myrmicas seem greatly to enjoy these attentions and from time to time reward their little riders with droplets of regurgitated food. But while the Leptothorax arrogate to themselves the right to mingle freely with the Myrmicas and to flatter them into regurgitation, they resent the intrusion of the latter into their own habitations and insist on bringing up their brood in perfect seclusion. Owing to the abundance of food thus received and possibly to the large amount of saliva it contains, the Leptothorax brood produces an enormous proportion of forms intermediate between the winged queen and regular worker caste. One of my former students, Miss Margaret Holliday (1903), from a careful study of one thousand specimens of *L. emersoni*, was able to distinguish no less than eleven different forms, grading down from the winged queen to the smallest worker, or micrergate. Dissections of the reproductive organs showed that " all eleven types possessed well-developed ovaries, containing mature eggs ; they had from two to four tubules on a side ; the number of eggs in each tubule varied both on the two sides of the same ovary and in the ovaries of different individuals of the same type, as also in those of different types. All the specimens dissected had better developed ovaries than the queen. Every individual dissected, except two, possessed the receptaculum seminis." Although similar transitional forms between queens and workers are known to occur in some non-parasitic species of Leptothorax, they are always rare.[1]

[1] Miss Holliday observed that " in *Lepththorax longispinosus* Rog. *L. curvispinosus* Mayr, *L. obturator* Wheeler and *L. canadensis* Prov.' the sexual phases are represented by males, queens, macro- and micro-ergates. No ergatoid females were found in these species. Among *L. canadensis* one tri-,one bi-, and one uni-ocellate worker were present. The queens and workers possessed a single tubule on each side. No receptaculum seminis was present ". Wasmann (1895), however,

Their extraordinary frequency in *L. emersoni* must, I believe, be attributed to the unusually favourable trophic conditions under which the larvæ are reared. Evolution in this case might be expected to lead eventually to a complete elimination of the sterile worker and a return to two very similar fertile types of female, one winged and migrant, or colony-founding and one apterous and combining the fecundity of the queen with a behaviour more like that of the worker. The foregoing examples will suffice to show that in the Formicidæ the female larva is much more plastic than that of the social wasps and bees. This is obvious, of course, also from the greater number and more pronounced differentiation of the female castes in many of the genera and species.

Before passing to other matters, attention may be called to Emery's (1904, 1921) and Viehmeyer's (1923) two principles, or laws of larval and prepupal development among ants and other metabolic insects. Emery formulated his law, which he calls that of the "antagonism between head and abdomen," as a result of his study of the mermithergates, as follows: "The principal function of the larva is to accumulate provisions in the form of the fat-body for the labour of metamorphosis and to prepare the rudiments of the imaginal disks and internal organs. These rudiments lend themselves to the competition which arises during the prepupal stage when the ant's body is differentiated into head, thorax and abdomen. The organs essential to life, notably the intestine, which is situated in the abdomen, have the preference and the head and thorax divide what remains (small-headed workers). If the provisions are more abundant, the imaginal disks of the head, which are most voracious, attract the surplus to themselves (large-headed workers,

---

found microgynes and numerous ergatogynic "Mischformen" in one colony, of *L. acervorum*, and Adlerz (1884), Wasmann and Stumper (1921) have observed ergatogynes in *Formicoxenus nitidulus*. The latter lives as a guest-ant in the nests of *Formica rufa*, and *L. acervorum*, like our North American *canadensis*, which is usually regarded as a subspecies of *acervorum*, often nests very near, i.e., in plesiobiosis, or parœcism with other larger ants.

soldiers). Finally, if the provisions are very abundant, the rudiments of the ovaries and imaginal disks of the wings intervene and the thorax and abdomen grow at the expense of the head (winged female.)" Undoubtedly this law has considerable value as a general description of the ontogeny of the various castes, not only in ants but also in the honeybee and the termites. It is, however, a special and rather striking case of the more general phenomenon which Child (1924) calls the " axial gradient " and its reversal. The tissues of the prepupa exhibit a differential in the localization of growth, which in the soldier and worker major is anteroposterior and in the queen posteroanterior. In both cases the decrement is steep. In Child's language, the axial gradient of the soldier and worker major is the reverse of that of the queen, and he would, perhaps, say that the feeding of certain larvæ may fail to alter the original gradient, in which case they would produce queens, or that it may be reversed, in which case, according to the strength of the determining factors, a soldier would result. In the small worker the anteroposterior decrement is feeble. All this, too, is merely descriptive and preliminary to ascertaining the actual metabolic processes of which we are still ignorant.

Viehmeyer's law might, perhaps, be regarded as an application of the gradient conception to phylogeny as well as ontogeny. He regards most of the organs of the female ant as being in a state of flux, or instability, but divisible into three categories, those which are relatively fixed, or stable, those tending to a maximum development and those which are on the road to atrophy. His law applies to the two latter organ categories which alone are influenced by larval feeding and asserts that there is a tendency to accelerate both of these processes in polymorphism. He says : " It seems to me that a long series of irregularities in the disappearance of female organs and in the acquisition of worker characters may be understood as an acceleration or retardation of their development. On the same principle we might also, perhaps, explain most of the differences, so far as they imply a nondevelopment of organs, between small and large workers and between true workers and soldiers." The specific examples cited by Viehmeyer are the very rapid disappearance of the vestigal mandibular teeth in male ants (Formica) and of the vestigial eyes of workers (*Lasius flavus* and *myops*) under unfavourable trophic conditions of the colony.

Ezikov (1926) has recently shown that the ovaries of worker and queen ants are more interesting plastic organs which react with acceleration or retardation of development to trophic stimuli applied during the early larval stages. It has long been known that worker ants have fewer ovarian tubules, or ovarioles than their cospecific queens and that the number of ovarioles is roughly proportional to the size of the individual worker. Thus in *Formica rufa* there are only two tubules in each ovary in small, four or five in medium and six in the largest workers. In *Camponotus herculeanus* the number varies from one to five in a corresponding series of individuals. In the young stage of the larva the higher number is indicated by divisions of the embryonic organ but artificial starvation may reduce it to the formation of a single tubule. This reduction, apparently, takes place only if starvation occurs before the ovary has progressed beyond a certain point in its differentiation. After this period the full number of tubules may be developed as in unstarved larvæ. In the workers of certain ants (*Tetramorium cæspitum*) the ovarial rudiments may be entirely suppressed by starvation, so that the insect is sexless. These results are significant in connection with Koschevnikow's observations on the ovaries of worker honeybees (see p. 199 note). In this case, however, the suppression of ovarioles occurs during pupal life, whereas in ants it is carried back into the larval stages.

Within recent years the active discussion of caste differentiation in the social insects has shifted from the Aculeates to the termites. Grassi concluded that hatching termites all have the same constitution and that the castes are the result of qualitative feeding. The feeding of saliva to the nymphs during their early instars accelerates the development of the reproductive organs so that fertile sexual forms of the first, second and third forms result. Those of the second and third forms are neotenic because the development of their somatic characters (wings, eyes, pigmentation, etc) are simultaneously inhibited The question of castration by the protozoa in the hind gut has proved to be an unnecessary complication, since it has been shown that they are not parasites but wood-digesting symbionts (Cleveland), which would naturally disappear from the intestines of individuals nourished with saliva only. The workers and soldiers are forms inhibited both somatically and sexually by a constant diet of wood, either crude or partially digested, and fæces. This conception of caste differentiation in the termites has been accepted by Feytaud, Wasmann (1908*a*), Heath, Escherich Jucci and others. It is also accepted by Holmgren (1909), but he believes the production of fatty exudates from the skin to be the stimulus that determines the differential feeding of the various individuals. All the larval stages and all the castes produce exudates but in very different quantities. The quantity, and possibly, also to a degree the quality of the exudates, " is the object of an amical selection. The individuals which produce more exudate, receive more food, and their exudate tissue therefore increases, since the adipose tissue stores the superfluous food and gives it off in the form of exudate. Now when the so-called indifferent larvæ hatch from the termite eggs, some of them may receive a little more food from the very beginning. These larvæ being better fed, secrete a somewhat larger amount of exudate than the others and this may signify for them a different development than that of the larvæ which are more poorly nourished, so that they appear in the next instar as so-called 'small-headed' larvæ, whereas the poorly fed individuals moult as 'large headed' larvæ. The 'small-headed' are the larvæ of the sexual, the 'large-headed' those of the sexless individuals." Holmgren believes that the further differentiation of the former into workers and soldiers and of the latter into neotenics and royal forms is also due to quantitative and qualitative differences in exudate stimulation.

Very different are the views of Bugnion, Miss Thompson and Snyder, and Imms, who regard the castes as blastogenic, or as predetermined in the egg. Bugnion's observations on Eutermes are, as we have seen, very doubtful, to say the least. Miss Thompson and Snyder admit that hatching termites are all alike externally but claim that they are internally distinguishable as two types differing in the size of the brain and reproductive organs and that their further differentiation as definitive castes is also due to intrinsic, or blastogenic and not to extrinsic, or nutritional factors. They therefore regard the castes as mutations, comparable to those detected by De Vries in Oenothera, and are inclined to interpret polymorphism as a Mendelian phenomenon. But it is obvious that the Mendelian behaviour of the castes can be proved only by experimental breeding. This should not be impossible in view of the fact that there are three fertile castes and possibly a fourth if we include the soldiers, which in the lower termites occasionally produce offspring.[1] Here the matter rests for the present and we can only cry, like the dying Goethe, for more light.[2]

[1] We have already seen (p. 184) that Heath's recent paper (1927) supports the earlier views of Grassi and calls for a re-examination of Miss Thompson's investigations.

Silvestri (1901) describes also fertile gynæcoid workers (possibly third form adults ?) in a South American termite.

[2] The contrast between the blastogenic and trophogenic interpretations of caste development in the termites is clearly expressed by Jucci in the following passage: "The researches of Miss Thompson have contributed greatly to our knowledge of the termites and have brought out two important facts: 'Two important facts, which apply to all the termites here described, should be borne in mind; first, that, together with certain external characters, such as wings, there is a correlation in the size, structure and degree of development of the brain, the eyes and the sex organs; second, that, in general, in whatever castes are represented, there is a gradation in the size of these organs, from the first form down to the worker or soldier (1920) . . . . this gradation of structure possibly representing different degrees of ancestral mutations. On the other hand, certain types possess characters peculiar to themselves, which may represent progressive

If my contention is correct that polymorphism originated and developed independently in the various groups of social insects, and if, as seems probable, they are of very different geological age, we should expect the castes which first arose as functional, i.e., behaviouristic and physiological results of a division of labour and only gradually acquired their morphological differentiation, to differ with respect to their hereditary determination in the germ plasm. While there seems to be little doubt that in the two recent groups, the social wasps and bees, the feebly differentiated queen and worker castes develop from eggs of the same kind, it is more probable that in the two more ancient groups, the ants and termites, the pronounced fertile and sterile castes may be blastogenic. The latter statement may not apply to the soldier caste which is evidently of very different age in the ants and termites. Among the termites it is completely differentiated as early as the Oligocene, and since winged specimens of Mastotermes occur in the Eocene there is every reason to suppose that the soldier was also present and as perfectly developed as in the living *M. darwiniensis*. The queen is evidently the oldest and most conservative ant caste and corresponds to the typical female of the solitary Aculeates. The worker, even if we assume that the ants originally had dimorphic fertile queens, is a later and secondary form, and the soldier, being merely a peculiarly

---

mutation, for example, the stouter legs of the second form, the thicker chitin of the head of the workers and soldiers, and the elongated head and mandibles of the latter'. Miss Thompson therefore sees in these two morphological facts a proof of the ontogenetic hereditary origin of the castes and of their phylogenetic origin through essentially retrogressive ancestral mutations. But on the contrary, I see in these two facts a proof of the differentiation of the castes during the course of their development (ontogenetic as well as phylogenetic) from a common, single morphological type; a physiological differentiation which has acquired its morphological expression and produced the castes directly through changing the type of metabolism and indirectly through the mechanisms of correlation. Leaving the neuters aside for the present, I am convinced that in the various sexual forms the so-called retrogressive ancestral mutations are characters inherent in neotenia, depending on arrests of development and that the eventual progressive mutations represent characters correlated with this neotenic arrest of development".

specialized worker, is clearly the most recent of the castes. Hence it more nearly corresponds phylogenetically, but not functionally, to the worker of the termites. The existence of ant soldiers in the early Tertiary is not satisfactorily established. Most of the Lower Oligocene genera that have survived till the present time have no soldiers and of others, like Aneleus, we know only the minor workers. If that genus has been correctly identified, a soldier should have been present, but even this is doubtful, because there are closely allied extant genera, like Solenopsis, a few species of which (*S. geminata* and *sævissima*) have both soldiers and pleomorphic workers, although the great majority have only small monomorphic workers. The Formicid soldier caste, therefore, may not have differentiated out of the larger worker till Midtertiary times. It seems, moreover, to have appeared independently in most of the genera in which it occurs and not yet to have acquired a definitive representation in the germ plasm. Starting from my attempt in former lectures to derive the ants from solitary or subsocial ancestors with dimorphic females and the termites from solitary or subsocial ancestors with dimorphic males as well as females, we might assume that the queens and workers of the Formicidæ and the royal forms and soldiers of Isoptera are really blastogenic, but that the other castes and especially the pleomorphic forms of workers and soldiers are trophogenic.

The foregoing considerations lead up to others which make a uniformly trophogenic origin of the castes in all social insects and especially in the ants and termites seem doubtful. Though rather general these considerations are at first sight impressive as is shown by Forel's (1902) recourse to some of them to support Weismann's blastogenic as opposed to the more trophogenic hypothesis advocated by Emery in several papers (1894, 1896, 1904, 1906, 1910, 1918, 1921). The considerations to which I refer are: first, the extraordinary stability of the typical castes among existing species of ants and termites; second, their extraordinary constancy during geologic time;

third, the difficulty of accounting for the gynandromorphic ants on the trophogenic hypothesis, and fourth, the difficulty of accounting for the adaptive characters of the soldiers and workers.

(1) The myrmecographer who is constantly handling ants from all parts of the world and minutely comparing thousands of specimens from many colonies cannot fail to be impressed by the extraordinary stability of the castes, notwithstanding the innumerable slight varietal, subspecific, specific, and generic differences. To be sure, the worker caste is absent in some parasitic genera and the queen in some others (Diacamma, etc.), but such cases are due to unusual ethological conditions and are so exceptional as merely to prove the rule. Even the greatest climatic extremes of temperature have no disturbing effect on the castes which are supposed to be so readily modifiable by food. And the difficulty of producing morphological modifications comparable to those of the castes by feeding in other insects and in our domestic birds and mammals only increases our scepticism.

(2) Even more extraordinary is the fixity of the ant and termite castes during geological time. I have been quite unable to distinguish the workers of the Oligocene *Formica flori* from those of living *F. fusca,* except by differences due to long preservation of the former in a resinous matrix, and the workers of other fossil species differ from their cospecific queens in precisely the same characters as do those of living species. The soldier and worker termites figured by von Rosen have the same modern aspect. We may say therefore that the same castes have been produced and reproduced for many millions of years and in the most widely separated portions of the globe.

These objections to the trophogenic hypothesis can be met by pointing to the fact that the very constancy of the castes implies that the species to which they belong have been similarly constant during long periods of Cœnozoic and Tertiary time in all parts of the world and that therefore the same must be true of the threptic medium in which the castes have necessarily been produced æon after æon. This may be granted, but it does not prove that the castes are entirely trophogenic because the blastogenist may contend that though the castes are predetermined in the egg they have to be reared in a particular environment and fed on particular foods in order to develop at all.

(3) A considerable number of gynandromorphic ants have been described. They are rare, anomalous individuals, with the body made up of left and right halves, of anterior and posterior portions or of a mosaic of parts belonging to different sexes (Wheeler 1903*b*). The finest examples are those having the left half of the body male, the right half female, or conversely (Fig. 55). There are really two types of both these "lateral" and the "anteroposterior" and "mosaic" gynandromorphs, according as the female component is that of a queen or a worker. The former I have called gynandromorphs proper, the latter "ergatandromorphs." The differences are nearly always sharply defined so that few cases are known in which there is any doubt about which female caste is combined with the male component.[1]

If we assume that gynandromorphs are determined in the egg, either by abnormal chromosomal behaviour during fertilization or cleavage (Boveri, Morgan), doubling of the egg nucleus (Doncaster), fusion of oöcytes (Dönhoff, Wheeler), or fusion of oötids (Whiting), we must regard the worker and queen components in the cases we have been considering as blastogenic. The gynandromorphs and ergatandromorphs would be sharply distinguishable from Goldschmidt's "intersexes", because the latter develop during the ontogeny through the acquisition by

[1] In 1919 I described a gynandromorph of *Camponotus* (*Colobopsis*) *albocinctus* Ashm. from the Philippines, with the right half of the head male, the left half soldier (Fig. 56). According to Emery (1924), the left half of this specimen exhibits merely "generalized queen-worker" instead of specifically soldier traits. This may be the correct interpretation, though it is strange that rudiments of ocelli are lacking. If Emery is right, the absence of a soldier component in any known ant gynandromorph would seem to support the view that the soldier is a very recent, purely trophogenic caste.

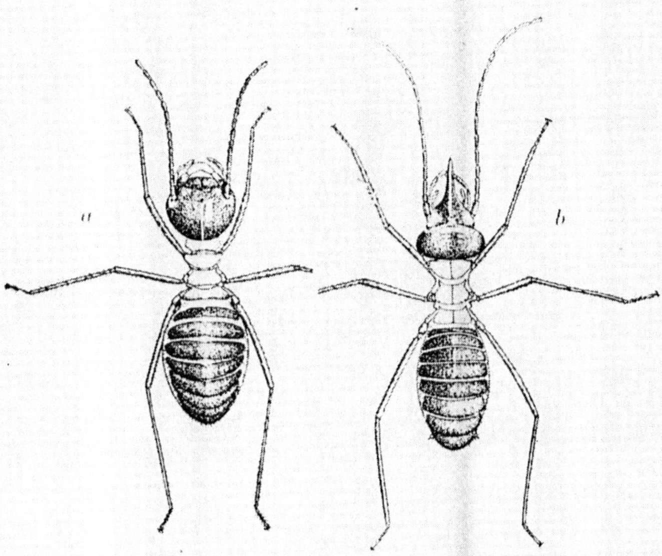

FIG. 53. a. Small, pigmented, worker-like form of *Nasutitermes* (*Constrictotermes*) *cavifrons* Holmg., which later changes into b, the mature, pigmented soldier (nasutus). (After Prof. A. E. Emerson.) (*See* p. 186.)

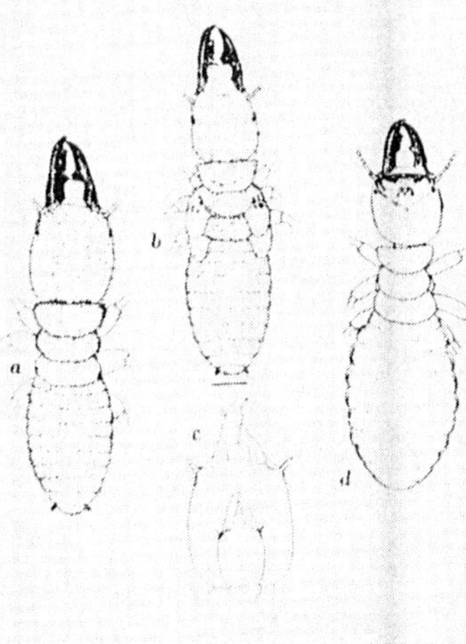

FIG. 54. a. Normal soldier of *Termopsis angusticollis*. b. Soldier with wing buds. c. Relative sizes of first and last soldiers produced in a colony. d. Fertile soldier of *T. nevadensis*. (After Prof. Harold Heath.) (*See* p. 185.)

[*face p.* 218

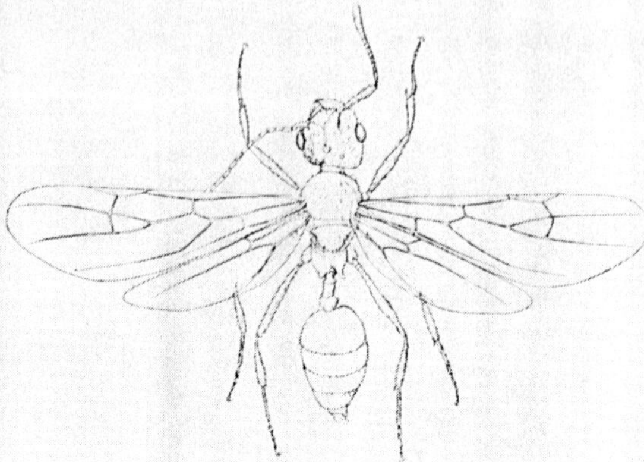

FIG. 55. Gynandromorph of *Epipheidole inquilina*; male on the left, female on the right side.

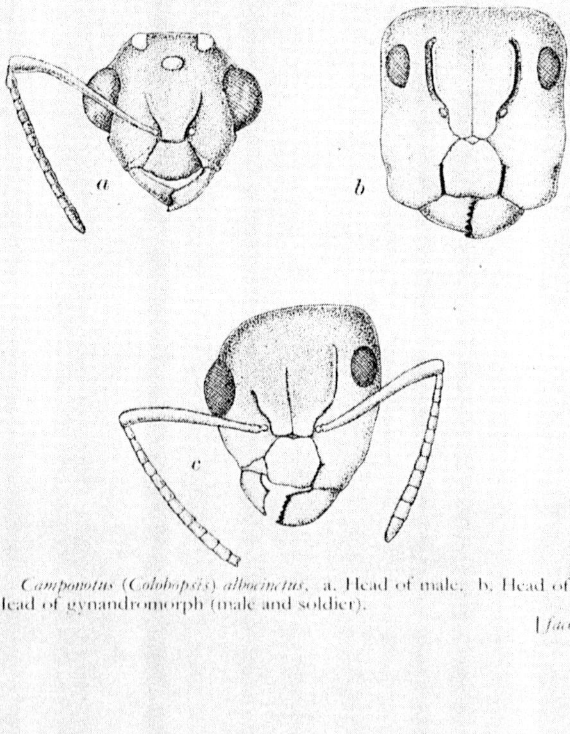

FIG. 56. *Camponotus (Colobopsis) albocinctus*. a. Head of male. b. Head of soldier. c. Head of gynandromorph (male and soldier).

[face p. 219

one sex of characters peculiar to the other. But these distinctions seem to be less absolute than has been supposed. Even if the male and female components in ant gynandromorphs are determined before or at the time of fertilization, it is conceivable that differences in larval feeding might affect the female component in such a way as to determine its caste in the adult, though we should expect the resulting form, so far as the female component is concerned, to be more frequently of an intermediate type between the worker and queen, i.e., ergatoid or pseudogynic and not so sharply of either the worker or queen type as we find in nearly all the anomalies hitherto described. Emery (1924), has indeed recently described two anteroposterior gynandromorphs of *Myrmica ruginodis*, which exhibit a peculiar mixture of worker, queen and male characters. They are, in fact, as he says, "pseudogynes with male genitalia." That the feeding of the larva may determine the caste of the female component of the gynandromorph is also indicated by the numerous cases of this anomaly among honeybees (von Siebold, 1864, Boveri, 1915, Fräulein Mehlis, 1915), since they seem all to have been reared in worker cells and to be ergatandromorphs. Recent investigations, however, have had a very unsettling effect on our views of the blastogenic origin of gynandromorphs in general. Although there is some doubt as to Kosminsky's definition of these anomalies (1924*a*, 1924*b*), both he and Emeljanoff (1924) claim to have produced gynandromorphs and intersexes in moths by submitting their larvæ and pupæ to abnormally low or high temperatures or by feeding the larvæ on unusual food. And Poulton (1927) has described the production by von Sommern of several gynandromorphs of the African butterfly, *Papilio dardanus*, by mechanical shock of the caterpillars at a "time during which the larval skin is being cast and the pupa is still soft and unset." If these observations are confirmed we may have to regard the distinction between gynandromorphs and intersexes as artificial and the former as special cases of the latter.

The most serious objection to the trophogenic hypothesis is the obviously adaptive or rather adapted structure of the various castes, especially of the soldiers and workers, as exhibited in the absence of wings, structure of head and mandibles, eyes, etc., and the correlated behaviour (difference of professions (Pricer, 1908, Miss Buckingham, 1911, Hingston, 1922, etc.)). And the difficulty is the greater, because the soldiers and workers in the most highly specialized species, which reveal these adaptations in their greatest perfection, are sterile, or if fertile, lay eggs only under exceptional conditions (abundance of food, high temperature, etc.). Such eggs in the ants, as in the other social Aculeates, regularly produce only males. Since the problem in the termites is very similar, it would seem that we are compelled to adopt one of the three following hypotheses :—

(1) The castes arise either as fluctuating variations or as mutations, which are sifted by natural selection acting not on the sterile castes but only on the mother queen and her queen offspring, as Darwin and Weismann maintained.

(2) There is some finalistic, metaphysical factor—some social entelechy—which initiates and guides the adaptive development of the sterile castes.

(3) The sterile castes are much more frequently fertile than has been supposed and can transmit their peculiar characters through their male offspring (social Aculeates) or male and female offspring (termites) and thus acquire representation in the germ plasm of the species.

The first hypothesis cannot be entirely rejected since the queen of most social Aculeates epitomizes the colony in its incipient stage of development, but natural selection has lost its value as an explanation of the origin of adaptive variations. The second hypothesis is avowedly vitalistic and is of no interest to the scientific biologist. That the third affords an opportunity for productive investigation is shown by a very suggestive paper by Verlaine (1926), who contends that the males of the social Aculeates in general are the offspring of the workers. While he admits that old honeybee queens whose spermathecæ are exhausted, may produce males, he is emphatically of the opinion that the drones, even in hives of the honeybee with young and vigorous queens, are all the progeny of workers. He shows that this is also true of the social wasps, and Plath (1922) has obtained the same results with humblebees. Emery (1918) reported that all the males in colonies of *Pheidole pallidula* are produced by the soldiers and not by the queens. It is, indeed, not improbable that the only queen ants which normally produce males are those of the parasitic genera (Anergates, etc.) which lack the worker caste. Verlaine's contention, if sustained by further investigation, will do away with several awkward problems which have long exercised the ingenuity of melittologists, such as the supposed alternate functioning and non-functioning of the spermatheca of the queen in adaptation to the worker and drone cells, and many of the hitherto inexplicable phænotypic peculiarities of hybrid honeybees.

No doubt, we have hitherto laid too much stress on the static, morphological aspects of polymorphism and too little on its dynamic, physiological and behaviouristic aspects. Many physiologists, too, are apt to overlook or at any rate to minimize the regulatory and adaptive peculiarities of biological processes and these are certainly very significant in the pœcilogyny and pœcilandry of the social insects. It has been suggested that the production of workers and soldiers in the colony is a kind of experimental teratogeny carried on by the worker nurses, but it is certainly strange that the monsters produced, e.g., the janitor soldiers of Colobopsis among the ants and the nasuti among termites, should be structurally and functionally so exquisitely adapted to their particular professions. And it strains our credulity to be told that such forms arise either from peculiar genes popping out of nowhere into the germ plasm or develop gradually under the guidance of natural selection from forms which, so far as we can see, must have an equal or even greater survival value. When we encounter such *impasses* as the foregoing, instead of embracing the Aristotelian Entelecheia, that belldame of more than two thousand summers, now so popular on the other side of the Rhine, or joining the apostles of the survival of the fittest and forever croaking "natural selection!", it is surely more commendable to sit down in the laboratory or in the field and say nothing but "*ignoramus*" till we have made a much more exhaustive behaviouristic and physiological investigation of the phenomena. "*Nullus sermo in his potest certificare, totum enim dependet ab experientiâ.*" (Roger Bacon.)

IX

THE SOCIAL MEDIUM AND TROPHALLAXIS

THE study of polymorphism has brought us to a realization of the significance of the social medium. In human society it is so complicated that probably few of us have a definite notion of what it is, though we all know that it is very important, but insect societies are so much simpler that they might be expected to yield a clearer conception of its nature. In one of my earlier lectures attention was called to the fact that the activities of social insects are merely modifications of the behaviour of their solitary ancestors. This is evident in the Aculeata and no doubt holds good also for the termites, although we know nothing of the habits of their solitary forbears, the Protoblattoidea. In the further discussion I shall therefore deal mainly with the Aculeata. The activities with which we are concerned are first, the nuptial or dissemination flight, second, nidification, third, foraging and storing food and its distribution among the members of the colony and fourth, defence. Among the solitary species these activities, with the exception of the nuptial flight or its equivalent, are carried on by the females exclusively. In the social species all four activities have become mass phenomena and are therefore intensified and modified, the nuptial flight but slightly, nidification and defence greatly, and most of all the trophic behaviour, which has to do with providing the adult and larval members of the colony with sustenance. The social medium, in very general terms, may be said to consist of these modified and intensified activities. Nidification and defence among the social Aculeates have had such a long and intricate evolution that their adequate presentation would require a special course of lectures. My time will permit me to consider only the trophic behaviour, to which all the other activities of the colony are really ancillary or contributory.

We know that there are two primary types of colony inception among social insects, either by single parental individuals (queens in Aculeates, royal couples in termites), or by swarming, i.e., by sexual forms accompanied by a certain number of workers from the parental colony. The former, again, may be either independent or dependent. In the independent subtype the queen or royal couple is able to establish a colony without worker assistance. This occurs among the Polistinæ in temperate regions, the Vespinæ, the Bombinæ in temperate and boreal regions, Halictinæ, Allodape and the great majority of ants and termites. The dependent subtype is represented by a few parasitic Vespinæ and Bombinæ and a number of mostly parasitic ants, the queens of which are quite unable to found colonies independently and must either be adopted by colonies of their own or other species or at least secure the brood of another species in order to provide themselves with the necessary worker aid in rearing their own young. With these cases I shall deal in a future lecture. The groups which establish their colonies by swarming are the Polybiinæ and some tropical Polistinæ among wasps, the tropical Bombinæ, Apinæ and Meliponinæ among bees and some termites (e.g., *Reticulitermes lucifugus* in Italy). It will be observed that swarming does not occur in temperate regions, except in the honeybee, which is really an Indian insect, and in some ants (*Formica exsectoides*, etc.).

Both the dependent and swarming methods may be regarded as modifications of the independent method of colony inception and are really more definite expressions of the dependence of the queen on the worker caste. But even the independent method is merely an interlude in the social life of the queens and royal couples that practise it, since the virgin sexual forms of all social insects remain for some time in the parental nest before their nuptial dissemination flight and then lead a very secluded life till they are in turn surrounded by their own worker offspring. The social insect is, therefore, like man, permanently social, the workers throughout their lives and the sexual forms of the independent type, except for a brief period while they are maturing their gonads. Naturally these conditions make it extremely difficult or even impossible to determine to what extent the behaviour of the individual insect is the result of the constantly acting social medium in which they are immersed, and to what extent it depends on inherited mechanisms. This difficulty is also responsible for the differences in the interpretation of behaviour by different investigators. The physiologist who studies social insects merely as individual organisms, experimentally isolated from their social medium, is apt to conclude that their behaviour is entirely reflex, or tropistic (Bethe, 1898, 1900, 1902, and Henning, 1916), whereas those who observe them in their social environment reach a very different conclusion, and while admitting that many of their activities are reflexes (" automatic " behaviour of Forel) feel confident nevertheless that they give unmistakable evidences of memory, appetites, emotion, imitation and a feeble intelligence, or ability to modify their reactions in conformity with previous experience and environmental changes (" plastic " behaviour of Forel). General agreement on these matters still leaves plenty of room for differences of interpretation in detail, according to the training, predilections and philosophic outlook of the investigator. Forel and Brun, trained as neurologists, are greatly impressed by organic memory and make constant use of Semon's " mneme " and his terminology in describing the behaviour of social insects. The Jesuit father, Wasmann, trained in the scholastic philosophy, operates with " instincts," virtues and faculties (Vermögen) in the manner of St. Thomas Aquinas, and with the adroitness for special pleading and ignoring of pertinent data, for which his order is so celebrated, wraps the whole subject of myrmecology and myrmecophily in a dense fog of teleology, "Fremddienlichkeit," amical selection, mimicry and theistical casuistry. Within recent years I have come increasingly to avoid the word "instinct" and to prefer "appetite," or "appetition" in Fouillée's sense (Wheeler 1921*c*). Of course, this is nothing new since the word was used with much the same signification by the scholastic philosophers (*appetitus sensitivus*, Wasmann's "sinnliches Begehrungsvermögen"). It is Aristotle's ὄρεξις and is synonymous with the "libido" and "craving" of modern psycho-analysts.

An adequate discussion of the central (cerebral) as contrasted with the peripheral (sensory, or receptor) factors involved in the behaviour of the social insects would require much time and be irrelevant to my immediate purpose, which is to call attention to the fact that the social medium is necessarily the creation of the worker caste. This caste is always produced in such numbers as to constitute nearly the entire colony, and all the important behaviour—nidification, defence, provisioning and caring for the queen and young, in addition to its own sustenance—is its sole prerogative. The activities mentioned are continuous, except during unfavourable climatic conditions, and extend not only throughout the nest but to a considerable area over which the insects forage for food. This territory, which has a circular circumference and may be called the "trophoporic field" is determined by the ranges of the foraging individuals and naturally increases with the growth of the colony. The trophoporic field of an incipient colony, with its few feeble minim workers, is very small, but as the foragers increase in number and size with succeeding broods, it gradually enlarges till it reaches its maximum area when the growth of the colony is completed. Except where they are few and sporadic, the trophoporic fields of colonies of the same or different species overlap more or less. The food supply is therefore limited in amount and the exploitation of the field is so intensive, that every available source of nutriment is scrupulously investigated and carefully utilized. According to the habits of the species the supply comprises all insects that can serve as prey, all those whose saccharine excreta can be collected (Homoptera), all plants whose nectar and pollen are accessible. Although the trophoporic fields of the social bees are much more extensive than those of the ants, the food supply is precarious and subject to great seasonal fluctuations. We find, therefore, that unlike many solitary bees, the Halictinæ, Bombinæ, Meliponinæ and Apinæ are decidedly polytropic, or not restricted to the flowers of particular species of plants (oligotropic). Certain foraging termites (Eutermes, Hodotermes, Termes, etc.) have rather extensive, others very restricted trophoporic fields. In the Doryline ants, which are so ravenously entomophagous that they can maintain no permanent nests, the trophoporic field shifts like that of a nomadic or hunting people.[1]

Since the size of the colony is a function of the amount of food obtainable from the trophoporic field, the latter determines and regulates the growth of the population. The workers are constantly endeavouring to obtain as much food and to rear as many young as possible. Hence extraordinary thrift and economy have to be practised, especially when the trophoporic fields of several colonies overlap. The sexual forms and especially the queens, which require a considerable amount of food during their larval stages, can be reared only during the most favourable season of the year. Unfavourable trophic conditions sometimes make economy so acute that the brood has to be devoured wholly or in part and portions of the adult population have to be sacrificed, as in the slaughter of the drones among honeybees or of the soldiers in certain ants of the genus Pheidole. Our North American *Ph. militicida* harvests seeds and its huge-headed soldiers are needed to crush them, but when the winter comes on and the services of the seed-crushers are no longer required, they are decapitated and their remains thrown out of the nest. Of course, the caste thus eliminated can be readily recruited from young larvæ during the ensuing spring and summer.

[1] What I have called the trophoporic field Le Dantec (1918) calls the "cantonal patrimony," the quantity of food which it furnishes, the "alimentary patrimony" of the species under consideration.

That the workers of all the social insects are really hunger forms is shown by their greedy behaviour, their normal lack of fecundity, their atrophic development and especially in the ants and termites by their rhachitic, wingless adult physiognomy. Yet the hunger and greed are those of a sterile proletariat (*sit venia verbo*), with little or no interest in producing but with a keen interest in rearing larvæ and in securing the food necessary for carrying on the business. This paradoxical proletariat, which takes upon itself the control and regulation of the numbers of individuals of all the various castes, its own included, has never ceased to excite the admiration of the paragon of animals. As de Gourmont (1924) says : " Of all sexual aberrations, chastity is perhaps the most singular. Not that it is unnatural, for nothing is unnatural, but on account of the pretexts which it serves. The bees, ants and termites present examples of perfect chastity and a chastity that is at the same time utilized and social. Involuntary and congenital, the neuter state in insects is an actual condition and a source of characteristic behaviour. In man it is a condition often merely apparent or transitory, adopted voluntarily or demanded by necessity, a precarious condition so difficult to maintain that around it have been erected all kinds of moral and religious and even true ramparts of real stones and real mortar. Permanent and voluntary chastity is nearly always a religious practice. In all ages men have been persuaded that a state of perfection is attainable only through such renunciation. This may seem absurd, but it is, on the contrary, supremely logical. The only way to avoid being an animal is to abstain from an act to which all animals yield themselves without exception. The same motive has suggested abstinence and fasting, but since it is impossible to live without eating and quite possible to live without mating, the former has remained a mere programme." De Gourmont might have gone further and called attention to the singular fact that throughout the ages the most diverse human societies have not only encouraged celibate castes to develop in their midst and lavished on them inordinate admiration, but have actually allowed them to behave like the workers of social insects in controlling and regulating the education and reproductive physiology of the fertile individuals as well as those of their own caste.[1]

The social medium obviously comprises not only the regular activities of the workers in the nest and mainly the collection of the food and its distribution among themselves, the queens, males and larvæ but also the relations to food-yielding insects or plants in the trophoporic field and to the various guests which as parasites *sensu latissimo* are sometimes present within the nest itself. The fact that the latter are often absent, that when present they are usually few in number and have relations with the workers only, and that these are usually sterile, constitutes an almost insuperable objection to Wasmann's attribution of the behaviour of the workers towards particular guests to specific blastogenic "instincts." One must either assume that there is an inheritance of acquired characters—and that with a vengeance!—through the males produced from workers' eggs, or that the behaviour towards the guests is a purely ontogenetic modification or perversion of the regular threptic behaviour exhibited by the workers towards one another, the queen, males and brood. I insist that the latter is the more probable and more economical assumption and the one in closest accord with the facts

[1] Perhaps voluntary human celibates are Mendelian recessives, and if one wished to resort to an Adlerian accessory hypothesis, their behaviour might be interpreted as inspired by the kind of "will to power" so conspicuous in individuals with pronounced "inferiority complexes". A similar interpretation seems to be implied in the following remark of Gerald Heard (1924): "Finally, history suggests that the invariable appearance at a certain stage of the priest-artist is because civilization necessitates specialization and real specialization postulates a genetic variety. Whenever a society is about to become a civilization, it begins by producing celibates. This "segregation of the best stocks" has always been a stumbling block to the old-fashioned Eugenist, but the fact that seems to have an inevitable naturalness about it must be faced, and perhaps, though this is not the place to discuss it, the "best stocks" might after all not prove to be the best for stock purposes. (Cf. Dr. Bateson's Herbert Spencer Lecture: "Biological Fact and the Structure of Society," p. 13). See also Towner (1923).

of organic behaviour in general. Even in the study of human behaviour which is far better known than that of any animal, the present tendency is to reduce the number of postulated " instincts " to a minimum or to eliminate them altogether from serious discussion.[1]

In a paper published in 1918 I attempted to stress the importance of the trophic and threptic behaviour of the social insects and to frame for the numerous and disparate facts a general conception to which I applied the term " trophallaxis " (from τροφή food, and ἀλλάττειν to exchange). The notion was favourably received by the physiologists and ethologists but was treated as a scandalous innovation by the Jesuit instinct mongers.[2]

We have seen that the insect colony or society may be regarded as a super-organism and hence as a living whole bent on preserving its moving equilibrium and its integrity.

[1] What really up-to-date laboratory behaviourists think of instincts and their inheritance may be gleaned from a perusal of Kuo's latest paper (1924). Though expressed in extreme language, this author's views are undoubtedly valuable in calling attention to the extraordinarily vague, unscientific and dogmatic assertions of Wasmann and others in regard to the hereditary basis of behaviour. See also Bernard (1924).

[2] That my efforts, like all such attempts, should be misunderstood or misinterpreted was only to be expected, but I did not realize the profundity of my ignorance of the biology of the Formicidæ and their guests. Father Wasmann, in a 176 page treatise (1920), promptly demonstrated my deplorable insipience, and as I had also been guilty of the mortal sin of pride, he very modestly placed as a motto on the cover of his polemic the well-known lines from " Faust " :

" Was glänzt ist für den Augenblick geboren,
Das Echte bleibt der Nachwelt unverloren ".

The first line refers, of course, to my daughter Trophallaxis, the second to Wasmann's own immortal progeny, the " Symphilic Instincts ". As it was feared that this howitzer in the camp of the Society of Jesus might miss its mark, I was treated with some machine-gun fire in the Mémoires of the Pontifical Roman " Accademia dei Nuovi Lincei " (1923). In the meantime Herr Reichensperger hurried to the rescue of his outraged mentor and with the loving Christian kindness of his ilk, exuded a little poison gas (1921) in the form of comments on my " recht kindliche Analogiebeweise ", etc. All of which shows how terribly dangerous it is to doubt even the biological dogmas of the reverend fathers and to permit oneself to joke about their pet lucubrations. Since I had not extolled the charms of Trophallaxis in every paper I had written since 1918, Wasmann concluded that I must have abandoned the poor thing as still-born. Here his perspicacity certainly forsook him, for I am so far from any such intention that I propose to devote the remainder of this and portions of the two following lectures to showing that she is an even healthier and more viable infant than I had supposed.

The individuals composing the colony must therefore be in communication with one another. The truth of this statement is indeed apparent from the observation of any insect society, but the methods of communication employed are so different from ours that their precise interpretation and even their detection may be matters of considerable difficulty. Undoubtedly the individual insects communicate by means of signs, i.e., by movements of the body and its appendages, especially of the antennæ (Forel, Wasmann, Crawley (1910*b*), Doflein (1920), etc.), by vibrations, or stridulation, Wheeler (1914), Santschi (1927), odors (von Frisch (1921)) and tastes. All these belong to the general biological category of stimuli and responses, and since the behaviour is very largely concerned with food and its distribution the stimuli are mainly olfactory and gustatory and the responses depend on the chemoreceptors and appetites. Many of the criticisms levelled against trophallaxis are at once dissipated by pointing out that in employing the word " food " I did not mean merely so much matter taken into the alimentary tract and assimilated but also its rôle as a stimulus, or excitation of the chemoreceptors. Obviously this latter role is of the greatest significance in social organisms. That it is taken for granted by physiologists is clear from the following quotations (Child, 1924) : " Even in most of the simpler animals the reaction to food involves excitation and the excitatory factor is undoubtedly concerned in the growth orientation of the roots and other parts of plants with respect to chemical and photic stimuli. . . . Like the relations between living protoplasm and its external environment the physiological relations between the different protoplasmic systems, parts, organs, etc., of an organism are either material, involving the mere transportation of substance, or dynamic, involving the transfer of energy. . . . In the higher animals and man the excitatory factor is unquestionably the primary factor between individuals, the material factor being significant only as it is excitatory. In fact all social integration is based upon the excitatory relation. . . In material or chemical correlation between parts, the part producing a substance which influences another part controls the latter to some extent. This is also true in cases in which chemical relations between the two parts are more or less mutual. Such cases are in some degree analagous to the social reaction system consisting of two groups of human beings between which mutual commercial relations exist. To take a rather primitive social system of this sort, one group for example inhabits the coast, the other an inland region. The first group exchanges fish, shell, salt for skins, game or perhaps for metal articles. The articles received constitute a factor in determining the further activities of the group in each case." I did not mention such obvious considerations in my paper on trophallaxis because I was writing for biologists and not for theologians. The latter, too, recognize two uses of food—one permissible, when it is employed merely as so much nutriment to keep body and soul together, the other sinful when it is so stimulating that it may be regarded as a " Leckerbissen" (tit-bit) and leads to " Naschhaftigkeit " (greediness), and this is the distinction imported by Wasmann into the discussion of insect behaviour as a telling argument against trophallaxis, after I had expressly stated (1918*a*) that the substances exchanged, e.g., in cases like *Formica sanguinea* and its guest Lomechusa, are not quantitatively equal but that the exudates of the guest, though produced in very small amounts, may nevertheless elicit intense reactions ! One would suppose that the food served in a Jesuit refectory must be even more tasteless than that served in a fifth-rate New York restaurant !

Other critics have imagined that I meant by trophallaxis not only an exchange of equal quantities of food *as mere nutriment* but an exchange *necessarily and immediately reciprocal*. Still others, like Morstatt (1922), believe that trophallaxis must be something superadded to the care of the young (Brutpflege), although I regarded this behaviour as the most primitive and characteristic manifestation of the activity. Kemner (1923) has also criticized trophallaxis on the basis of some interesting but incorrectly interpreted observations on a Javanese myrmecophilous moth. A discussion of this case may be postponed till the next lecture. Among entomologists Brun (1923*b*) alone seems to have grasped the true meaning of my paper. His knowledge of psychiatry enabled him to show that trophallaxis involves the " pleasure " as opposed to the " reality principle " and to call attention to the determining rôle which the former plays even in the lives of the social insects. This being the state of opinion in regard to trophallaxis, I may be pardoned for briefly reviewing the relevant facts. It will then be advisable to add some considerations which, I trust, will not only meet the really serious objections but extend the heuristic and explanatory usefulness of the principle.

Trophallaxis is most easily observed among the wasps, ants, and termites. Du Buysson (1903) and Janet (1903) were, I believe, the first to show that the larvæ of Vespa and Polistes, after being fed by the workers or when their mouth-parts are stimulated, secrete from their salivary glands drops of a thin, sweetish liquid which are at once greedily swallowed by their nurses. I have repeatedly observed the same behaviour in our North American *Polistes metrica*. It was Roubaud (1916, 1924), however, who first made a detailed study of the reciprocal feeding in certain African wasps (Belonogaster) and grasped its social significance. He called it " œcotrophobiosis ", a word which seemed to me to be rather awkward and inadequate. I therefore substituted the term " trophallaxis ". In his latest paper (1924) Roubaud has adopted the word " trophœcism " and therefore calls the wasp-colony a " trophœcium ". In my study of ant-larvæ (1918*a*) I cited a number of phenomena which can only be interpreted in the same manner as the trophallaxis of the Vespidæ. In certain Ponerinæ (Pachycondyla) which are primitive, wasplike ants, the larvæ when fed produce a salivary secretion which is licked up by the nurses. Some Myrmicine larvæ (Pædalgus) have enormous salivary glands, the secretion of which very probably has a similar function since it is not used in spinning a cocoon, an envelope never produced by any species of the subfamily. In the larvæ of all Pseudomyrminæ the mouth is surrounded by a cluster of singular papillæ or appendages which are clearly exudatoria very much like the exudatoria of some termitophiles (Spirachtha). These structures are greatly developed in the very young larvæ (trophidia) of two species of the African genus Pachysima (*æthiops* and *latifrons*) (Figs. 57), but decrease rapidly in size in the later stages when the salivary glands develop and apparently usurp their secretory function (Fig. 58). Moreover, all ant larvæ store in their bodies an enormous amount of fat, some of which exudes through the thin chitinous cuticle which is frequently licked by the worker ants. The surfaces of ant-larvæ are, in fact, usually covered with a lipoid, hydrofuge substance which is probably derived from the blood or fat-body. That such substances can actually pass even through rather thick chitin is shown by the observations of Pantel (1909), Roubaud (1924) and others on the young larvæ of Tachinid flies. Although these larvæ are completely enclosed in a chitinous sac developed from the tracheal intima of their hosts, they nevertheless obtain their food by filtration or osmosis from the blood of the latter.[1] All ant larvæ, therefore, probably produce small quantities of exudates which are appreciated by their nurses. But even if this be doubtful, it must be admitted that trophallaxis is of general occurrence among ants, since the queen, after bringing up her first brood, is fed with regurgitated food by the latter, that is by her offspring in their *adult* stage. That the workers of the higher ants (Myrmicinæ, Dolichoderinæ, and Formicinæ) almost universally practice this form of reciprocal feeding is a fact too well known to require further substantiation.

[1] Eidmann (1922) has proved that both alkaline and acid solutions will diffuse through thin layers of chitin in which pores are invisible under a high magnification, and though his results were obtained with the intima of the crop and hind gut of the cockroach, there can be no doubt that the thin chitinous integument of insects is similarly permeable.

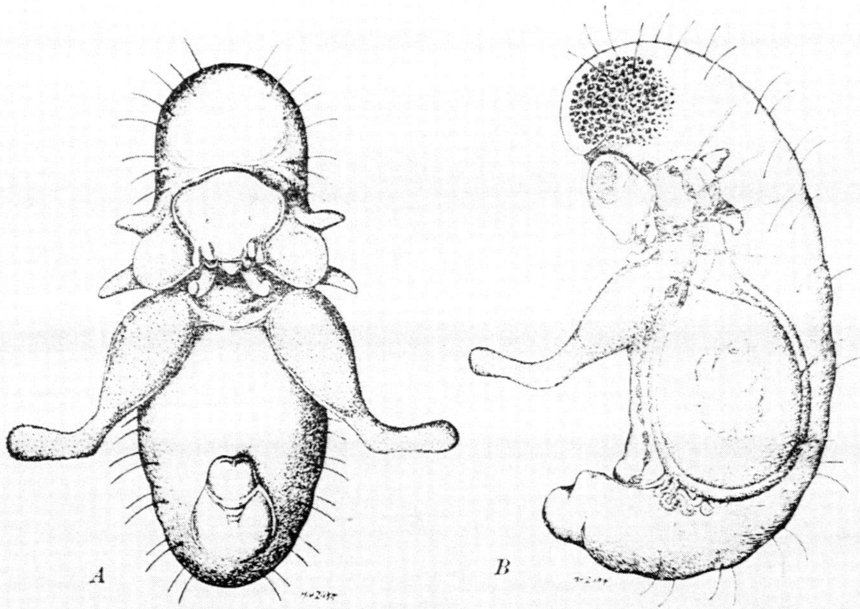

FIG. 57. A. Ventral. B. Lateral view of the first larval stage ("trophidium") of the Ethiopian *Pachysima latifrons*, showing the peculiar appendages ("exudatoria") surrounding the head. These belong to the three thoracic and the first abdominal segments.

[face p. 234

Plate XXXVI.

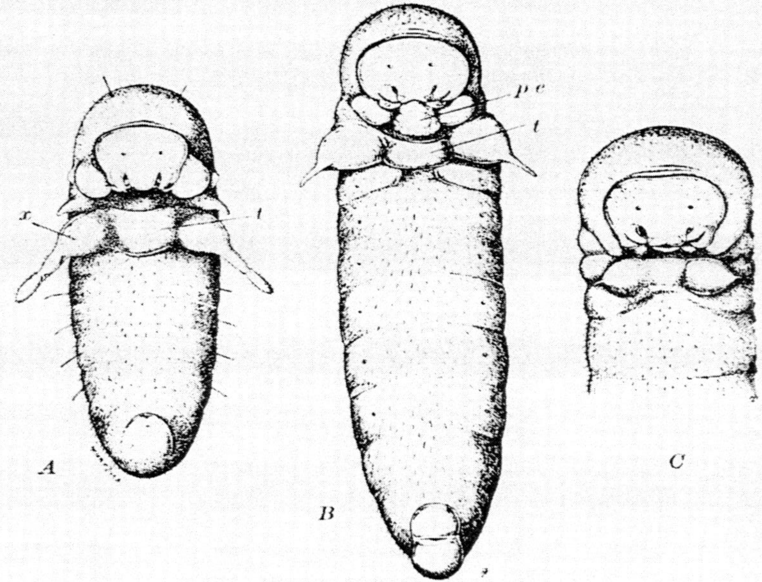

FIG. 58. Second, third and fourth (adult) larval stages of *Pachysima latifrons*, showing the gradual dwindling of the exudatoria. A. and B. show the trophothylax (t) and B. also shows the food pellet pe., which is the pellet formed in the infrabuccal pocket of the worker nurse. x. Exudatorium. (*See* Figs 62 and 63.)

[face p. 245

I regard it as certain, furthermore, that adult ants also produce fatty exudates or attractive glandular secretions and that the mutual licking in which these insects indulge is attributable to the presence of such substances on their bodies and not to affection or to a desire for cleanliness. Nor are the toilet operations of the individual ant due to such a desire but probably to a reflex response to the irritation of foreign matter on the surface of its body and appendages.

The mutual exchange of liquid foods finds its most spectacular expression in the honey ants (Figs. 59 and 60), which, as I have shown (1908*b*, 1914, 1923), belong to a number of genera and species confined to desert or xerothermal regions where liquid food is very scarce. Usually certain workers store the food in the crop, which is so enormously distended that the gaster becomes very large and spherical (Myrmecocystus, *Camponotus inflatus*, *Melophorus bagoti*), but in many species of Pheidole, Pheidologeton and Oligomyrmex the soldiers may act as honey-pots and the gaster is much less conspicuously enlarged. Emery, many years ago (1898), observed in the workers of certain ants from the deserts of Asia Minor and North Africa (*Camponotus fedschenkoi* and *atlantis*) a peculiar enlargement of the abdominal fat-body. He noticed that this "adipogastry", as it may be called, was best developed in nocturnal species and that the fat represents food stored as sustenance not only for the individual worker but also, after metabolism into salivary secretion, for the larvæ. A very similar adipogastry is often noticeable in many hypogæic ants, especially in our North American Lasii of the subgenus Acanthomyops, which obtain nearly all their sustenance from root-aphids and -coccids. The yellow nocturnal forms of *Camponotus castaneus* in New England and of *C. sansabeanus* in the deserts of Arizona may also exhibit the same "Fettleibigkeit". As Ottramare (1919) has shown, there is evidently a significant connection between absence of light and the accumulation of fat. This is also apparent in the fatty physogastry of termites and their guests.

Grassi long ago described the extensive development of trophallaxis among termites and all succeeding investigators have confirmed his observations. The members of the colony feed one another on saliva, partially digested and regurgitated (stomodæal) foods and fæces, or scybalum (proctodæal food). Mutual infection of the workers with the symbiotic, wood-digesting protozoa is a result of this singular coprophagy. As previously stated, Holmgren (1909) has shown that all the termite castes, but especially the queens, have voluminous exudate tissues, consisting of the peripheral layers of the abdominal fat-body. The trophocytes of these layers do not contain fat but numerous minute granules which are discharged into the blood and thus convert it into an exudate, which passes through numerous fine pores or lacunæ in the chitinous cuticle to the surface where it is licked up by the members of the colony. The development of the exudate tissue differs considerably not only in the different castes but also in their various developmental stages, and we have seen in the preceding lecture that Holmgren traces the differentiation of the castes to exudate hunger. Escherich (1911) gives a more vivid, not to say more spectacular account of the greed of termites. So eager are the workers of the Ceylonese *Termes redemanni* for the exudate of their huge physogastric queen that they actually tear little strips out of her cuticle in order to get at the liquid more readily. Escherich noticed that old queens sometimes have their white abdomens dotted with little brown scars of the wounds thus inflicted by their progeny. Here the feeding behaviour of the mother and offspring is the reverse of that in incipient ant-colonies, since the queens are fed with regurgitated food by the workers and feed the latter with their exudates, but in all probability the same is true of established ant colonies, after the workers have matured and the queen no longer feeds the brood.

It is more difficult to find evidence of trophallaxis among social bees, but the case is perhaps not as hopeless as I had supposed (1918*a*). Adult worker bees feed their

Plate XXXVII.

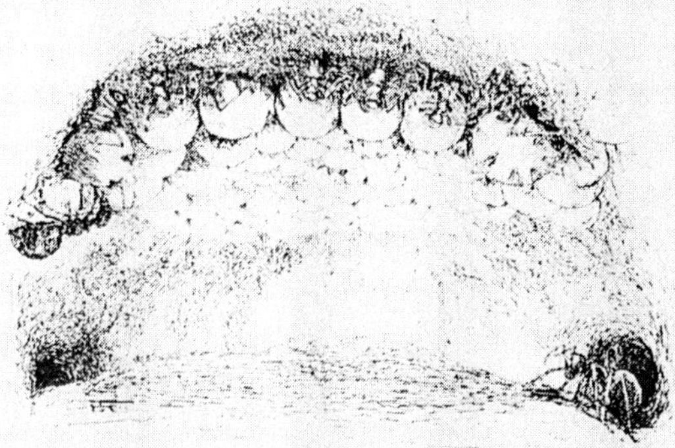

FIG. 59. Repletes of honey ants, *Myrmecocystus horti-deorum*, suspended from roof of nest chamber. (After H. McCook.)

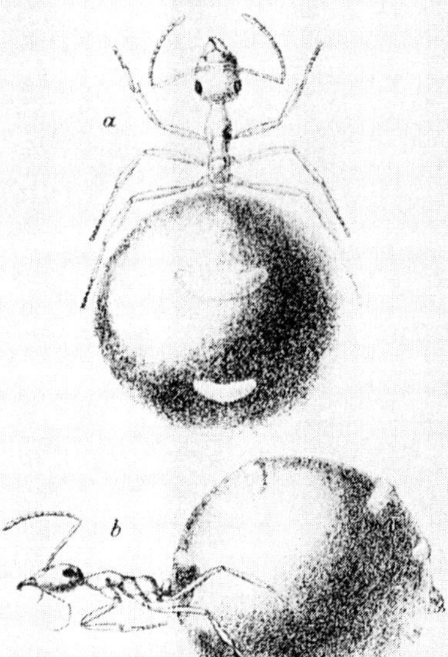

FIG. 60. Repletes of the honey ant, *Myrmecocystus horti-deorum*. (Dorsal and lateral views.)

[*face p.* 236

Plate XXXVIII.

FIG. 61. *Atelura formicaria* about to snatch the droplet of food that is being regurgitated by one *Lasius mixtus* worker to another. (After C. Janet.) (*See* p. 254.)

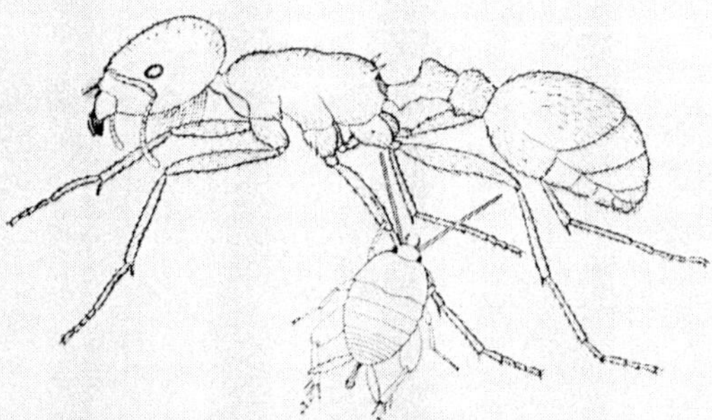

FIG. 62. A small cricket, *Myrmecophila nebrascensis*, gnawing at the tibia of the harvesting ant, *Pogonomyrmex molefaciens*. (*See* p. 254.)

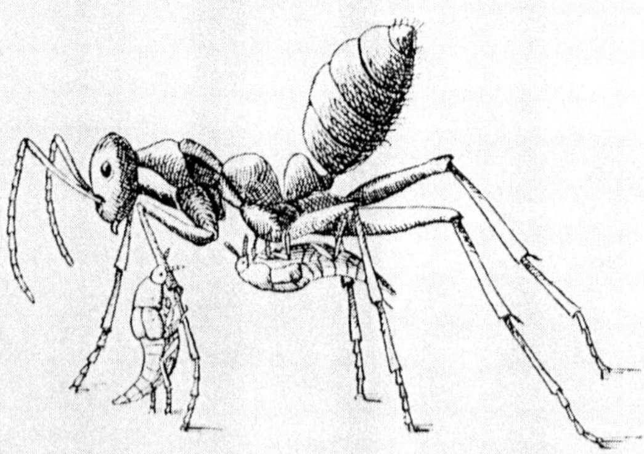

FIG. 63. Staphylinid beetles, *Oxysoma oberthueri*, feeding on surface secretions of *Cataglyphis viatica*. (After Prof. K. Escherich.) (*See* p. 254.)

[*face p.* 237

queen and one another with regurgitated honey (von Frisch, 1921), but whether this ever occurs among the Bombinæ and Meliponinæ seems not to have been observed. Lineburg (1924) has made a careful study of the honeybee larva to ascertain whether there is reciprocal feeding as in the wasps and ants. " Reciprocal feeding ", he says, " has never been observed in the honeybee, but since the food given to the older larvæ is not complex in character and probably requires no great amount of work on the part of the nurse bees for its elaboration, it is difficult to explain the great amount of time spent by the nurse bees in the cells merely on the basis of the belief that food is being fed to the larva all that time. That the usual method of feeding is directly to the larval mouth is clear from the fact that no significant amount of food is ever found in the base or on the sides of the cell of an older larva. In the determinations of the weight of food residue in the cell it is noteworthy that for each age it is quite uniform. If one dared to assume reciprocal feeding, this might perhaps account for the peculiar development of the worker bee in the repression of the sex organs and their morphological modifications ". These remarks show that reciprocal feeding between worker honeybees and worker larvæ is not altogether improbable. That the worker eggs and larvæ have some powerful attraction is shown by the number of visits they receive and the time workers spend in the cells. " Averaging the results found for the eggs and larvæ of all ages, it is seen that on an average more than 1,300 visits are made in twenty-four hours. On the last day before capping no less than 2,855 visits are made by the nurse bees to a single cell. On this last day before capping approximately 4¾ hours are spent by the nurse bees within the cell. . . . Over 10,000 visits are made to each developed bee during the eight days from the time the egg is laid until the cell is capped."

In a previous lecture I called attention to the secretion of wax in social bees and of a similar substance in certain solitary species (Tetralonia, Eucera, etc.). Wax, which is very largely myricyl palmitate, is really a lipoid exudate related to the fats (Mathews, 1921). Its production would seem to be a result of the great amount of sugar consumed by bees, since similar exudates are produced by Coccids, Aphids, Fulgorids, etc., which live on the sap of plants and ingest so much sugar that they have to rid themselves of much of it as fæces. Curiously enough, many of the Coccinellid larvæ (e.g., Brachyacantha) which feed exclusively on these Homoptera also produce abundant wax exudates (Wheeler, 1911*b*). According to Friese (1923), the fatty exudates of solitary bees "evidently represent the excess energy derived from the food which tensely fills out the body of the larva in the form of the whitish corpus adiposum before pupation. This stored chemical energy, which in the trophocytes is used to maintain the life of the pupa and for some years that of the imago, is usually not entirely consumed but produces the above mentioned fatty exudate in freshly killed solitary bees. Such exudates are also represented by the "cushions of an apparently fat- or wax-like substance", found between the four median segments on the dorsal side of the abdomen of the long-horned bee, *Tetralonia ruficollis*".[1] Although wax is now used exclusively in constructing the cells in which the larvæ are reared and the food is stored, the question arises as to whether it was not, among the immediate ancestors of the Bombinæ, Meliponinæ and Apinæ, employed to some extent as a larval food. Two of the older and at least one of the modern observers of humblebees (Swammerdam (1737) Réaumur (1742) and Haerter (1890)) have, in fact, claimed that the "pollen paste", or mixture of pollen and wax, in which the queen Bombus lays her eggs, is devoured by the larvæ. The majority of recent investigators, however,

[1] In connection with the fact that only the young honeybee workers secrete wax, Friese remarks : "We see therefore that little attention has been given to the organic nexus between the fat-body of the larva and the wax-production of the adult honey bee. I am of the opinion that the wax exudation is entirely dependent on the supply of fat energy of the larva. From this point of view the condition of older honey-bees with respect to wax production as well as the phylogenetic origin of the wax scales and wax-glands may be readily explained".

have been unable to confirm the observation. Since the larva of the well-known wax-moth (*Galleria melonella*) can digest wax, although it cannot complete its development without additional substances (Dönhoff (1882), Lieber and Metalnikow (1904), Biedermann (1911)), it is not improbable that humblebee larvæ may be able to produce some ferment capable of rendering the wax assimilable. Elaboration of the suggestion that the ancestors of the three higher groups of social bees may have fed their larvæ on fatty exudates would be pure speculation, but our knowledge of the intimate behaviour of these insects is so rudimentary that there can be no harm in calling attention to the possibility.

Extensive as is the scope of trophallaxis, or exchange of food in the various social insects, the principle is sufficiently elastic to cover an even greater number of phenomena, if we include besides the substances that are socially excitatory through the taste receptors also those that affect the other chemical sense, namely olfaction. At first sight, this seems to be inadvisable, if not absurd, but I believe that the following remarks will show that there is abundant justification for taking a different view of taste and smell in insects from that commonly held in regard to these senses in human and mammalian physiology. In mammals the gustatory and olfactory receptors are clearly separate in structure and position, though we are all familiar with the fact that olfaction enters very largely into what we call our taste sensations, and it is known that in certain fishes the taste-buds are scattered over the surface of the body.[1] In both receptors

[1] "In fishes Herrick (1903) lists over thirty-five species in which taste-buds are known to occur on the outer surface of the animal as well as in the mouth. The catfish Amiurus is remarkable in this respect in that its whole outer surface is provided with these organs which are most abundantly present on the barbels. When a piece of meat is brought in contact with the barbel of one of these fishes, the animal will immediately seize and swallow the morsel. The same is true when the meat is brought in contact with the side of the fish". (G. H. Parker, 1922). Herrick has proved that the responses to meat-juice discharged from a pipette against the fish's side are due to taste, not touch, and Parker (1912) has shown that when the seventh cranial nerve which innervates the taste-buds in the flank of Amiurus is severed, the gustatory responses no longer occur.

the nerve terminations are affected by chemical substances dissolved in a liquid or mucous layer overlying them. It is usual to transfer our notions of distinct taste and smell receptors to insects although even in vertebrates taste is not a unitary chemical sense and in insects cannot be distinguished from smell by the structure of the sensillæ.[1] We have merely acquired the habit of regarding chemoreceptors (sensillæ) on the mouth parts as gustatory and those on the antennæ as olfactory. But lately Minnich (1921, 1922*a*, 1922*b*, 1924, 1926) has demonstrated by carefully conducted experiments that butterflies and Muscids taste with their feet (with the four terminal tarsal and distal portion of the basitarsal joints of the second and third pairs of legs). This form of chemoreception is probably of more general occurrence among insects and other Arthropods. Even the distinction between distance receptors for smell and contact receptors for taste does not help us, since insects use their antennæ in both ways as well as for tactile sensations.

The study of the chemoreceptors of insects leaves us confused and baffled with the variety of the sensillæ and their wide distribution over the body. And the confusion is increased by the difference of opinion in regard to their structure, which grades all the way from sensillæ which may be either tactile or olfactory, through a great variety of presumably olfactory and gustatory to the Hicksian, or campaniform sensillæ which have been variously interpreted as organs of pressure, temperature,

[1] Parker says of the vertebrates: "It is true that gustation is a strikingly unified operation, but when this unity is looked into, it is seen to depend on simultaneousness of action rather than on independence of activities. Smell is related to taste in much the same way that one taste is related to another. On the whole it would seem more consistent with fact to speak of the sour sense, the saline, the sweet, and the bitter sense than of the sense of taste. Just as the sense of feeling in the skin has been shown to consist of at least four senses, touch, pain, heat and cold, so taste must be regarded as composed of at least four senses. That these act together and in everyday experience produce a unified effect upon us is no more reason for classing them as one sense than in the case of the integumentary senses. The sense of taste must, therefore, be regarded as a generic term under which at least four true senses are gathered: sour, saline, bitter and sweet (Öhrwall, 1891, 1901)".

humidity or vibration. While most authors believe that the olfactory sensillæ are confined largely or exclusively to the antennæ, McIndoo (1914*a*, 1914*b*, 1914*c*, 1915, 1916, 1917, 1918, 1920), for some inscrutable reason, finds olfactory organs on nearly all parts of the body, except the antennæ, and believes he has secured experimental evidence in support of his contentions. Berlese (1909) in a remarkably lucid account of the olfactory sensillæ, describes them as always containing glandular in addition to sensory cellular elements. He calls attention to the earlier papers of Erichson (1847) and Saulcy (1891), who found the antennæ of insects to be covered with a thin film of liquid. According to Berlese, this film is the secretion of the glandular elements and forms with the alfactory substance a solution which acts on the terminations of the sense cells. It is, indeed, difficult to conceive how olfaction can occur without such a solvent. But other investigators (Hauser (1880), vom Rath (1894), Röhler (1905), Vogel (1911), Hochreuter (1912), Deegener (1912), Demoll (1917), McIndoo, etc) interpret Berlese's glandular elements as sense-cells and say very little or nothing about an olfactory liquid or its source In an excellent review of the insect sense-organs, Snodgrass (1926) comments as follows on the matter : " The chief objection to the idea that a liquid exudes upon the surfaces of sense organs is the lack of any observations on the presence of such a liquid. Yet, the vacuole which surrounds the distal processes of the sense cells in many organs that have been regarded as chemoreceptive, suggests a possible source of a solvent liquid. Though Berlese's attempt to show that one of the elements of the insect sensillum (*sic*) is always a gland cell has not been generally accepted, it is not unreasonable to suppose that one of the cells might take on a secretive function in certain organs. The cell which contains the vacuole, when a vacuole is present, however, is the basal enveloping cell, which is the trichogenous cell and not a special gland cell ". In certain respects the physiological accounts are more satisfactory than the morphological and leave little doubt that the chemoreceptors are widely distributed over the antennæ, mouth-parts, feet, and possibly other portions of the integument, and that responses to chemical substances either from a distance or on contact are far and away the most important sensory reactions of insects. This is notably true of the social species, and numerous investigators have shown that much of their behaviour is determined by such reactions. In ants, bees, and termites, individual, caste, colony, species and nest odours have been distinguished experimentally by Forel (1910), Bethe (1898), Brun (1917), Miss Fielde (1904, 1905*b*), and others. The odour of the ant-brood, especially of the pupæ and cocoons, is clearly perceptible even to our blunt human olfactories. Many adult ants have an intense odour, often quite unlike formic acid, e.g., Acanthomyops, Crematogaster, Eciton, Megaponera, many Pheidole, Tapinoma, Atta, etc. The same is true also of bees and termites. The strong odours of Meliponinæ are often very pleasant ; some of the neotropical Trigonas smell like cocoanut, others like lemons, others like atta of roses, etc. The seat of the odour of honeybees seems to have been definitely located in Nassonoff's gland (Sladen, 1902, von Frisch, 1923) between the fifth and sixth abdominal segments. In ants the odour perhaps arises from the peculiar epinotal glands described by Janet (1898*a*, 1898*b*) which are present in all the castes and are admirably constructed for producing, conducting to the exterior and volatilizing odouriferous secretions. Other glands, such as the poison, or formic acid glands of the Formicinæ and the anal glands of Dolichoderinæ (Tapinoma, Azteca, etc.) produce strongly odoriferous secretions and, according to Forel, the same may be true of certain glands in the head. The fact that odoriferous substances are readily absorbed and retained by fats has long been known. Every housewife is familiar with this property in butter and from the earliest times unguents have been used as vehicles for perfumes. Fats and paraffins are still employed in "effleurage", a process of extracting the perfumes from flowers. The lipoid coating on the surfaces of larval, pupal and adult ants and the dense hairy investment of bees seem to be particularly well adapted to retaining both the individual and the adventitious, or nest odours and retarding their diffusion.[1]

There is no doubt that the glandular secretions of social insects are emitted in greater volume at times of excitement, but since even the persisting individual, caste, colony and nest odours are important means of recognition and communication, there is no reason why the odours should not be included with the gustatory stimuli as trophallactic. In a very interesting book von Frisch (1921) has recently shown that honeybees actually employ as means of communication the adventitious odours of nectar and pollen adhering to their bodies. Nectar-laden bees on returning to the hive perform a peculiar dance ("Rundtanz") which attracts the attention of unemployed workers and the same is true of bees returning with pollen. The latter perform a different dance ("Schwänzeltanz"). In both cases the unemployed individuals are informed of the kind of flowers that are yielding nectar and pollen in abundance and stimulated to fly out in search of them. Bees also discharge among the flowers the odours of Nassonoff's gland and thus enable the foraging individuals

[1] According to Haupt (1922) "it would seem that both nectar and ambrosia denote fragrant fat, especially the nidorous smell of the sacrifices ascending to heaven. The fragrant steam of a burning sacrifice was the nourishment of the gods". The ancient Hebrews had the same notion about Javeh, and the "personal" God of the Christians still receives incense. Haupt regards the ambrosia used in anointing and shampooing the Homeric heroes (Iliad, XVI., 670 and 680, XIV., 170) as a "scented massage cream". "The ancients had no scents dissolved in alcohol but perfumed greases, solid or liquid fats, charged with odours". That smelling and drinking may become synonymous is clear from the saying quoted by Haupt from Ben Johnson ("the most divine tobacco I ever drunk") and the Arab expression "to drink smoke" instead of "to smoke tobacco". Sunamitism is also interesting in this connection. Henning (1916) says: "The belief that the breath and perspiration of young human beings have a tendency to prolong the lives of the aged is well known from the relations of David to Abisag of Sunem (Kings I. 1-4). This magical superstition propagated its waves down to our own time and is responsible for many curiosities (J. H. Cohausen, Von der seltenen Art sein Leben durch das Anhauchen jüngerer Mädchen bis auf 115 Jahr zu verlängern. Stuttgart, 1847)".

to direct their flight more readily. The late F. W. L. Sladen, a well-known authority on Bombinæ, once showed me in Canada how the males of some of our humblebees find the virgin females. The latter fly into the recesses between the roots of large trees and there discharge a peculiar and even to the human olfactories distinctly perceptible odour, which is so heavy and diffuses so slowly that the males are able to follow the scent from tree to tree and thus overtake their mates. Similar behaviour is also exhibited by the males and females of solitary bees, with the substitution of flowers or foliage as places of deposition of the odour instead of the recesses of tree-trunks. Every collector of female Andrenas must be familiar with their powerful odours, which apparently emanate from the large glands of the facial foveæ to which I referred in a former lecture.

I admit that in my paper of 1918 I failed adequately to emphasize the rôle of olfaction in the reciprocal behaviour of the social insects. It would seem, however, from the foregoing discussion that the question as to whether an ant or bee smells or tastes its food, a larva, pupa or another ant or bee with its antennæ, is largely academic, or at any rate of no very great physiological significance. Since the words " taste " and " smell " are charged with anthropomorphism and the stimuli in both cases are chemical, it would be better to use the word " chemorecept ". And since, moreover, the food stimuli are necessarily chemical, I can see no reason to change the term " trophallaxis " because it happens that much of the behaviour of social insects is what we have been calling " olfactory ". Nor is the fact that some of this behaviour depends on stimuli other than food a valid argument against trophallaxis, since I have never asserted that it includes all the social activities. I believe, nevertheless, that it constitutes the most essential characteristic of the social medium. If we compare the distribution of food in the colony regarded as a superorganism with the circulating blood current (" internal medium ") in the individual insect or Vertebrate, trophallaxis, as the reciprocal exchange of food between the individuals of the colony, may be compared with the chemical exchanges between the tissue elements and the blood and between the various cells themselves. As stated in the passages quoted from Child (p. 231) this process necessarily involves not only the transportation and distribution of nutritive materials but also the transmission of stimuli, or excitations among the living elements. And inasmuch as the diverse alien organisms that have taken up their abode in the nests of social insects in many cases really constitute integral parts of the colony, much as some of our domestic animals form integral parts of human society, the guests and parasites of the social insects are also to be included in the trophallactic circuit.[1] The facts and arguments that support this statement will be presented in the two following lectures.

[1] Cf. De Grange (1923): "Sociological phenomena are never exclusively human. Lower animals were often an integral part of the primitive group. The dog, e.g., was in no sense a competitor of the savage, subjugated by cunning or force, nor a mere object of luxury, but a veritable ally, a *socius*".

X

THE EVOLUTION OF THE GUESTS AND PARASITES OF THE SOCIAL INSECTS

THE singular success of the social insects in creating for themselves a peculiar medium has exposed them to the invasion of a host of greedy guests and parasites, which it is now our task to consider. Nearly all of these intruders are Arthropods, and the great majority are Hexapods belonging to the most diverse families and orders. Why so many alien species should come to live in the nests or trophoporic fields surrounding them is obvious when we think of the many attractions and opportunities afforded by the social medium. In the first place, the galleries and chambers of the nests furnish very convenient hiding places and are, moreover, owing to the congestion of their population and especially during cold weather, of a somewhat higher temperature than the surroundings. Secondly, the refuse, the weak and dying or even the able-bodied workers, and particularly the inert and defenceless larvæ and pupæ represent an abundant supply of food. And thirdly, since the social insects have a more or less pronounced tendency to repel aggressive enemies of all kinds, all small organisms that can elude the hostility of the owners of the nests may share in the general protection. The disadvantages are less obvious, but they are real nevertheless, because the workers, even of the same nest, often vary considerably in their toleration of intruders. And even when most of the workers of a colony are willing to tolerate or adopt a strange insect, the latter may be eventually ejected or destroyed if there be present only a few persistently hostile and intolerant workers, since they must sooner or later come in contact with the intruder. Hence unanimous toleration or adoption on the part of the colony has to be secured by any insect if it is to become a regular guest, and this is more difficult than would be supposed. It is therefore surprising to find so great a number of satellites in the nests of all the various groups of social insects. We can most naturally divide all these creatures into two groups, namely, other social insects (social parasites), and sporadic, solitary, non-social species. The latter I shall discuss here and devote the next lecture to a consideration of the various social intruders.

The attraction exerted on alien insects by the colonies of the social species is interesting also in connection with certain matters to which Cuénot (1911) has called attention. He was of the opinion that uninhabited areas, the "places vides" as he called them, tend to attract faunal elements more strongly than areas already stocked with organisms. But Rabaud (1911, 1917, 1922*a*) and Picard (1919) have shown, on the contrary, that the more densely inhabited an area, the more it tends to attract additional species, whereas the "places vides" exert no such attraction and often remain for indefinite periods altogether uninhabited. Since the nests of the social insects, though among the most densely inhabited areas in nature, nevertheless constantly tend to withdraw and retain additional alien inhabitants from the general environment, they may be cited as a striking confirmation of the truth of Rabaud's and Picard's contention.

It is by no means easy to classify the Arthropod invaders of the social medium. Perhaps the most convenient general grouping would be into sphecophiles, melittophiles, myrmecophiles and termitophiles, according as they are associated with the wasps, bees, ants or termites. We might also classify them according to the types of relationship which they have developed towards their hosts. Wasmann, after many years of painstaking investigation, has adopted the following five categories:

(1) Synechthrans, or predators, i.e., species which prey on the social insects or their brood and are not, as a rule, tolerated but persecuted.

(2) Synœketes, or commensals and scavengers, which are ignored or indifferently tolerated. They are evidently either odourless or possess neutral odours which fail to arouse the curiosity or animosity of their hosts.

(3) Trophobionts, or insects such as the Aphids, Coccids, Membracids, etc., which excrete honey-dew, and the larvæ of certain Lepidoptera (Lycænidæ), which produce sweet secretions, and may therefore be attended or even cherished by their hosts. These trophobionts live mostly outside the nests in the trophoporic field, but may occasionally occur within the nests (on the roots of plants, on the lower surfaces of stones, etc.).

(4) Symphiles, or true guests, which have established the most intimate and amicable relations with their hosts and exhibit special structural adaptations (trichomes, exudatoria, etc.) to their peculiar mode of life.

(5) Parasites (*sensu strictiore*), which live either on or in the individual social insect and may therefore be classified as external or internal.

In general Wasmann's groups represent categories of increasing intimacy of the intruder with the host. The symphiles and parasites have evidently been developed from synechthrans or synœketes, and in so far as they are Hexapods, the parasites are mostly parasitoids in the sense assigned to that word in an earlier lecture. Although Wasmann's classification is the best hitherto devised, it can hardly be regarded as altogether satisfactory or definitive. Not only do some forms exhibit the distinctive behaviour of different categories in their larval and adult stages, but several combine the behaviour characteristic of more than one category in a single stage of their life-history. Still others exhibit such aberrant behaviour that it is difficult to place them in any of Wasmann's categories. At least one of the categories, that of the symphiles, occurs only among the myrmecophiles and termitophiles, and though bees and wasps may feed on the honey-dew of Aphids and Coccids, this excretion is very rarely solicited directly.[1] Hence the sphecophiles and melittophiles are all either predators or parasitoids which attack the brood, or scavengers (synœketes) which feed on the materials of the nest or on refuse. Certainly the most interesting and far and away the most numerous intruders live with the ants and termites. Unfortunately no one has endeavoured to treat the sphecophiles and melittophiles comprehensively, but enough is known about them to make it certain that any young investigator who is willing to devote the same patient study to them that Wasmann, Donisthorpe, Silvestri, Mann, Reichensperger and others have devoted to the guests of ants and termites, will be amply rewarded for his pains. As I shall have to give considerable attention to the myrmecophiles and termitophiles in this lecture, my account of the guests of wasps and bees will have to be brief.

The sphecophiles belong to several orders, one of the most singular of which, the Strepsiptera, comprises internal parasites of both solitary and social Vespids. In North America the solitary species of Odynerus, Eumenes and Zethus are occasionally stylopized by a peculiar genus, which is now being studied by one of my students, Dr. George Salt. The species of Xenos, both in Europe and North America, are peculiar to the social wasps of the genus Polistes. Occasionally as many as 25 per cent. of the wasps in a given locality, may be infested by these Strepsiptera (see Wheeler 1910*b*). According to Roubaud (1924), the African social wasps of the genus Belonogaster are only rarely stylopized, evidently by some peculiar Strepsipteran genus. Other singular sphecophiles related to the Strepsiptera belong to the Coleopteran family Rhipiphoridæ, of which Chobaut (1891) described one species, *Emenadia flabellata*, as feeding in the larval stage on the brood of *Odynerus simulator*, and Chapman (1870), Murray (1870) and others have described the very similar development of *Metœcus*

[1] Belt (1874) gives an account of a Nicaraguan wasp which attends plantlice and solicits their honey-dew in the same manner as ants.

paradoxus in the nests of Vespa. There are also several peculiar Hymenopterous sphecophiles. The parasitism of Trigonalids on the larvæ of various wasps was mentioned in my second lecture, and Bequaert has recently taken an interesting Chalcidid, Polistomorpha, in the nests of certain Polybiine wasps in Honduras. E. André, who published a valuable review of the European sphecophiles as early as 1881, cites the Ichneumonid *Tryphon vesparum* as a well-known parasite of *Vespa germanica* and *vulgaris*, and several other species (*Crypturus argiolus, Mesostenus gladiator* and *Ephialites extensor*) as parasitizing the larvæ of Polistes. *Mutilla brutia* and *littoralis* also invade the nests of the latter wasp. Brauer (1869) records the Neuropteron Trichoscelia, which is allied to Mantispa, as developing in the nests of a South American honey-storing Polybia. The larvæ of a small Tineid moth, *Melissoblaptes anellus*, is known to destroy the nest-materials of Vespa. Among the Diptera, there are at least three remarkable groups of sphecophiles, namely, various species of Volucella (*zonaria, inanis*, etc.) some of which were long ago exhaustively studied by Künckel-d'Herculais (1869, 1870, 1875-81, 1895), certain species of Conopidæ and of the Tachinid genus Anacamptomyia. Roubaud (1924) has published a beautiful and suggestive paper on this last genus, which comprises internal parasites of Belonogaster larvæ. Various authors have called attention to a number of beetles which are either predators, like the Staphylinid *Quedius dilatatus* in nests of European Vespæ and *Triacrus superbus* in nests of *Polybia vicina* in Brazil (Wasmann, 1904), or merely scavengers living in the refuse of the nest (small species of Staphylinidæ, Cryptophagidæ, Dermestidæ, Phoridæ, etc.). The larvæ of the Syrphid Microdon, though more frequently associated with ants, have also been found in Vespa nests.

No less multiform are the melittophiles, which are associated with all the groups of social bees, except, perhaps, the South African species of Allodape. The Halictinæ, which, as we have seen, nest gregariously, are infested by many parasites and scavengers (flies, beetles, Mutillids, etc.) as Melander and Brues (1903) have shown in the case of North American *Chloralictus pruinosus*. The Bombinæ have long been known to be parasitized by Mutillids, Conopids, Tachinids, Meloids and the peculiar nematode worm *Sphaerularia bombi*, which has been studied by Réaumur (1742), von Siebold (1836), Lubbock (1861), Schneider (1883, 1885), Leuckhart (1887), Dimmock (1886) and Frison (1926). Their waxen cells are devoured by the larvæ of moths, *Aphomia colonella* (Hase, 1926) in Europe and *Vitula edmondsi* in North America (Packard, 1865, Frison, 1926), while the refuse of the nest is eaten by the larvæ of beetles of the genus Antherophagus (*pallens*, *silaceus*, *ochraceus*, etc.), which in the adult stage, visit flowers and there attach themselves to the legs or antennæ of the bees so that they can be carried into the nests (see Wheeler, 1919*c*, Frison, 1926). The genus Volucella furnishes a remarkable series of humblebee brood parasites, which are often coloured like their hosts. They are common in Europe, but nothing has as yet been written on the habits of our American species. In 1924, Gabritchevsky published an interesting paper on the genetics and mimicry of *V. bombylans*, *hœmorrhoidalis*, *flava*, *plumata* and *caucasica*. He finds that these five forms are merely varieties of a single species, since they hybridize readily and produce offspring in Mendelian proportions. The same investigator has more recently (1926) studied our North American forms of *Volucella bombylans* and finds that they show striking colour-resemblances to the various species of humblebees of the same geographical areas. Even the Meliponine bees have their peculiar guests. Wasmann (1904) has recorded a predaceous Staphylinid, *Belonuchus mordens*, and four species of the Silphid genus Scotocryptus, taken by Goeldi in the nests of some six species of Melipona in Brazil, and a Silvanid beetle, *Nausibius clavicornis* and its larvæ as living in a nest of *Trigona ruficrus*. Quite a number of insects have long been known to infest honeybees (see Assmus, 1865). The most famous are the common wax-moth, *Galleria melonella* (Phillips, 1907, Hase, 1926), which destroys the combs, the Clerid beetle, *Trichodes apiarius*, and certain species of Meloe which in Europe prey on the brood, the tiny, wingless and aberrant Dipteron, *Braula cæca*, the " bee louse ", which is an ectoparasite, and *Mermis albicans* which is an ento-parasite, of individual bees. Of course, the bacteria, moulds, and Microsporidia, to which various well-known bee-diseases are due, should also be included among the parasites of the Apinæ (see White, 1906). " Isle of Wight disease " has been shown by Rennie, Harvey and White (1921) to be caused by a minute mite (*Tarsonemus woodi*) which enters the thoracic spiracles of the bee and reproduces in the trunks and larger branches of the trachæ. The mite, feeding on the tracheal walls, renders the bee useless as a working unit, " disorganizes the social system and eventually shortens the bee's life. Further, these vital effects are accompanied by pathological conditions in the tissues. The most obvious of these is a blackening and thickening of the tracheal wall. The thickened trachæ become progressively hardened and brittle in texture, and certain muscle fibres become atrophied ".

The number of myrmecophiles seems to be very great. At least 2,000 species have been described up to the present time and careful scrutiny of ant colonies, especially in the tropics, is continually bringing to light new and remarkable forms. Although the order Coleoptera contributes most of the species, nearly every order of insects and several of other classes of Arthropoda are represented. The behaviour of these organisms is so unusual that whole volumes would be required for their adequate description. Here I can present only some selected examples to show the diversity of relations which they bear to their hosts. The synechthrans or persecuted intruders are mostly Staphylinid beetles which lurk in the less frequented galleries of the nests, avoiding contact with the ants as much as possible and preying on the brood or enfeebled adults. As this category is of no unusual interest I will cite only a single example, a rather common beetle in the nests of our North American *Formica exsectoides*, *Megastilicus formicarius*, which somewhat resembles its host in coloration and form. When confined with the ants in an artificial nest, it is invariably killed within a few hours, but in the natural nests it adroitly eludes its host in the same manner as the European species of Myrmedonia, for when an ant tries to seize it, it raises the flexible tip of its abdomen and emits a whitish fluid, which causes the ant to start back as if a flask of ammonia had been suddenly uncorked in its face, a manœuvre which, of course, gives the beetle time to escape.

A very large proportion of the myrmeophiles are synœketes, or tolerated inquilines, but in behaviour they are so diverse as to defy classification and in some cases merge into the group of symphiles. Many belong to a group which I have called neutral synœketes, because they live quite unnoticed by the ants on the refuse or nest materials. Typical examples are the small white panmyrmecophilous Podurans of the genus Cyphodeira, the slow-moving, snow-white European Isopod, *Platyar thrus hoffmannseggi*, many small beetles, mites, Phoridæ, Lepidopteran and Dipteran larvæ and the singular larvæ of Microdon. The food of the Microdon larvæ long remained unknown till Donisthorpe (1912, 1921*a*) showed that it consists of the small pellets (" boulets de nettoyages ") cast out of their infrabuccal pockets by the ants. More recently Borgmeier (1923) has found that the larvæ of a Brazilian species actually devours the larvæ of its host, the common fire ant, *Solenopsis sævissima*. The moth *Wurthia aurivillei*, which Kemner (1923) describes as living in the silken walls of the nests of *Polyrhachis bicolor* in Java and regards as representing a new type of myrmecophily, " hyphænosymphily ", is clearly a synœkete like so many other Tineidæ. Among the synœketes must also be included a large number of Staphylinid and Histerid guests of the Doryline ants, Eciton in the New and Dorylus in the Old World tropics (Wasmann, 1910*a*, 1915*c*, etc.). These beetles probably devour portions of the prey collected by their hosts and carried into their temporary nests. Some of the Staphylinids are supposed by Wasmann to be protected from their hosts by their ant-like form and the peculiar texture of their integument, which closely resembles that of the Dorylinæ, others by their broad, convex bodies and hard or slippery surfaces. Another group comprises the myrmecocleptics, the best examples of which are the many species of the Thysanuran genus Atelura, of world-wide distribution. *Atelura formicaria* (Fig. 61), first observed by Janet (1896), might be regarded as a synechthran, since it is sometimes threatened by the ants which are unable to hold it fast on account of its slippery integument and rapidly tapering abdomen. It moves cautiously about among the ants (*Lasius mixtus*) and snaps up the droplets of food while they are being regurgitated by one ant to another. An even more interesting myrmecocleptic is the larva of a small Phorid fly, *Metopina pachycondylæ*, which I observed in Texas (1901*a*) in the nests of the Ponerine *Pachycondyla montezumia*. The larva of this ant is fed by its worker nurses with fragments of insects placed on the broad, dish-like ventral surface of its abdomen and the Metopina larva, which encircles the neck of the ant-larva like an Elizabethan collar, reaches forward with its head and partakes of the feast. When both ant-larva and commensal are full-grown, the former spins a cocoon and within it the latter forms its puparium at the posterior pole. The adult ant emerges first, leaving at the anterior end of the cocoon an opening through which the fly can escape.

Finally there is a group of synœketes which spend much of their time licking or nibbling at the surfaces of the adult ants and evidently deriving some nutriment from the secretions or exudates covering their bodies. The nearly blind crickets (Fig. 62) of the genus Myrmecophila (Wheeler (1900*b*), Wasmann (1901), Schimmer (1909, 1910)) and the Staphylinid beetle *Oxysoma oberthüri* (Fig. 63) of North Africa (Escherich, 1902) belong to this group ; also the small subapterous roaches of the genus

Attaphila which live in the fungus gardens of the leaf-cutting Attini. I described the type of the genus, *Attaphila fungicola*, from the nests of *Atta texana* in 1900 and have recently taken the same species in British Guiana in the fungus gardens of *A. cephalotes* and in those of *Acromyrmex octospinosus* in Panama. In the meantime a few other species of the same genus have been described from the nests of other Attini in South America. Shelford (1906, 1907) has described an allied roach, *Sphecophila polybiarum* from the nest of Polybiine wasp, *Polybia pygmæa*, in British Guiana and a second species, *S. termitium* from East African termitaria.

In this place I may also consider several peculiar synœketes and synachthrans which have been observed more recently and are of unusual interest because they either live entirely outside the ant nests in the trophoporic field or are actually imported from the latter by the ants into the nest and there exhibit quite a different type of behaviour. They are the more remarkable because they belong to the orders Diptera, Lepidoptera and Heteroptera, orders which we should hardly expect to acquire such specialized relations to ants.

(1) Certain paleotropical Calliphorinæ belonging to the genera Bengalia and Ochromyia are known to station themselves near files of foraging or migrating ants and to rob them of their prey or brood. These flies differ markedly from other Calliphorinæ in the structure of the proboscis, which is stiff and chitinized, with strongly toothed tip, directed forward. The recorded observations on these flies, which were first studied by Nangle (1905) in India and E. E. Green in Ceylon (1906, 1908) have been recently reviewed by Bequaert (1922), who has observed them in the Congo.[1]

[1] The flies observed by Nangle and Green were *Bengalia obscuripennis* and were hunting winged termites flying at night. J. W. Yerbury saw the same species "trying to take her burden from a large ant (Lobopelta species)". F. W. Thompson made the following observation with regard to the Indian *B. jejuna* (Fabricius): "I always noticed specimens of this species on the ground, or on a stone or leaf near an ant's nest. On watching I saw them swoop down on an ant carrying an 'egg' or larva, take it from the ant, carry it

(2) *Ptilocerus ochraceus* is a very interesting Javanese Heteropteron, which preys on the ants in the trophoporic field. Its structure and behaviour have been described by Kirkaldy (1911) and Jacobson (1911*a*). On the middle of its ventral surface it possesses a peculiar gland with a flame-coloured tuft of hairs (trichomes). Stationing itself along the foraging file of a common East Indian ant, *Hypoclinea bituberculata*, it at once becomes alert on the approach of an ant and raises the anterior portion of its body so as to display its trichomes. Their odour attracts the ant and incites it to lick and nibble them. The bug quietly submits, merely folding its forelegs over the ant's head as if to make sure of its prey. The ant often milks so avidly with its mandibles at the trichomes that it moves the bug up and down. But the secretion of the gland has a toxic and paralyzing effect on the ant. As soon as the poor creature draws in its legs and is about to topple over, the bug seizes it with the forelegs, runs its proboscis through one of the thoracic sutures or preferably into an antennal insertion and sucks out the contents of its body. That the paralysis is due to a substance imbibed by the ant from the gland and not by a thrust of the bug's proboscis is, according to Jacobson, " proved by the fact that a great number of ants, after having

---

away a short distance and proceed to suck it ". *Bengalia latro* de Meijere in Java lurks in the neighbourhood of the column of *Pheidologeton diversus* (Jerdon) ; when a worker ant comes along carrying its prey, the fly dashes into the moving ant column, quickly steals the prey from the carrier and returns to its perch where it devours its catch at leisure (Jacobson 1910, de Meijere 1910). Lastly, C. R. Dutt, in his interesting " Life Histories of Indian Insects " (1912), writes of *Monomorium indicum* Forel as follows : " One morning I observed the inmates of a nest marching out with young ones. Close to the nest was sitting a Muscid fly (*Ochromyia* species) which attacked from time to time the larvæ and pupæ which were being carried by the workers. The fly never snatched the victim from the grasp of the ant, but simply " licked " it from its place with the proboscis, which when withdrawn left the larva or pupa quite shrivelled up ". The African Bengaliæ evidently have much the same habits as their Indian congeners. According to W. A. Lamborn (1913, 1919), " *Bengalia depressa* (Walker) in Southern Nigeria, regularly follows the marauding armies of *Dorylus nigricans*, to rob them of their prey." See also Surcouf and Guyon (1912), Cragg (1917-1918) and Farquharson (1919).

licked for some time the secretion from the trichome, leave the bug to retire to some distance. But very soon they are overtaken by paralysis, even if they have not been touched at all by the bug's proboscis. In this way a much larger number of ants is destroyed than actually serves as food to the bugs, and one must wonder at the prolificacy of the ants which enables them to stand such a heavy draft on the population of one community ". Both the nymphal and adult Ptiloceri attack the ants indiscriminately, and though the bugs are local, they may infest the ant-trails in considerable numbers.

(3) A much more benign exploitation of the ants in the trophoporic field is exhibited by the beautiful dark-coloured, silver-spangled paleotropical mosquitos of the genus Harpagomyia. Their distribution and taxonomy have been investigated by Muir (1918), Edwards (1922) and others, their metamorphosis by de Meijere (1909, 1911), and their remarkable behaviour by Jacobson (1909, 1911*b*) in Java, James (1914) in Ceylon, Banks in the Philippines and Farquharson (1918, 1922) in Nigeria. Harpagomyia breeds in the water-filled cavities of tree trunks and in the same situations frequents the trails of various species of Crematogaster, importuning the ants for regurgitated food and receiving it without difficulty. The feeding process has been carefully described by Jacobson and Farquharson. I cite the latter's account of *H. farquharsoni* (1918, p. XXXIV). The Crematogasters while running up and down the tree trunks are constantly " making little regurgitory exchanges " of food with one another and it is this trophallactic behaviour which the mosquito exploits. The insect " hovers an inch or less over the line of ants (at times resting on the stem and dodging out of the way when necessary), till it sees what is presumably a likely ant. If the ant is running downwards the mosquito drops down (in flight) also, keeping a little in front of the ant—as near as possible without touching it. The ant tries to evade it, but the mosquito as a rule declines to be put off, and the ant at last stops. The mosquito quickly settles and the usual rapid exchange

begins, the mosquito thrusting out its proboscis—which when not in action is carried bent under the body much like the rostrum of a Reduviid bug—so that the swollen end is practically within the ant's jaws. I have seen the ant's palpi (*not* the antennæ) vibrating on it. The swollen portion of the proboscis is undoubtedly capable of independent movement. I may say that the mosquito is not ' nervous ', and I have had no difficulty in several cases in standing with my eyes sufficiently close to the two insects to make out the ant palpi and the proboscis movement in the mosquito. The ant raises its head slightly when the exchange takes place ". Farquharson (1918, 1922*a*) has also described the behaviour of other small Nigerian flies (*Milichia argyratoides, præctes, prosætes* and *dectes*) which solicit and receive food from the ants. Another species, a Cecidomyid, *Farquharsonia rostrata* (Collin, 1922), was seen flying over the carton nests of *Crematogaster alligatrix*, approaching the ants engaged in feeding one another and adroitly snatching portions of the regurgitated droplets, thus exhibiting a type of behaviour like that of the Atelura described by Janet.

(4) Even more extraordinary is the behaviour of other midges described by Jacobson (1909) and Farquharson (1918, 1922*a*). The former observed a Milichiid, *Prosætomilichia myrmecophila* (de Meijere, 1909) which runs about after the ants (*Hypoclinea bituberculata*) and licks the tips of their abdomens, evidently feeding on the anal secretions. Occasionally this fly was also seen to imbibe the excretions (honey dew) of Membracid nymphs. The fly observed by Farquharson is an Ephydrid, *Rhynchopsilopa apicalis* (Collin, 1922). It was first seen (1918) feeding at the anus of dead Crematogaster workers, but later (1922*b*) Farquharson observed it running after and licking the tips of the abdomens of living ants, like the Javanese Prosætomilichia. Both observers describe the ants as resenting the persistent and impudent attentions of the midges.

The three following cases refer to Lepidopteran

caterpillars which acquire relations to the ants after being carried by them into their nests :—

(5) F. P. Dodd (1912) found that the first-stage larva of a small gray Queensland moth, *Cyclotorna monocentra*, is ectoparasitic on a Jassid Homopteron which feeds on certain trees and is attended and "milked" by *Iridomyrmex sanguineus*. The ant carries the parasite but not the Jassid into its nest. There the former spins a temporary cocoon and later emerges from it as a peculiar, flat, bright red, second stage larva, with two long tails. In this stage it subsists "solely on the ant grubs by sucking out their juices", but as in the case of Lomechusa in the nests of the European *Formica sanguinea*, the ant is partially recompensed for the destruction of its brood. Dodd says : "Reference has been made to the caterpillars raising their terminal segments, even the small ones from the cocoons doing so. This was quite sufficient to warrant investigation. Consequently at various times I have placed them with ants and grubs under glass, in order that they could be seen to advantage and without risk of disturbance. When the anal parts are protuded, an ant generally soon becomes aware of the fact and will be seen to pay these great attention. I soon noticed that a liquid, often perfectly transparent (it looks so on the blue-green ground, probably was pale bluish), is emitted, and that it is greedily drunk up by the ants. Over and over gain, with and without a lens, I have seen this issue, and the ants speedily absorb it. Some ants, perhaps hungry or more enterprising than others, would take in a supply from a second caterpillar. If an ant is not satisfied with the quantity given out, she deliberately seizes the protruding parts and gives them a gentle nip, the mandibles can plainly be seen to press upon the juicy flesh ; if the hint is not immediately acted upon a more vigorous squeeze is given, and the tails may be gripped and pressed. This is very comical, the ant's meaning is unmistakable and the caterpillar so thoroughly understands it, too, for a second hint never fails. This liquid, though frequently quite clear, is often mixed with yellowish

matter, and at times some jelly-like substance is extruded; the latter the ants do not care about, for after the moisture is licked up this is in their way, and if they have not been imprisoned too long, will seize and tug at it until it comes off, and carry it to a spot set apart for waste matter, such as their own pellets and pupal skins, etc., are kept in". Dodd also observed the ants licking and cleaning the caterpillar and the caterpillar cleaning the ants! When mature the caterpillar leaves the nest, travels to the nearest tree in company with the foraging ants, spins its cocoon in a crevice of the bark and pupates. In about twenty days the moth emerges.

(6) The second caterpillar is that of *Lycæna arion*, which has been studied by Chapman (1915*a*, 1915*b*) and Frohawk (1915) in England. The butterfly lays its eggs on thyme and other plants. On these the larva feeds, and is often attended by ants as it possesses a honey-gland like many other larval Lycænids. When it has reached the third, or last stage it crawls down to the ground and on encountering a foraging worker of *Myrmica lævinodis* or *scabrinodis* hunches up the anterior segments of its body in a singular manner. Frohawk interprets this behaviour as a "signal" which induces the ant to seize the caterpillar and carry it into the nest. Chapman was able to observe that after the caterpillar was taken into the nest it fed on the Myrmica larvæ. During this period of its life it was not seen to yield the secretion of its honey-gland but was treated by the ants as what Wasmann would call an indifferently tolerated guest, or synœkete.

In Argentina Bruch (1926) has found that a butterfly caterpillar, *Hamearis epulus signatus* (Fig. 64), belonging to the family Erycinidæ, has habits resembling those of *Lycæna arion*. It feeds on the foliage of a vetch (*Vicia graminea*) and is visited by one of the fiercest South American ants, *Solenopsis sævissima richteri*, which it provides with a greenish secretion from two glands (their orifices are shown as white spots in the figure) situated on the eighth abdominal segment. When full grown the caterpillar is found in the Solenopsis nest and there passes

Plate XXXIX

FIG. 64.—Myrmecophilous caterpillar and chrysalis of a butterfly, *Hamearis epulus signatus.* (After C. Bruch.)

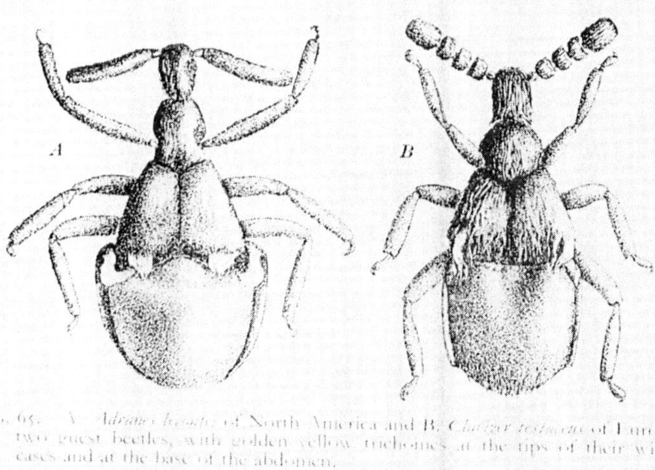

FIG. 65.—A, *Adranes lecontei* of North America and B, *Claviger testaceus* of Europe, two guest beetles, with golden yellow trichomes at the tips of their wing cases and at the base of the abdomen.

[*face p.* 268]

Plate XL.

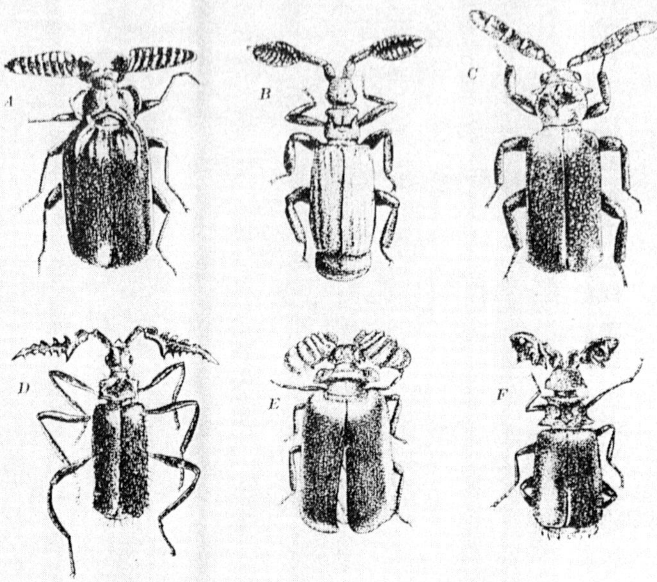

FIG. 66. Various species of Paussidae. A. *Pleuropterus brevicornis*. B. *Paussus bora*. C. *Pentaplatarthrus natalensis*. D. *Paussus dama*. E. *Lebioderus goryi*. F. *Paussus spiniceps*. (After E. Wasmann.)

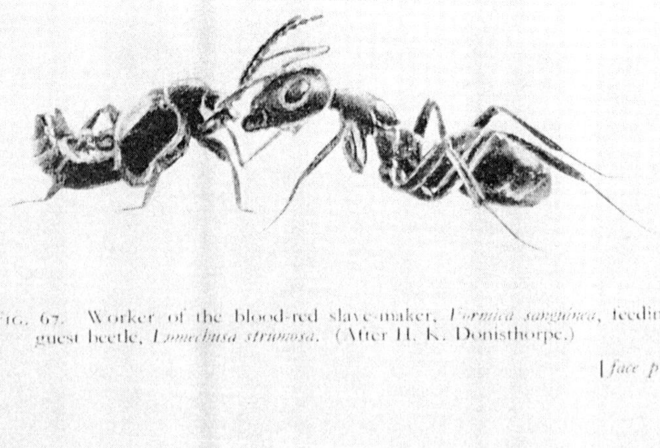

FIG. 67. Worker of the blood-red slave-maker, *Formica sanguinea*, feeding its guest beetle, *Lomechusa strumosa*. (After H. K. Donisthorpe.)

[*face p.* 261

its pupal stage during the winter. According to Bruch, the chrysalis is provided with certain papillary (myrmecophilous) organs which seem to preserve it from attack by the ants.

I have dwelt on these cases—and many more might have been cited—because they show that the predators and synœketes do not necessarily confine their activities to the nests and because they illustrate the truly astonishing diversity of methods employed in exploiting the normal behaviour of the ants. The Bengalias are, of course, frankly predatory, the Harpagomyias and Milichiids might be designated as myrmecocleptics or symphiloids, the Ptilocerus is predatory, but its narcotic trichomes recall the conditions in the true symphiles, to be considered presently. The Cyclotorna is a parasite of a Homopteron in its first and a predatory symphile of the ants in its second larval instar, while the *Lycæna arion* caterpillar is a trophobiont in its first and a predatory symphile in its later larval stage.[1]

The true guests, or symphiles, comprising only some three or four hundred species, mostly Coleoptera (Lomechusini, Clavigeridæ (Fig. 65), Paussidæ (Fig. 66), Histeridæ, etc.) constitute the élite among the myrmecophiles and not only show behaviouristic adaptations to their hosts, like the predators and synœketes, but also singular structural adaptations. These are, generally speaking, the development of a peculiar red, oily integument, the presence on various parts of the body of golden-yellow trichomes and glands which produce pleasant ethereal secretions, modifications of the mouth-parts, especially a reduction of the tongue and number of palpal joints and modifications of the antennæ, due to a fusion of their joints so that they often become clubbed,

[1] The literature on the Lycænid caterpillars associated with ants is too extensive for citation in this place. I cannot refrain, however, from referring to Lamborn's important paper (1914) on the West African Lycænids and other trophobionts, because it presents such an instructive picture of the number and diversity of insects attended by the ants in a circumscribed tropical area. See also the interesting accounts by Dodd (1902*a*, 1902*b*), and Chapman (1902) of *Liphyra brassolis*.

and adapted to stroking the host (Clavigeridæ) or broad and compacted so that the ants can use them as handles by which to drag or carry the guests about in the nests (Paussidæ). Wasmann has made a detailed study of the European symphiles and especially of the Lomechusini (1915*a*), but our knowledge of the behaviour of tropical ants guests is still very meagre.[1] I referred to the Lomechusini in a previous lecture, in connection with the production of pseudogynes in colonies in which these beetles breed. Brief statements in regard to *Lomechusa strumosa* (Fig. 67), one of the most typical and certainly the best known of symphiles, may here be added. In its larval stage this Staphylinid is really a predatory parasite in the nests of *Formica sanguinea*, devouring the ant larvæ. The fat tissue of the Lomechusa larva probably supplies the ants with an agreeable exudate, and the adults certainly furnish an agreeable secretion from their abdominal trichome glands. When the larvæ, which are evidently treated as if they were ant larvæ, mature, they are buried in the soil, in the same manner as the ant larvæ, in order that they may pupate. The pupæ are also unearthed like the ant pupæ, after they have spun their cocoons, but this treatment is fatal to the parasites and only those that have been overlooked or forgotten (?) and left in the soil are able to develop into adult beetles. After the greater part of the ant-brood has been destroyed by the Lomechusa larvæ, many of the neglected queen larvæ, as we have seen, develop into pseudogynes, or pathological forms intermediate between workers and queens and unable to perform the functions of either of the castes which they imperfectly represent. This is in its essential outlines the history of Lomechusa and I must refer for the details to Wasmann's many papers (see bibliography in his papers of 1915*a* and 1920).

Lack of space also prevents me from considering the parasitic myrmecophiles. Some of the more interesting of them, such as the Eucharidæ and Mermis, which produce distinct structural modifications in their hosts, have been described in preceding lectures, and many other

[1] The reader will find an excellent account of British myrmecophiles in Donisthorpe's *Guests of British Ants* (1927).

forms, such as the mites, parasitic fungi, etc., are so similar to those which infest solitary insects that they offer little of interest for the present discussion. We may therefore turn to the termitophiles, which may be grouped very roughly in the same categories as the myrmecophiles, i.e., as synechthrans, synœketes, symphiles and parasites. The category of trophobionts is not represented, since the termites have never learned to solicit the saccharine excretions of Homoptera and Lycænid caterpillars.

Although the termitophiles have been less studied than the myrmecophiles, nevertheless between 600 and 700 species are known. At least 70 per cent of them are Coleoptera ; the remaining 30 per cent belonging to a number of insect orders. Warren (1919) has drawn up a table to show the distribution of termitophiles known in 1919. Since that time some representatives of other orders have been added, notably Braconid Hymenoptera, of which Cushman (1923) has described two species of Ypsistocerus from Bolivia and Brues (1923*a*) a species of Termitobracon (*T. emersoni*) from British Guiana. In his list Warren cites the number of Ethiopian termitophiles (319) as greatly exceeding that of any other region, but so many species have been recently discovered in South and Central America that we shall probably have to regard the neotropical region as possessing the largest number of forms. This is true also of other organisms ; the exuberance of the insect fauna, especially, of the American tropics is almost inexhaustible. Very few termitophiles are known from the nests of primitive termites (Calotermitidæ), the great majority having been taken from the very populous termitaria of the Termitidæ. Similarly myrmecophiles are rare in the small colonies of the primitive subfamilies of ants (Ponerinæ, Cerapachyinæ and Pseudomyrminæ) and numerous in species and in individuals only in those of the more specialized Dorylinæ, Myrmicinæ, Dolichoderinæ and Formicinæ. In other words, the guests and parasites are most readily tolerated and adopted in the colonies of the most highly socialized species of ants and termites—a suggestive fact, which has not been sufficiently stressed in the literature.

Owing to the fact that nearly all the termites are tropical, little is actually known of the behaviour of the termitophiles. Living as they do in dark, cool galleries, thronged with their hosts, they certainly have a more highly specialized and less diversified medium than the myrmecophiles. In nearly all cases the relations with the hosts have been inferred, with more or less probability, from the structure of the termitophiles and not from actual observations or experiments. Warren (1919) has so clearly summarized our present knowledge of these relations that I quote his remarks :—

" It is clear from the condition of the mandibles that many termitophiles attack the young of the termites and suck the fluids of the body, and thus should the termitophiles in a nest be numerous, the community would be injured by their presence. It is accordingly necessary that the termitophile should adopt some means for cajoling the termite-community, since it is quite obvious that no termitophile could withstand a combined attack by a number of soldiers or workers. The various means adopted may be now summarized :

(1) By the assumption of a simple, rounded contour the body may be rendered inaccessible to easy attack, and further, it may be flattened and very smooth. This is the defensive type (" Trutztypus ") of Wasmann.

(2) Less frequently the body is provided with stiff bristles, which would render approach by the termites somewhat troublesome or disagreeable.

(3) It is inherently probable that some termitophiles can protect themselves from attack by the power of producing a scent which is disliked by the termites, and they accordingly avoid the intruders as much as possible.

(4) On the other hand, some termitophiles are not shunned in the least by the termites, and they live in the closest communication with them. These termitophiles may possess no obvious exudatory organs, but they may

have well-developed scent glands, and it is to be supposed that the odour is pleasing to the termite community.

(5) Other termitophiles are provided with conspicuous outgrowths from the body, and their microscopic structure indicates that fluids exude from them. From analogy with myrmecophiles it is certain that these fluids either evaporate and form an agreeable perfume or are licked up by the termites.

(6) In a few cases the bodies of termitophiles carry clusters of bright yellow hairs. Such yellow hairs are frequently present on myrmecophiles, and the ants of the nest have been seen to lick them with great assiduity. Dr. Brauns has drawn my attention to the occasional occurrence of these hairs in termitophiles, and we may suppose that they have a similar attraction to the termites.

(7) Sometimes the bodies of termitophiles are greatly inflated, or physogastric, as it is called, and there may result a marked mimicry of the young or adult termites. This general resemblance to the termites would assist in rendering the termitophiles less liable to attack ; but it is doubtful if the origin of physogastrism is to be explained wholly on these lines."

It is obvious that the course of termitophile evolution parallels that of the myrmecophiles. In both cases there is the same origin from synechthrans and synœketes, which, as they become more intimate with their hosts, gradually take on the peculiar characters of symphiles. Warren does not consider the internal and external parasites or the important symbiotic protozoa mentioned in a previous lecture. He calls attention, however, to the interesting fact that the peculiar adaptive modifications in all seven of his termitophile categories are met with as independent lines of development in each of the termite regions in which the guests have been adequately studied, namely, the Indomalayan, African and South American regions.

The most striking characteristic of many termitophiles is physogastry, or the excessive enlargement of the abdomen, which closely resembles that of the hosts and especially that of the old queens. It is due to several causes, such as a great increase in the fatty or exudate tissue or enlargement of the alimentary tract or reproductive organs (Wasmann (1903), Trägårdh (1907 *a* and *b*) and Warren (1919, 1920)) and is especially noticeable in certain Coleopteran larvæ, adult Staphylinids of numerous genera and the strange Diptera of the genera Termitoxenia, Termitomyia, Ptochomyia, Termitomastus, etc. In some predatory Coleopteran larvæ, like those of the Carabid *Glyptus punctulatus*, which live in the nests of the African *Termes bellicosus*, the physogastry is less extreme, but in the Staphylinids it occurs in all degrees and culminates in most remarkable, one may truly say monstrous, forms. Warren has constructed a diagram (Fig. 68) to illustrate the three independent lines of phylogenetic development in the physogastry of these beetles in the three great geographical regions, from an ordinary predaceous Staphylinid like Myrmedonia. Among a considerable number of forms we may select as examples of highly developed physogastry *Corymbogaster miranda* from British Guiana, *Corotoca melantho* and *phylo* from Brazil, *Œdiprosoma mirandum* from Africa and *Thyreoxenus pulchellus* of British Guiana. The two last forms are so aberrant that only an expert Coleopterist would recognize them as Staphylinids. But there are even more bizarre types, which are not only extremely physogastric but possess in addition long exudatorial appendages, e.g., the viviparous *Spirachtha eurymedusa* (Fig. 69) long ago described by Schiödte (1853, 1856) from Brazil and the even more extraordinary form, *Spirachtha mirabilis* (Fig. 70), recently discovered by Emerson in British Guiana and described by Mann (1923). The peculiar structure of its three pairs of exudatoria has been elucidated by McIndoo (1923). I happened to be with Dr. Emerson when he found a number of specimens of this termitophile in the nest of *Nasutitermes* (*Constrictotermes*) *cavifrons* in

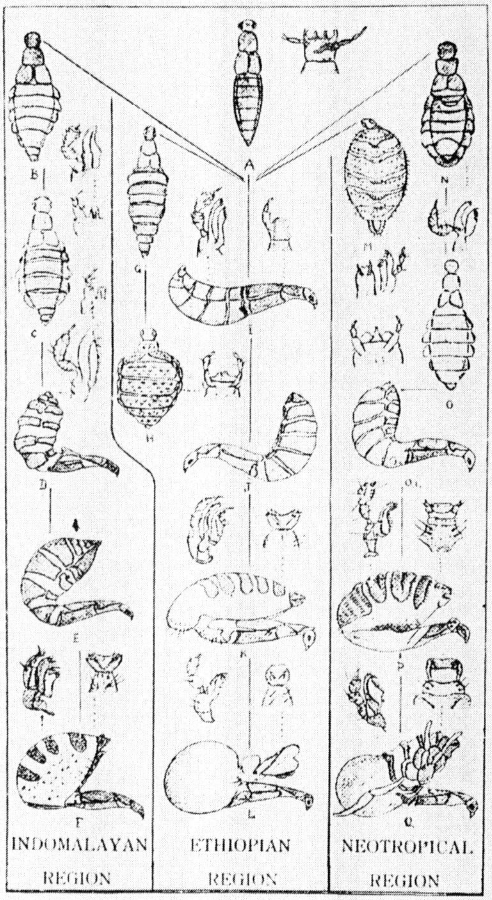

FIG. 68. Diagram illustrating three parallel and independent lines of phylogenetic development of physogastric Staphylinid termitophiles from an undifferentiated type. A. *Myrmedonia* (unmodified condition). B. *Jacobsonella termitobia* Silv. C. *Discticta capritermitis* Wasm. D. *Asticta butteli* Wasm. E. *Termitotima assmuthi* Wasm. F. *Termitoptochus indicus* Silv. G. *Termitopaedia kohli* Wasm. H. *Termitobia physogastra* Wasm. I. *Termitotecna braunsi* Wasm. J. *Idiogaster escherichi* Wasm. K. *Paracorotoca akermani* War. L. *Termitomimus entendveniensis* Trag. M. *Timeparthenus regius* Silv. N. *Termitophya heyeri* Wasm. O. and O1. *Xenogaster inflata* Wasm. P. *Corotoca melantho* Schiödte. Q. *Spirachtha eurymedusa* Sch. (After E. Warren.)

[*face p.* 266

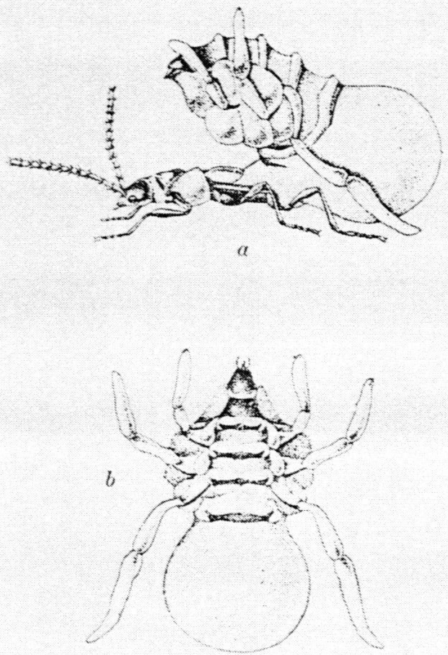

FIG. 69. *Spirachtha eurymedusa*, a physogastric termitophilous Staphylinid beetle from South America. a. Lateral view of the whole insect. b. Dorsal view of abdomen, showing the three pairs of appendage-like exudatoria. (After J. C. Schiödte.)

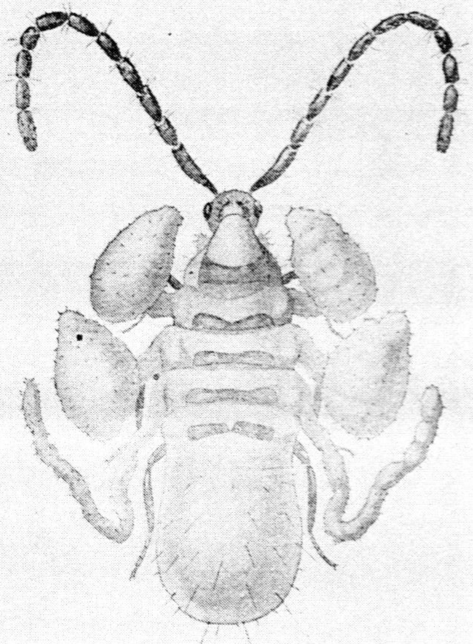

FIG. 70. Dorsal view of *Spirachtha mirabilis* Mann, a physogastric termitophilous Staphylinid, from British Guiana, with three pairs of huge exudatoria. The abdomen is bent upward and forward and covers the thorax. (After W. M. Mann.)

[*face p.* 267

the superb forest at Kalacoon. When the termites and their guests were brought into the laboratory we saw the former carrying the beetles about and frequently licking not only their exudatoria, which seemed to have sticky or glutinous surfaces, but also other portions of their bodies. Comparison of the exudatoria of this strange insect with those of the larvæ of the African ant Pachysima (Figs. 57 and 58), described in a former lecture, is very suggestive.

The physogastric Diptera above mentioned are in certain respects even more aberrant than the Staphylinids, because they also have much reduced wings and eyes (Fig. 71). One of the genera, Termitomastus (Fig. 71C), belongs to the suborder Nematocera and was discovered by Silvestri (1903) in South America. The others, Termitoxenia (Fig. 71A), Termitomyia and Ptochomyia (Fig. 71B) occur in the Oriental and Ethiopian regions and have been studied by Wasmann (1900-1901*a*), Assmuth (1913), Bugnion (1913*a*), Brues (1918), and Silvestri (1920). The great similarity of the physogastry, subaptery, microphthalmy, etc., of the flies and beetles here briefly considered to the termites themselves would seem to be most naturally explained as the result of adaptation to the peculiar trophallactic habits of the termites and to the same physical and trophic conditions, namely the crowding in dark cavities poor in oxygen and of rather low temperature compared with the surrounding air and soil, the abundance of food, etc. These very conditions are known to produce quite similar results in other insects. (Compare e.g. the physogastry and accumulation of fat, microcephaly and absence of eyes in the parasitic larvæ of Hymenoptera, Diptera, etc., the aptery and blindness of many hypogæic and cave insects, etc.)

In conclusion something must be said concerning the opinions that have been advanced to account for the extraordinary behaviour and morphological idiosyncrasies of the myrmecophiles and termitophiles on the one hand and the reciprocal behaviour of their hosts on the other.

Unfortunately the general consideration of the subject has been wrapped in clouds of speculation, largely through Wasmann's unceasing efforts to remove it from the field of legitimate biological inquiry to that of the neothomistic philosophy which is now so assiduously cultivated by the Jesuits. He has elaborated his opinions in a long series of articles, the bibliography of which may be found in his works of 1910, 1915 and 1920. Since his latest article is devoted in great part to a polemic against my views of 1918 and since I remain quite unconvinced by his arguments, or rather assertions, I shall here repeat some of my former statements.

Having found that particular symphiles live only with particular ants and termites, Wasmann concludes, first, that the latter have during their phylogeny acquired particular symphilic instincts as differentiations or modifications of their original nursing and adoptive instincts, and second, that the true ant and termite guests have been developed by these symphilic instincts through a process called "amical selection", which he likens to the conscious artificial selection employed by man in perfecting the numerous, often bizarre varieties among his domesticated animals and plants. Escherich (1898, 1902, 1911), Schimmer (1909, 1910) and I (1910*a*) have never accepted this view, and I am still unable to see that Wasmann has successfully disposed of our arguments. The whole matter comes down to the answer to two questions: Do ants and termites possess special symphilic instincts? and: Is the assumption of amical selection necessary to account for the facts? In my opinion both questions are to be answered in the negative.

It is unnecessary to consider all the various symphiles which Wasmann has so long and so diligently studied. A brief account of *Lomechusa strumosa*, his chief battle horse and according to his own statement one of the most typical of symphiles, has already been given and will suffice as an illustration. He believes that *Formica sanguinea*, the host of this insect, has acquired during its phylogeny a special symphilic instinct which impels it to foster Lomechusa to the detriment of its colonies and therefore to the detriment of its species, and regards the case as furnishing a crucial argument against natural selection and an incontestable proof of the existence of amical selection. The same reasoning is, of course, made to apply, *mutatis mutandis*, to *Claviger testaceus* which lives with *Lasius flavus*, the Paussids which live with different species of Pheidole, the physogastric Staphylinids which live with various termites, etc.

The argument looks plausible till we examine it more critically. When we ask how the particular instinct to foster Lomechusa became established, i.e., hereditary, in *sanguinea*, we see that Wasmann has taken a great deal for granted. Of course, we really know nothing about the phylogeny of *sanguinea* in its relation to Lomechusa. The *sanguinea* queen and her fertile female offspring in colonies that are old enough to be infested by the beetle, pay no particular attention to the parasite and could therefore acquire such an instinct as Wasmann postulates only by inspiration. The workers, which do look after the beetles, rarely reproduce and probably never reproduce in infested colonies and would therefore not be in a position to transmit even if they acquired such an instinct. And as the *sanguinea* brood is either largely devoured or converted into infertile pseudogynes, so that the whole colony tends to die out, we have anything but a favourable environment for engendering and transmitting an instinct so specialized as to be concerned with a particular symphile.[1] Add to this fact that Lomechusa is a sporadic or local parasite and Wasmann's case becomes quite hopeless. Of course, he might contend that Lomechusa was once a universal *sanguinea* parasite, or, at any rate, much more abundant and more uniformly distributed

[1] According to Lomnicki (1921), males produced in *sanguinea* colonies infested by Lomechusa may also be affected, becoming dwarfs, or "micraners" only about half the length of normal individuals. Admitting that such males may be the offspring of the workers, they are certainly not favourable material for transmitting symphilic instincts acquired by their mothers.

than at present, but if this had been the case, how could *sanguinea* have survived, if its ravages are as great as he asserts, especially when we consider that this ant is itself, as will be shown in the next lecture, a parasite on another ant, *Formica fusca* and therefore dependent on a host ?

The perusal of Wasmann's paper leaves one with the impression that he is bent on showing that symphily is something biologically unique and that for every peculiarity in ant behaviour we are bound to postulate a specific instinct. To repeat what I said in my " Social Life Among the Insects ", if three of my maiden aunts are fond of pets and prefer cats, parrots and monkeys, respectively, I am not greatly enlightened when the family physician takes me aside and informs me sententiously that my aunt Eliza undoubtedly has an ælurophilous, my aunt Mary a psittacophilous and my aunt Jane a pithecophilous instinct, and that the possession of these instincts satisfactorily explains their behaviour. It is only too apparent that the physician has merely called the stimuli that severally affect my aunts by Greek names, *plus* a suffix denoting, " fondness ", assumed their existence as entities in my aunts' minds and naïvely drawn them forth as " explanations ". One wonders how many more centuries will be required to eradicate this scholastic mode of thinking so exquisitely satirized by Molière when he adduced the " virtus dormitiva " as an explanation of the soporific effects of opium.

The facts and considerations recorded in this and the preceding lecture seem to me to constitute a formidable argument against the existence of specific symphilic instincts, for in the first place, if the relation between parent and offspring or between the nursing workers and the brood, is trophallactic, it is essentially the same as the relation between host and symphile. It becomes unnecessary, therefore, to assume that in the ants and termites the primitive nursing instinct or behaviour, which is a mutual feeding, has been specialized or modified during the phylogeny in adaptation to particular symphiles. It is difficult to see why there should be any more reason to assume that the responses of socal insects to particular symphiles have specific genetic determination than that their tolerance for particular ecto- and ento-parasites should be similarly determined. Slight onto-genetic modifications well within the limits of the plastic, or " intelligent " behaviour of the ants and termites, as responses to the specific organization and behaviour of the symphiles, seem amply sufficient to account for the phenomena.

But the symphilic instincts and amical selection are only two of the peculiar notions which Wasmann has introduced into the general consideration of the ant and termite guests. He also operates with other dubious, or at any rate highly speculative, conceptions such as mimicry,[1] hypertely, " Fremddienlichkeit " and the relations of his personal Deity to the ants and myrmeco-philes. I have no desire to discuss such opinions but would merely point out that the biologist who falls under their spell must sooner or later become involved in fancies which far transcend and necessarily tend to deform the facts of observation. That Wasmann has been unable to avoid this pitfall is apparent from his account of Termitoxenia and its allies, the peculiar subapterous, physogastric Phorids to which I have referred. When he first discovered them he advanced the sensational claim that they were hermaphroditic, that they did not lay eggs and have larvæ like other Diptera, but were viviparous, or rather imaginiparous, producing young of the adult form, that their aborted wings were used as organs of equilibration or as handles to enable the termites to carry them about, that they fed as parasites on the termite brood, etc. After these suppositions had been repeated by Wasmann, Assmuth and others for some years and had even been cited as facts in various

[1] For a discussion of Wasmann's account of " mimicry " among ant-guests and his incorrect use of the term, see Heikertinger (1919, 1925, 1925-1926).

popular works, they have now been proved to be without foundation. Brues (1908*b*), Bugnion (1913) and Silvestri (1920) have shown that Termitoxenia is not hermaphroditic, and Kemner (1922, 1926) has found that it lays eggs, has larvæ of a normal Dipteran type, which feed on the termite eggs, that the adult is in all probability fed by regurgitation, that the wings are not employed as balancers or handles, etc. In view of the collapse of this edifice of sensational inference, erected on a collection of alcoholic specimens, can we be censured if we suspect that some of the other elaborate constructions of the learned Jesuit may, on further examination, prove to be quite as insecure ?

XI

THE EVOLUTION OF THE SOCIAL PARASITES

We come now to a consideration of the social insects which have themselves adopted a life of parasitism in the colonies of other social insects, a subject which has received much attention from a number of investigators during the past quarter of a century. Cases of such social parasitism have been observed among all the groups, wasps, bees, ants and termites, but are most frequent and have been most carefully studied among the ants. The relationships between the social parasites and their hosts are very diverse both in their character and in the degree of their intimacy. A parasite and host which are closely allied structurally and ethologically and represent species of the same genus can obviously enter into much more intimate relationships with one another than distantly related species, or those belonging to different subfamilies or orders. Thus although even colonies of ants and termites may contract relations with one another, they are never very intimate and are due either to predatism on the part of the ants or to mere parœcism, or indifferent toleration. From the viewpoint of evolution it is interesting to note that the numerous cases of social parasitism now known are all sporadic and convergent developments, and if we omit the termites as we shall have to, owing to insufficient knowledge and the limits of our time, and confine our attention to the Aculeates, we shall find that the infestation of their colonies by other Aculeates is in many of its features quite distinct from any other known type of parasitism. The only similar phenomena are the parasitism of human groups and races by other human groups and races, i.e., parasitism existing between societies of the same species, a subject which has not received the attention it deserves by sociologists since attention was called to it many years ago by Massart and Vandervelde (1893). That the erudite Giard was fully aware of the significance of human parasitism is apparent from the fact that he gave the Belgian investigators' work a prominent place in the journal which he founded, the " Bulletin Scientifique ".[1]

The parasitic social wasps and bees may be briefly considered before we pass to a review of the cases among the ants. Even the parasitic solitary wasps are rare and confined, as far as known, to the species of Ceropales, Nysson and a few species of Pompilus (*pectinipes*, *campestris* and *aculeatus*) and Stizus (*unicinctus*). Among the social species we know of only two parasitic forms, *Vespa austriaca* and *V. arctica.* The former has long been known in Europe, where it lives in the nests of *V. rufa* (Morawitz (1864), Robson (1898), Carpenter and Pack-Beresford (1903), etc.). Recently Bequaert (1916) and Sladen (1918) have found *austriaca* in the United States, British America and Alaska, but its North American host is still unknown, though believed to be *V. consobrina. V.*

[1] Cases, like those recorded by Massart and Vandervelde in the following paragraphs, and many others, will occur to any unbiased observer of our existing human societies, or to any student of history :

" The disturbances caused by social parasitism, which is characterized by appropriation of the means of existence, are less serious than those caused by the nutritional parasitism whose effects we have described. Nevertheless, social parasites may sometimes multiply so greatly or attain such a degree of nocuity as to exhaust the workers that support them. Spencer reports that the rapacity of Oriental monarchs has sometimes led them to despoil their subjects to such an extent that they were obliged to return enough produce to enable them to sow their crops.

" Spain, at the end of the past century, may be regarded as the type of a community reduced to a kind of economic phthisis through the development of parasites and insufficiently productive workers. Under Philip III. there were 988 convents for women and 32,000 mendicant friars ; the number of monasteries had tripled during the fifty years preceding 1624, and the number of monks had increased in even greater proportion. The census of 1788 gave a total of 1,221,000 priests, soldiers, sailors, nobles, lawyers, government clerks, students and servants in a population of about 3,800,000, from which we should still have to deduct a mass of beggars, vagabonds, etc. (Rocher, ' Principes d'Econ. Pol., II., pp. 144 and 145. Transl. by Wolowski.)"

arctica, as Fletcher (1908), Taylor and I (1921) have demonstrated, lives in the nests of our common *V. diabolica.* Both *austriaca* and *arctica* have completely lost the worker caste, so that they are represented only by males and fertile females. They were at one time undoubtedly non-parasitic like their present hosts, but are now reared and fed by the workers of the latter like their own more favoured sexual forms. As a result of such nurture, what were once independent social insects, with two distinct female castes, have actually reverted to the status of the solitary insects, with only a single type of female.

Since many genera of parasitic solitary bees are known (see Wheeler, 1919*a*), it is not surprising to find that the social bees also have Apid parasites. In fact, all the subfamilies, except the Apinæ, have such satellites, and in each case it is evident that the parasite is descended from the same genus as the host. Brauns (1902) has described a species, *Eucondylops konowi*, as a parasite of Allodape, but has made no observations on its habits. The common bees of the genus Sphecodes have been studied by a long series of investigators: Walckenaer (1817), Wesmael (1835), Lepeletier (1841), Spinola (1851), Taschenberg (1866), F. Smith (1851), Sichel (1865), Breitenbach (1878), Pérez (1884), Perkins (1887, 1889), Morice (1901), Ferton (1890, 1898, 1905), Marchal (1890, 1894), Sladen (1895), Nielson (1903), Friese (1923), etc., but we still lack a definitive account of their relations to their hosts, the bees of the genus Halictus. A paper by Stöckhert, which promises to elucidate this matter, has not yet been published. From the observations of many of the authors above mentioned it would seem that the old Halictus females, which guard the nest entrances, violently resist the intrusion of the Sphecodes. Ferton (1905) saw a *S. subquadratus* breaking into a nest of *Halictus malachurus*: "Unable to seize the sentinel that obstructed the entrance with her head, she burrowed towards the bees' gallery and thus succeeded in seizing and killing the guardian and cast her backward and out of the tunnel. A second and then a third Halictus endeavoured to replace the first but met with the same fate". After gaining an entrance to the nest the Sphecodes apparently destroys the eggs in the completed cells of the Halictus and lays her own eggs in their place, so that her larval offspring do not have to compete with the lawful owners of the stored pollen and nectar. Sphecodes is therefore a predatory parasite. Some recent authors (e.g. Heselhaus (1922)), however, still maintain that this bee constructs and provisions her own cells in the Halictus nests.[1]

The most interesting parasites of bees are undoubtedly the species of Psithyrus (Apathus) which have long been known to live and breed in the colonies of Bombus. The European species have been studied by F. Smith (1885), Hoffer (1881, 1888), Friese (1888, 1923), Sladen (1899, 1912), Wagner (1907), etc., some of the North American species by Sladen (1915), Frison (1916, 1921) and Plath (1922). The genus Psithyrus is evidently an offshoot of Bombus and like the parasitic Vespas has lost its worker caste. In all probability, the species of the genus are polyphyletic, i.e., derived from several different species of Bombus. As in other parasitic bees, the females fail to develop the collecting apparatus of the hind legs, but in other respects closely resemble their hosts. In some species this resemblance is remarkably close, especially in colouration, and has been interpreted as mimetic, but on the theory of a polyphyletic origin, it may be really genetic (Plath, 1922). The female Psithyrus has some difficulty in entering a Bombus nest, and after entering, will not tolerate the presence of another Psithrus female, According to Sladen, the parasite kills the host queen, but Smith, Hoffer and Plath have shown that this is not always, nor perhaps generally, the case. Plath finds that our North American *Ps. laboriosus*, after gaining admission to the nest of *Bombus vagans*, intimidates the workers by

[1] According to Robertson (1926) the North American bees assigned to Chloralictus, which he regards as a genus, are probably parasitized by the species of Paralictus. The taxonomic characters of the latter genus suggest that it is derived from the host genus.

roughly rolling them about under her body, but does not sting them. "After this period of 'getting acquainted' the members of the *vagans-laboriosus* colony show no more hostility towards each other than the members of an uninfested colony." The parasitic queen does not always, if ever, lay her eggs in a pollen mass already containing the eggs or larvæ of her host, but like the Bombus queen, constructs her own egg-cells and like the latter attaches them to one or more of the existing cocoons. Nevertheless, according to Plath, the Psithyrus female dsestroys the eggs and young larvæ of her host and permits the Bombus workers to rear her own brood. Here, too, the parasite behaves like a predator.[1]

Friese (1923) has published on a predatory neotropical Meliponine bee a few brief notes of considerable interest. He regards the species, formerly called *Trigona limao*, as the type of a new subgenus, Lestrimelitta (Friese, 1903), because the workers lack the corbula on the hind legs, a condition which proves that they do not collect pollen. He maintains that *limao* is really a robber bee, which overpowers the colony of some industrious species of Trigona, drives away or kills the rightful owners of the nest, settles down in it and feeds on the stored pollen and honey, and then moves on to capture the nest of some other peaceful colony.

These very brief accounts of the parasitic social bees show very clearly that they are all more or less predatory. The Lestrimelitta is, in fact, so highly predatory that it can hardly be regarded as a parasite. Psithyrus also reveals decidedly predatory traits in its behaviour on entering the nest, in compelling adoption and in destroying the eggs and larvæ of its host. The behaviour of Sphecodes, so far as known, seems to be similar. That of *Vespa austriaca* and *arctica* has not been studied, but they, too, can hardly be adopted by their hosts without a conflict.

The literature on the social parasites among ants has

[1] For more recent valuable observations on Psithyrus see Frison (1926).

grown to be so voluminous that I am unable to cite all of it. Much of it is listed in my books (1910*a*, 1923) and in the first volume of Wasmann's "Gesellschaftsleben der Ameisen" (1915*c*). As would be expected, the conditions become peculiarly complicated when a social superorganism, such as an ant colony, becomes intimately associated with another colony. Wasmann long ago (1891) divided all these consociations into two categories, the "compound nests" and "mixed colonies". Although these are still convenient divisions, the distinction, as I shall show in the sequel, is not as sharp as Wasmann supposed. The compound nests are associations between two or more entire colonies of ants, of ants and termites, or of termites, living in close contiguity but really in different nests and rearing their respective broods separately. The mixed colonies of ants (none is known among termites), represent a fusion of the personnel of two or rarely more colonies, which occupy the same nest and rear their broods in common. With very rare exceptions, only one of the colonies is complete, i.e., has a fertile mother queen as well as workers. The ants which form compound nests may be divided into several minor categories, the cleptobiotic, lestobiotic, parœcismic (plesiobiotic), parabiotic and xenobiotic, or myrmecophilous. Together these subgroups comprise a number of cases, ranging from mere contiguity, based on the active predatism of one of the components, through indifferent toleration or a feeble coöperation to a definitively symphilic condition. The series, of which I describe some examples, therefore parallels the progressive development of the myrmecophiles and termitophiles, from synechthran, through synœketic to symphilic relations with the host.

(1) Different colonies of ants even of the same species are so hostile to one another that their mere existence in contiguity implies that one of the species is to some extent exploiting its neighbour. Certain small but aggressive species, which secure at least a portion of their sustenance by waylaying the foraging workers of another

Plate XLIII.

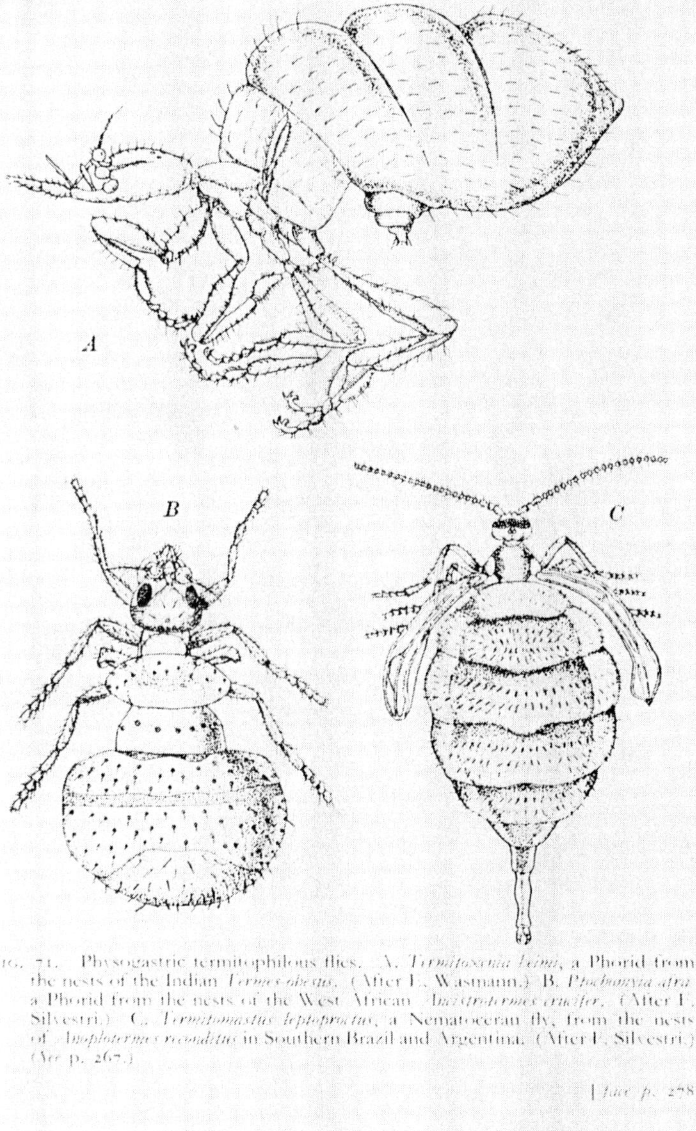

FIG. 71. Physogastric termitophilous flies. A. *Termitoxenia heimi*, a Phorid from the nests of the Indian *Termes obesus*. (After E. Wasmann.) B. *Ptochomyia afra* a Phorid from the nests of the West African *Ancistrotermes crucifer*. (After F. Silvestri.) C. *Termitomastus leptoproctus*, a Nematoceran fly, from the nests of *Anoplotermes reconditus* in Southern Brazil and Argentina. (After F. Silvestri.) (See p. 267.)

[*face p.* 278

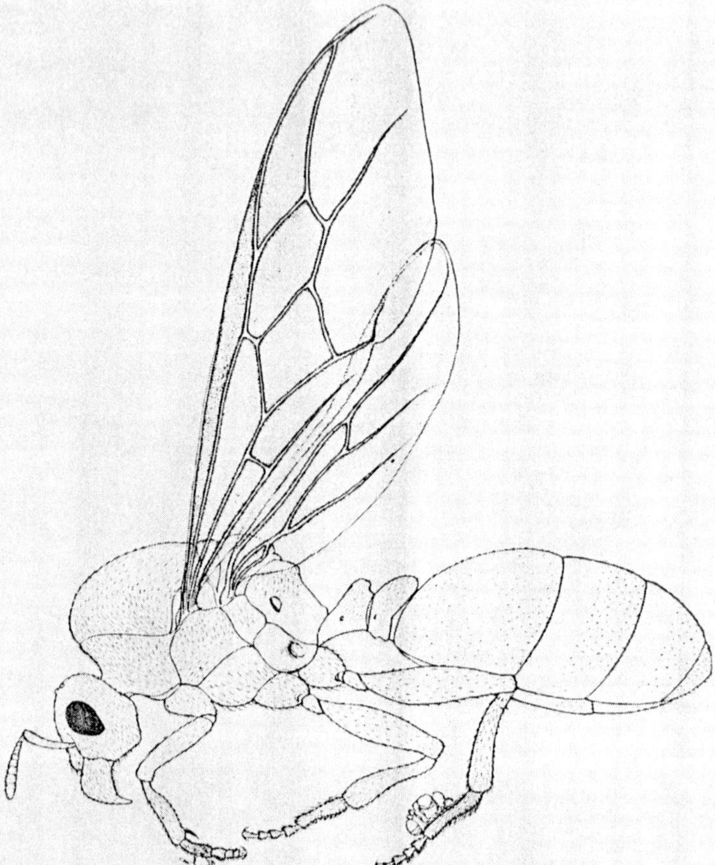

FIG. 72. Winged queen of *Carebara vidua* carrying the minute, blind workers of her own species attached to her tarsal hairs.

[*face p.* 279

species and snatching away their prey, deserve the name of brigands (cleptobiosis). Such ants naturally make their nests near those of the species they plunder. Thus *Dorymyrmex pyramicus* in our southwestern states often constructs its nest in the clearing surrounding or even on the large mounds of harvesting ants of the genus Pogonomyrmex.

(2) In the cases of lestobiosis, or thieving, the exploitation is more subtle and efficient. The thief-ants, all of which are subterranean and have very small workers, nest in the earthen walls of populous formicaries or termitaria. The chambers of the two nests are connected by extremely tenuous galleries permitting the thief-ants to invade the nests and devour the brood of their large neighbours, but preventing the latter from entering the nests of the robbers, who are either ignored or overlooked on account of their diminutive size and neutral odour. The abundance of food which they secure enables them to rear very large and often deeply coloured queens and males, but the workers themselves are condemned to perpetual dwarfhood and a pale colouration by their criminal mode of life. The most remarkable thief-ants belong to the genus Carebara and are found in the huge termitaria of the tropics. The workers of these ants are minute, pale yellow and blind, the queens and males deeply coloured and several thousand times as large as the workers. Arnold (1916) has recently suggested that these extraordinary differences in size must make it impossible for the queen to feed her first brood of workers and hence to establish her formicary in the typical independent manner of other ants. For this reason, when she leaves the parental nest to take her nuptial flight, she carries, attached by their mandibles to the tufted hairs on her feet, several of her minute workers, which thus accompany her till she has made her cell in some termite mound and then take charge of rearing her first brood (Fig. 72).

(3) Plesiobiosis, or parœcism ("neighbourliness") is a very common relationship between two or more rarely

three or even four species of ants living in nests, often with interdigitating but never with intercommunicating galleries, under the same stone or in the same log. Usually the ants of the different colonies, when they meet, are more or less hostile. If one of the species is small and weak it undoubtedly derives some protection from merely living within the immediate trophoporic field of a large and powerful neighbour, or the feebler may feed to some extent on the refuse of the larger form. When both species are large and aggressive they may perhaps find it advantageous to present a combined hostile front to the attacks of other ants.

(4) What Forel (1898) calls "parabiosis", a word which may be translated as "tutelage", seems to be a more definite relation of mutual or unilateral protection. In a typical case, which I recently observed in British Guiana (1921), we have two species, a small black Crematogaster (*C. parabiotica*) and a large brown Camponotus (*C. femoratus*) together inhabiting a large ball of earth which they build around the branch of a tree. In this ball, which Ule calls an "ant-garden", because it supports numerous epiphytes, the Crematogaster inhabits only the superficial, the Camponotus the central portions. When the nest is slightly disturbed the small black ants rush out to attack the intruder, but a more serious disturbance at once brings out the battalions of the much larger and extremely pugnacious brown species. The Crematogaster seem therefore to act as a skirmishing line for the Camponotus. Although the galleries of both species open freely into one another, and though the workers of both forage in long common files on the surrounding vegetation, they nevertheless keep their broods rigidly separated.

(5) The relationship between the ants occupying compound nests becomes still more intimate in the cases of xenobiosis, of which *Leptothorax emersoni* and *Myrmica canadensis* furnish the best example. As we have seen in a preceding lecture, this Leptothorax has really become a true guest, or symphile, feeding on the surface

secretion of its host and being fed with regurgitated food by the latter (Wheeler, 1901*c*, 1903*c*, 1907*c*). Although both species bring up their broods separately, it is possible in artificial nests containing no earth, to compel them to mingle their broods and form a mixed colony. To the same group of guest-ants belongs another North America species, *Symmyrmica chamberlini*, which lives in the nests of *Manica mutica* in Utah, and has wingless males (Wheeler, 1904*c*), and the European *Formicoxenus nitidulus*, which lives in the mounds of *Formica rufa* and *pratensis* and also has wingless, but decidedly ergatomorphic males.

Although *F. nitidulus* has been studied by Adlerz (1884), Wasmann (1891, 1915*c*), and Stumper (1918*a*, 1918*b*, 1921), none of these investigators has been able to determine the precise nature of its food. Stumper believed that it feeds on the rain water charged with vegetable matter after filtering through the materials of the Formica nests, but Stäger (1925) has been able to show that it obtains its food from its hosts in much the same manner as *Leptothorax emersoni*. This is proved by the observation which I quote: "Two *pratensis* workers were feeding one another in the customary manner, squatting and stroking one another's heads with their antennæ till one of them regurgitated a drop which the other imbibed. This was nothing unusual, but during the performance one of the guest ants that had been hanging around the feeding pair, comes up in haste, climbs from the ventral side to the mouth of the regurgitating individual and partakes of the drop. I can scarcely believe my eyes. The two *pratensis* workers separate, and the guest ant dismounts from her host. Then in another part of the artificial nest two ants begin the act of feeding. Instantly a Formicoxenus (this time) mounts the back of the regurgitating ant, reaches forward over her head and snatches its share of the meal. The *pratensis* worker that is being fed moves away, perhaps disturbed by the intrusion. The regurgitating ant, however, does not budge, but remains squatting with wide open mandibles and permits the guest to keep on with its meal. The latter remains perched on the head of the *pratensis* and with rapidly vibrating antennæ licks the drop from her mouthparts. Not until somewhat later does the unequal pair separate". Though the regurgitated food, which it steals or even secures directly by persistently stroking the heads of workers not engaged in feeding their sisters, seems to form an essential part of its diet, Formicoxenus has not lost the ability to feed independently on sugar, injured ant larvæ, etc. Stäger finds that the hosts are often unwilling to tolerate the guests and may shake them off or even attack them, but that when seized, the latter, by "feigning death" and slyly defending themselves with their stings, succeed in compelling their hosts to release them.

I recently found in the small clearing made for the erection of the tropical laboratory on Barro Colorado Island, in Gatun Lake, Panama, a new and very peculiar guest ant, of which I published an account in 1925. During the rainy season at least fourteen species of fungus-growing ants of the tribe Attini, representing nearly all the described genera, may be found on the island, many of them within the small clearing. Of one diminutive form, *Sericomyrmex amabilis*, I found more than a dozen nests within a few steps of the laboratory, each containing one or more fungus-gardens in small chambers and often suspended from the roots left intact by the ants during their excavations. There were from one hundred to three hundred workers in each nest, a single sluggish queen and a quantity of brood scattered over in the fungus gardens, which were only two to three inches in diameter. In nearly every nest I encountered a colony of a guest-ant belonging to the singular neotropical genus Megalomyrmex but representing a new subgenus and species, which I have called *Cepobroticus symmetochus* (Fig. 73). It is a slender red ant, smaller and much smoother than its velvety, brown host. Its colonies, too, are smaller, comprising only about forty to fifty individuals, with a single queen not much larger than the workers. Both

Plate XLV.

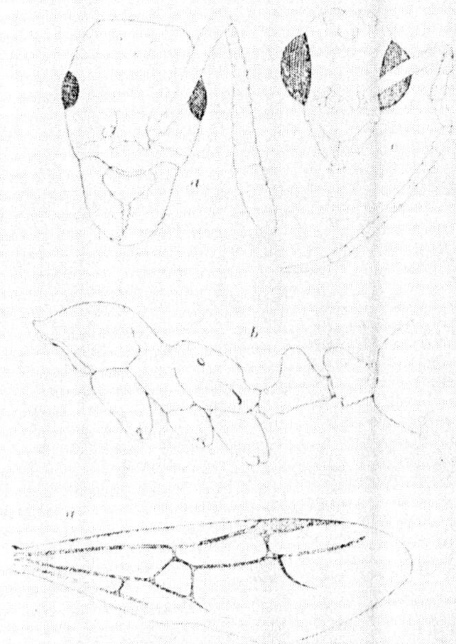

FIG. 73. *Megalomyrmex (Cepobroticus) symmetochus* Wheeler. a. Head of worker, dorsal aspect. b. Thorax and pedicel of same, lateral aspect. c. Head of male, dorsal aspect. d. Fore wing of female.

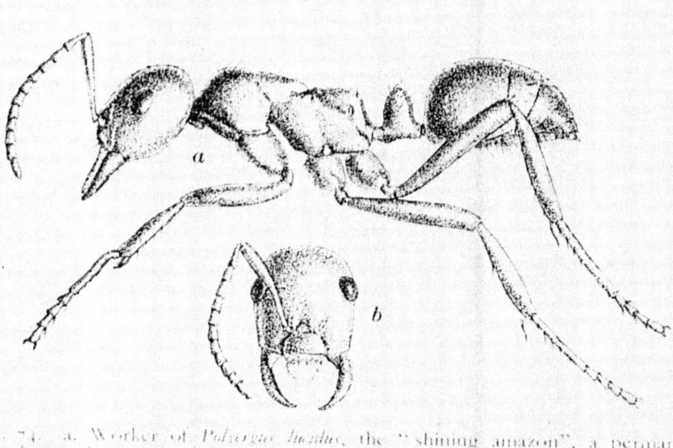

FIG. 74. a. Worker of *Polyergus lucidus*, the "shining amazon", a permanent slave-maker of the Eastern United States. b. Head of same, showing the sickle-shaped mandibles.

[*face p.* 282

Plate XLVI.

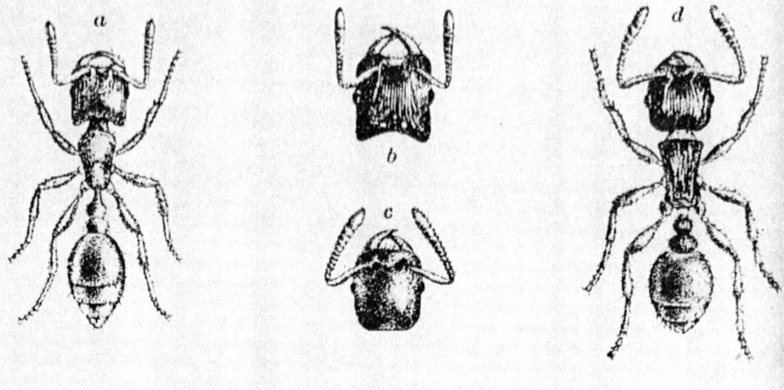

FIG. 75. a. Worker of *Strongylognathus testaceus*, a degenerate slave-maker of Europe. b. Head of female of same. c. Head of worker of *S. huberi*, an allied species. d. Worker of the pavement ant (*Tetramorium caespitum*), the host of the two species of Strongylognathus.

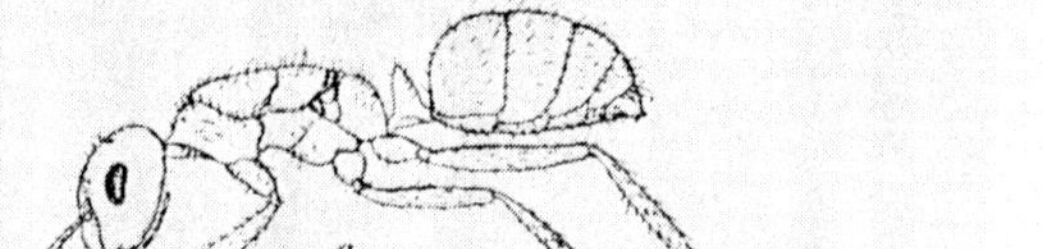

FIG. 76. A temporary social parasite (*Formica microgyna*). a. Queen, with wings removed. b. Large worker drawn to same scale. c. Head of same. d. and e. Petiole of worker and queen seen from behind. (*See* p. 292.)

[*face p.* 283

species live amicably together and occasionally lick one another but exchange no other attentions. When transferred to artificial nests, the workers of both species were seen to live and feed on the fungus-gardens which are cultivated exclusively by the Sericomyrmex. The queens, which were often found resting side by side on the fungus, scatter their eggs among its tufted hyphæ. The larvæ and pupæ are therefore mingled together, but so far as I was able to observe, each species cares only for its own progeny. This case is of unusual interest first, because the guest ant belongs to a peculiar genus, the other known species of which are not mycetophagous, and second, because it evidently exhibits a condition clearly intermediate between the compound nests and mixed colonies of Wasmann. I have called this new and unique form of social symbiosis " mycetometochy ".

The ants that live in the various compound nests are not closely allied but belong to different genera or even subfamilies, a fact which may explain why, except in the case of Sericomyrmex and Cepobroticus, they occupy separate nests and do not bring up their broods in common, for the rearing of the brood is a very delicate operation and would be apt to differ considerably in different species. We are therefore prepared to find that mixed colonies are formed only by closely allied species, i.e., either by those belonging to the same genus or to closely allied genera, and this proves to be the case. But before considering the various types of mixed colonies, two facts must be emphasized : First, many ants are fond of kidnapping and devouring the larvæ and pupæ belonging to other colonies of their own or other species. One whole subfamily, the Dorylinæ, exhibits this rapacious habit very clearly. I have often witnessed such raids by our smaller species of Eciton (Acamatus), and have observed more recently the same behaviour on a more extensive scale in the larger species (*E. burchelli* and *hamatum*). Moreover, as I observed in Australia, the members of at least one genus of Cerapachyinæ, Phyracaces, make similar forays on other ants and carry home their brood as prey. Mr. J. Clark, of Perth, West Australia, writes me that he has recently witnessed such a raid of an undescribed species of Phyracaces on a colony of *Iridomyrmex gracilis.*[1] Frequently the kidnapped young are devoured, but in well-nourished colonies they may be permitted to complete their development and the emerging workers may be adopted as *bona fide* members of the colony, even if they belong to a different species. It is therefore possible to produce a mixed colony experimentally by giving the colony the mature brood of some other species. In this manner Miss Fielde (1905) succeeded in inducing species belonging even to very different subfamilies to live together in perfect amity. It is also interesting to observe that ants thus reared in the colony of an alien species may be very hostile to their own sisters that have been left to grow up in the parental nests.

Second, the mixed colonies found in nature are not, in the first instance, produced by the mere kidnapping of the brood of an alien species, but by the young queen of a parasitic species that is unable to start a colony independently, invading the nest of another species, which then becomes the host. The behaviour of the invading parasite and the host colony differs in different species, but in nearly all the observed cases the host queen, if present, is eventually killed and her place is taken by the alien intruder. Since the queen ant is really the reproductive organ of the colony considered as a superorganism, the host colony may be said to be castrated and its sterile worker personnel is constrained to devote all its energies to rearing the brood which is forthwith produced by the fecund parasite. With these general statements in mind

[1] It will be remembered that Darwin derived the slave-making propensities of *Formica sanguinea* from just such a condition, and I still believe, despite all of Wasmann's arguments to the contrary, that there is considerable truth in the great naturalist's phylogenetic hypothesis. *F. sanguinea* is primarily a predatory ant, in all probability descended from forms which behaved very much like Eciton, though it has now come to confine its raids to certain abundant and prolific species of its own genus (varieties of *F. fusca, cinerea, neogagates, pallidefulva*, etc.). It still devours a large portion of its larval and pupal prey.

we may turn to the types of mixed colonies, of which three may be distinguished, those of the slave-makers, of the temporary and of the permanent social parasites.

(1) The peculiar phenomenon known as slavery, or dulosis, which occurs in two genera of Formicinæ, Formica and Polyergus, and two genera of Myrmicinæ, Strongylognathus and Harpagoxenus, represent three phylogenetic stages: a primitive stage in *Formica sanguinea* (Fig. 51), a culminating stage in Polyergus (Fig. 74), and a degenerate or evanescent stage in Strongylognathus (Fig. 75*a*-*c*) and Harpagoxenus. The blood-red slave maker, *Formica sanguinea*, is a common but rather local red ant, with black or brown gaster, and is represented by numerous subspecies and varieties ranging over northern Europe, Asia and North America. It can be readily distinguished from the other species of the genus at least in the Old World and the Eastern United States, by the pronounced notch in the clypeus. *Sanguinea* is one of the most intelligent of ants and therefore one of the most interesting to keep in artificial nests. Its habits were first studied more than a century ago by Pierre Huber (1810), the son of the blind François Huber, and have ever since commanded the attention of myrmecologists, because its armies of workers make periodical forays on the colonies of the common black *Formica fusca*, carry the worker larvæ and pupæ into their nest and permit many of them to emerge and become members of the colony. Thus the colony is mixed, and the black individuals, on account of their colour and provenience, have been called slaves. It is evident, however, that this term is inappropriate, for a slave is "a man who is the property of another, politically and socially at a lower level than the mass of the people, and performing compulsory labour" (Nieboer), and none of these distinctions applies to the *fusca* workers in the *sanguinea* nest. They might more properly be called "auxiliaries" (Hilfsameisen). At least one of the subspecies of *sanguinea* (*aserva*) does not make slaves, and the colonies of some of the other forms give up the habit after a time,

for the *sanguinea* colony, when once established, is quite able to lead an independent life. Darwin and others offered various explanations of the peculiar slave-making habit of *sanguinea*, but its meaning remained obscure till I discovered (1905, 1906, 1907*b*, 1908*c*), that it had its origin in the behaviour of the young queen. She is quite unable to found a colony independently and therefore, after her marriage flight, may adopt one of three courses: she may return to the nest in which she was reared or enter some other *sanguinea* nest, or she may invade a nest of *Formica fusca*. As the first and second courses are sometimes adopted by other ants and do not lead to the formation of mixed colonies, they need no further consideration in this place and we may confine our attention to the last. As soon as the *sanguinea* queen invades a *fusca* colony, she becomes greatly excited and interested in the brood, seizes and collects in a pile as many pupæ as she can snatch up and mounts guard over them. She slays any *fusca* workers that are bold enough to attempt to regain their property and is therefore soon left in undisputed possession of her plunder. Eventually *fusca* workers emerge from the cocoons and at once assume a friendly attitude towards the queen, feed her by regurgitation and behave towards her as if she were their own mother. She begins to lay eggs and the resulting larvæ are fed and reared by the black workers, so that when the *sanguinea* workers emerge a mixed colony is established. These workers show that they have inherited their mother's proclivities by kidnapping the brood of neighbouring *fusca* colonies, but they do this as an army and carry the *fusca* brood to their nest. In some colonies, as I have stated, this kidnapping, or slave-making proclivity may disappear after a time, and in *aserva* it seems to lapse very early or perhaps is not even inherited by the workers. In such cases, therefore, the personnel of old colonies may be made up entirely of *sanguinea* after the batch of *fusca* workers kidnapped and reared by the queen has died of old age.

The behaviour which I have observed in the young

queens of the American subspecies of *sanguinea* was later shown to be exhibited also in the typical European form by Viehmeyer (1908, 1909, 1910*a*, 1910*b*, 1910*c*), Wasmann (1908*b*, 1909*c*, 1910*b*) and Donisthorpe (1910, 1911). It is evident that slavery is at bottom a form of predatism and has its origin in the inability of the young queen to establish a colony without the aid of workers. Unlike the great majority of ant-queens, she has been unable to store enough food in her body to stand the strain of long fasting and nourishing her first brood. In another sense, she is, of course, a parasite and the *fusca* workers represent the host. Owing to the fact that the colony may eventually cease to increase its worker personnel by the kidnapping of *fusca* brood, we may call this type of slavery temporary, acute, or facultative.

(2) The species of Polyergus, or "amazons", as they were called by Pierre Huber (1810), have much the same distribution as *sanguinea* and have the same species of Formica as slaves, but their method of securing the latter is more highly perfected. The amazons are very beautiful red ants (except the Japanese *P. samurai*, which is black) and their mandibles are slender and sickle-shaped and perfectly adapted to fighting, but of no use for digging in the earth or capturing food. Hence these insects are unable to make nests or even to feed themselves, or to care for their own young, but are absolutely dependent on their slaves. Like *sanguinea* the amazons make periodical forays, which for some reason are always carried out in the afternoon, but their armies show a more highly perfected tactical organization and the subjugation and plundering of the *fusca* colonies are effected with much greater dispatch and precision—one might say, with the most consummate *éclat*. At the approach of the amazons the *fusca* workers usually flee in dismay, but if they offer any resistance the amazons pierce their heads with the sickle-shaped mandibles. The young workers on emerging from the kidnapped pupæ excavate the nest, feed the Polyergus and bring up their brood but do not accompany the armies on their

raids. The initial stages in founding the colony have been studied by Emery (1908, 1909*a*, 1909*b*, 1911*a*, 1911*b*), who found that the young Polyergus queen secures adoption in some small, weak *fusca* colony after killing its queen by piercing her head. She then produces her brood which will later make the slave raids on the *fusca* colonies. Since this raiding proclivity never lapses even in old colonies, Polyergus is to be regarded as a permanent, chronic, or obligatory slave-maker.

In Europe there are several species of the interesting genus Strongylognathus (Fig. 75*a-c*), which have sickle-shaped mandibles like Polyergus, though they belong to a very different subfamily, the Myrmicinæ, and always live in colonies of the common pavement ant, *Tetramorium cæspitum* (Fig. 75*d*). Our fragmentary knowledge indicates that we have here some of the degenerate or evanescent stages of slavery. The workers of *S. rehbinderi* and *huberi* seem still to make forays on Tetramorium colonies and to carry home their brood (Forel, 1900), and Kutter (1920, 1923) has recently shown that *alpinus*, a subspecies of *huberi* which I discovered some years ago (1909) near the headwaters of the Visp, within sight of the Matterhorn, makes nocturnal slave-raids and is accompanied by its slaves, which do most of the fighting and carry home the brood of their own species. In this case the slaves are really the masters and seem to use the Strongylognathus merely as a means of disconcerting or terrifying the colonies of *cæspitum* whose brood they are bent on kidnapping. Finally, *S. testaceus*, the best-known species of the genus, no longer makes forays and is tending to lose its worker caste. Wasmann (1891, 1915*c*), Mrázek, Forel and I (1909), have found that colonies of *cæspitum* infested by this species may retain the host queen. In order to establish her colony, therefore, the young *testaceus* queen probably associates herself with a young, nest-founding *cæspitum* queen. In the mixed colonies of other species of Strongylognathus and Tetramorium the host queen appears to be eliminated as in the colonies of Polyergus.

Yet another Myrmicine genus, Harpagoxenus, comprising only two species, *H. sublævis* of Central and Northern Europe, and *americanus* of the Eastern United States, must be included among the dulotic, or slave-making ants. The European *sublævis* has been studied by Adlerz (1896) in Sweden, where it has only ergatoid queens, and by Viehmeyer (1906, 1912, 1921) in Germany, where it also has winged queens. Both of these investigators found that the ergatoids and workers break into nests of *Leptothorax acervorum* or *muscorum*, seize the brood and carry it to their nest. The behaviour of the Harpagoxenus, according to Viehmeyer, is similar to that of Polyergus and indicates a like absolute dependence on the slaves. He describes the method of colony formation, which resembles that of *sanguinea*, as follows: " As in all parasitic ants the founding of the colony is of the dependent type. It probably takes place during the slave-raids. The ergatoid females participate in these forays and remain behind in the plundered nest in order to rear for themselves a first batch of auxiliaries from a few of the Leptothorax pupæ that have been left or have fallen to their share. Twice I found such newly established colonies (Dresden, 1908 and Oberrittergrün, 1915). They consisted of a solitary worker-like female, with in the one case a single, in the other two Leptothorax workers and some brood of the auxiliaries. Moreover, it is proved by experiment that an isolated female may be successful in plundering a nest and robbing pupæ. But in the case of ergatoid females this type of founding a colony is improbable since the outcome must be easily frustrated compared with the founding of a colony during a foray. The two colonies, furthermore, had the appearance of having been established in plundered nests ".

Harpagoxenus americanus was discovered by Pergande in a nest of *Leptothorax curvispinosus* at Washington, D.C. and described by Emery in 1895. In 1906 I published a few notes on three mixed colonies of the same ants found in elder twigs at Bronxville, N.Y. In 1925, Sturtevant discovered a mixed colony in an oak gall on Naushon

Island, Massachusetts. More recently (1927) this investigator and Creighton (1927) have been able to make so many observations on *H. americanus* that we have suddenly acquired a better knowledge of its behaviour than of that of the European *sublævis*. In two localities in New Jersey, Sturtevant discovered some seventeen colonies, in most of which the yellow *L. curvispinosus* was the host, but occasionally this rôle was assumed by the black *L. longispinosus*. His experiments show that the behaviour of the parasite is very much like that of *sublævis*. The queens are winged and ocellate ; ergatoid individuals are rare. Creighton's field observations on a number of colonies were made at Tuxedo, New York. In this locality the common host of *americanus* was *L. longispinosus*. though one unusually large colony contained both hosts, as I had found in a single colony at Bronxville. He summarizes his conclusions on the method of colony formation and the slave raids, which as in Polyergus are conducted during the afternoon hours, as follows :

" While much remains to be done it is now possible to sketch the ethology of *H. americanus* ; a hasty sketch to be sure, with many missing features hypothetically supplied, but at least a beginning. In this picture we see the fertilized Harpagoxenus queen entering a Leptothorax nest by force. Having driven away or killed the original owners of the brood she appropriates this, tends it and is in turn tended by the resulting Leptothorax workers, until in time there arises a mixed colony. When the colony is well established the dulotic instinct manifests itself in the Harpagoxenus workers. These gain entrance to some Leptothorax nest after a long struggle in which their greater hardiness and superior size finally enable them to kill or intimidate the Leptothorax workers. If the external conditions are favourable, the raid is carried to a successful conclusion and the brood of the pillaged nest is carried back to the mixed colony. If the return of the raiders is rendered impossible then a fragmentary Harpagoxenus-Leptothorax colony results.

"The slave raids of Harpagoxenus show none of the organization and spirit so characteristic of those of Polyergus. They fall far short even of the lesser degree of co-operation exhibited by *F. sanguinea*. Nevertheless they are of great interest, since they appear to be the last manifestation of a vanishing character. They show the decay of dulosis and foreshadow a state of abject parasitism. During the progress of a raid the Harpagoxenus workers and their slaves cluster about the entrance of their nest in a manner suggestive of Polyergus but with this the similarity ends. There is no rapidly moving phalanx of raiders, no concentration about the entrance of the raided nest, no frantic activity to enlarge the entrance. The Harpagoxenus leave their nest singly and amble awkwardly and uncertainly to the nest of their victims. Only once did I see more than one Harpagoxenus leave the nest at the same time. On one occasion a column of three departed for the Leptothorax nest. However, this column broke up almost at once and was, I believe, purely fortuitous. On arriving at the Leptothorax nest, the Harpagoxenus wastes no time in preliminaries but enters at once. Having secured a larva or pupa it emerges as quietly as it entered and returns with its burden to its own nest. Quite often they lose their way and I have seen a number of them captured by small spiders while raiding. The action of Harpagoxenus after it enters the raided nest could not, of course, be followed in the field. However, observation of ants in artificial nests shows that the brood is obtained by force and not by stealth, although indeed, the action of the Leptothorax during a raid furnishes ample evidence that this is the case."

(2) In 1904 I detected another method of forming mixed colonies, which I called temporary, although it might have been called acute, or "protelian", to use Caullery's term (1922), social parasitism. It is practised by a number of species belonging to at least three of the subfamilies of ants; several Formicæ and Lasii among the Formicinæ, certain species of Bothriomyrmex among the

Dolichoderinæ and a few species of Aphænogaster and Crematogaster among the Myrmicinæ. I first detected it in *Formica consocians*, a member of a group of species (*microgyna* group) represented only in North America and characterized by having very small queens (Fig. 76), often smaller or no larger than the largest workers and sometimes of a very different colour and pilosity—red or yellow, with long golden hairs. These characters are very significant since they are precisely those of symphilic myrmecophiles. The young queen of these ants enters the nest of another Formica belonging to the *fusca* or *pallidefulva* group and is very apt to be adopted, probably on account of her small size and attractive odour. The fate of the host queen in such invaded nests has not been ascertained, but she is probably killed by her own workers. The parasite then proceeds to produce her brood, which is reared by the host workers, and a mixed colony results. As there is no inclination on the part of the queen's offspring to plunder other nests of the host species, and as all the host workers naturally die off in the course of a few years, a pure colony of the parasitic species is left behind and may grow to be very populous and aggressive —a beautiful analogue of some human institutions, which after starting in humble and cringing parasitism have come to acquire during the centuries a most exuberant and insolent domination. Subsequently this same type of temporary parasitism was detected in two other groups of which *F. rufa* and *exsecta* are the types. These ants (*F. rufa*, *pratensis*, *obscuripes*, *exsectoides*) often make mound nests of considerable size in Europe and North America. Their method of colony formation has been studied by Wasmann, Viehmeyer (1908), Donisthorpe (1909, etc.), Brun (1910, 1912*a*, 1912*b*, 1913*a*, 1913*b*, 1924), Crawley and Donisthorpe (1913), Kutter (1913*a*, 1913*b*, 1917), Tanquary (1911), myself (1908*c*, 1909) and others.[1]

[1] More than twenty species of Formica of the *rufa*, *exsecta* and *microgyna* groups, with a large number of sub-species and varieties may now be regarded as temporary social parasites on species and varieties of the *fusca* and *pallida-fulva* groups.

It has also been found that the various subspecies and varieties of *Lasius umbratus* in Europe and North America are temporary social parasites of *L. niger* and its varieties (Crawley, 1909, Crawley and Donisthorpe, 1913, Wasmann, 1909*b*, 1923*a*, Tanquary, 1911, Wheeler, 1917*a*, Rüschkamp, 1924) and, what is more remarkable, the aberrant European *Lasius* (*Dendrolasius*) *fuliginosus* has been proved to found its colonies with the aid of *umbratus* or *bicornis* workers (Donisthorpe, 1897, 1922, etc., Crawley, 1910, Crawley and Donisthorpe, 1913, de Lannoy, 1908, Forel, 1908, Emery, 1908*a*, Wasmann, 1909*b*, Bönner, 1915, Lomnicki, 1922). *L. fuliginosus* is, therefore, a temporary social hyperparasite. Tanquary observed the adoption of a *L. umbratus minutus* queen by a colony of *L. niger* var. *americanus* and I was able in 1917 to observe numerous young queens of *L. umbratus subumbratus* in the act of hunting out colonies of *L. niger* var. *neoniger* and *sitkaënsis* and securing adoption in them at an altitude of 9,000 feet at Cloudcroft in the Sacramento Mountains of New Mexico. The female *subumbratus* is covered with long, yellow pile and has a delicate and peculiar odour. A similar pilosity is also observed in the female of *L. ciliatus* of Cashmir, which is therefore in all probability a temporary social parasite. Our North American *L.* (*Acanthomyops*) *latipes* and *murphyi*, the females of which have a dense golden trichome-like pilosity, are, perhaps, temporary parasites of hyperparasites on some of our forms of *L. niger* or *umbratus*.

The precise fate of the host queen in the mixed colonies of Lasius has not been ascertained, but the *niger* queen is probably killed by the *umbratus* queen. In the case of *Bothriomyrmex decapitans* the massacre of the host queen has actually been observed by Santschi (1906, 1920) in Tunis. The young queen, on descending from her marriage flight, wanders about on the ground till she finds the nest of a *Tapinoma nigerrimum* colony, when she permits herself to be seized and " arrested " by its workers. These then proceed to drag her into their burrow by her legs and antennæ. After entering the nest the parasite may be attacked from time to time by the workers, but she takes refuge on the brood or on the back of the large Tapinoma queen. In either of these positions she seems to be quite immune from molestation, probably because her own odour is overlaid by that of the brood or the host queen. Santschi observed that the parasite often spends long hours on the back of the Tapinoma queen and that while in this position she busies herself with sawing off the head of her host ! By the time she has accomplished this cruel feat, she has acquired the nest-odour and is adopted by the Tapinoma workers in the place of their unfortunate mother. The parasite thereupon proceeds to keep them busy bringing up her brood. They eventually die of old age and the nest then becomes the property of a thriving, pure colony of *Bothriomyrmex decapitans*.

(3) No less than fourteen genera and seventeen species of ants, from various parts of the world, may be classed as permanent, or chronic social parasites. They have all completely lost the worker caste so that in this respect they closely resemble *Vespa austriaca* and *arctica* and the species of Psithyrus. We may distinguish two types among these permanent parasites, those in which the male and female are similar in size and structure and of normal habitus, i.e., but little modified (Epipheidole, Sympheidole, Wheeleriella, Epœcus, etc.) and those in which one or both sexes show distinct degenerative modification (Anergates, Anergatides, Bruchomyrma). The young queens enter the nests of other ants and secure adoption, like the queens of the temporary social parasites. The host queen seems to be regularly assassinated by her own workers. At least this has been observed by Santschi (Forel, 1906) in the case of *Wheeleriella santschii* (Fig. 77), which lives in the nests of the common North African *Monomorium salomonis*. After fecundation the *Wheeleriella* queen roams about over the surface of the soil in search of a Monomorium nest. When near the entrance of one of them she is "arrested", to use Santschi's

Plate XLVII.

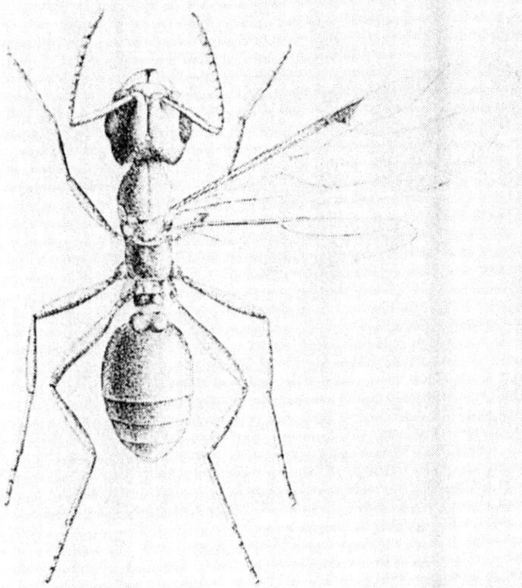

FIG. 77. Female of a workless permanent social parasite, *Wheeleriella santschii* Forel from North Africa.

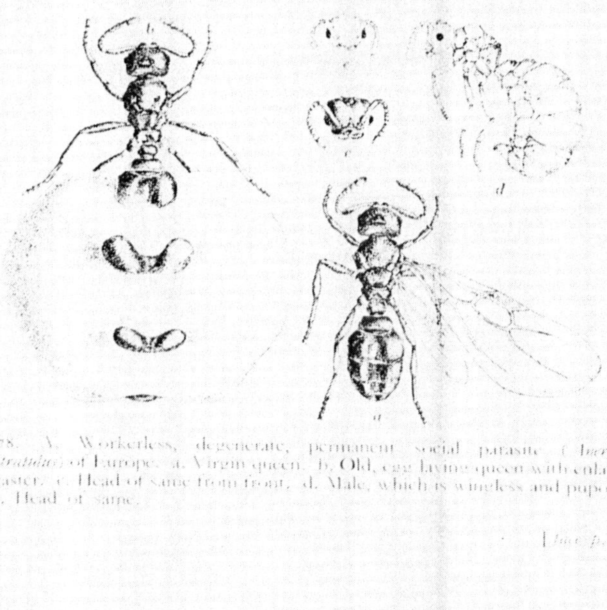

FIG. 78. A. Workerless, degenerate, permanent social parasite (*Anergates atratulus*) of Europe. a, Virgin queen. b, Old, egg laying queen with enlarged gaster. c, Head of same from front. d, Male, which is wingless and pupoidal. e, Head of same.

[Face p. 204

Plate XLVIII.

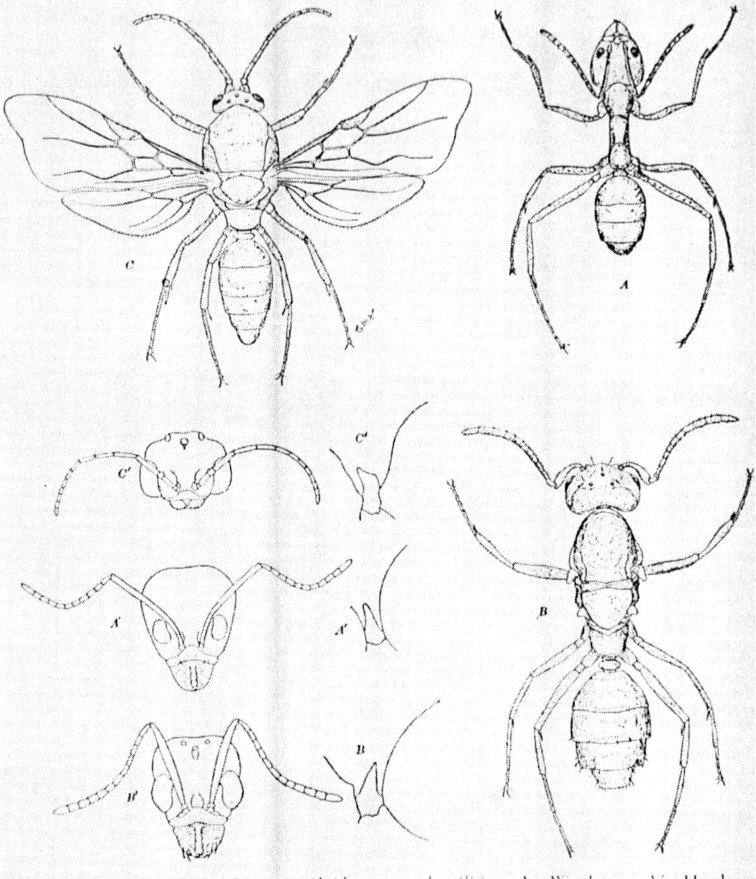

FIG. 79. The Argentine ant (*Iridomyrmex humilis*). A. Worker. A'. Head. A''. Petiole of same in profile. B. Dealeated female. B'. Head. B''. Petiole of same. C. Male. C'. Head. C''. Petiole. (After W. Newell.) (*See* p. 320.)

[*face p.* 295

expression, by a band of Monomorium workers, which tug at her legs and antennæ and draw her into the galleries. Sometimes she may be seen to dart suddenly into the entrance of her own accord and is arrested within the nest. There are no signs of anger on the part of the Monomorium, and she is soon permitted to move about the galleries unmolested. The workers then begin to feed and adopt her and in the course of a few days she lays her first eggs, which are accepted and cared for by the host. The parasite pays no attention to the much larger Monomorium queen, but the latter is eventually assassinated by her own workers. Other species, like the famous *Anergates atratulus* (Fig. 78) of Europe, the recently discovered *Anergatides kohli* (Wasmann, 1915*b*) of the Congo and *Bruchomyrma acutidens* of the Argentine (Santschi, 1922) are much more highly modified and represent the last stages of parasitic degeneration. In Anergates, which lives with *Tetramorium cæspitum* and has been studied by a number of investigators, the queen is small and winged, but after deälation and adoption her gaster swells enormously with eggs till she somewhat resembles an old termite queen. The male is wingless and pupa-like and unable to leave the nest. Mating therefore takes place between brothers and sisters ("adelphogamy" of Forel). The conditions in Anergatides and Bruchomyrma which live in the nests of Pheidole species, are somewhat similar. In all these workerless parasites the offspring of the intrusive queen are, of course, all males and females and are produced during the lifetime of the host workers. The colonies are therefore mixed throughout their existence which is necessarily terminated by the death of the host.

Although the foregoing account of the various forms of social parasitism has been greatly condensed in order to bring it within the compass of a single lecture, but little space remains for a few concluding remarks. It will be noticed that all the hosts are very abundant, widely distributed and therefore dominant species. This is particularly true of *Formica fusca*, *Tetramorium cæspitum*, *Tapinoma nigerrimum*, *Monomorium salomonis*, *Leptothorax acervorum*, etc. Nearly all the hosts, moreover, produce large colonies because they have prolific queens and workers that are very successful in rearing the brood. The parasites themselves, on the contrary, are very rare or sporadic. This is noticeably the case with the degenerate slave-makers and workerless permanent parasites, some of which are among the rarest of ants. During more than forty years Forel succeeded in finding only a few colonies of *Anergates atratulus* in Switzerland, which has a rather exuberant ant fauna, and the species was only recently discovered in Great Britain. The North American *Epœcus pergandei* was taken once many years ago, and neither Mann nor I have been able to find it again, though we have examined hundreds of colonies of its host, the common *Monomorium minimum*. Nearly all the other permanent parasites are known only from specimens taken in single colonies of abundant hosts. It will also be noticed that some of the latter, e.g., *Formica fusca* and *Tetramorium cæspitum* are parasitized by several species of ants. It is also true that only the ants which form vigorous colonies and belong to dominant species harbour myrmecophiles in any number or variety. This peculiar susceptibility to parasitic infection on the part of vigorous species of Formicidæ is interesting in connection with the similar conclusions of Raines (1922) on the relation between vegetative vigour of plant hosts and their susceptibility to diseases. He says: " While the claim that increased vigour of the host means greater susceptibility to an infection may appear somewhat anomalous from the point of view of current theories regarding the infectious diseases, observations such as form the subject of the present paper are readily understood when we consider the infectious diseases in the light of the larger class of biological phenomena of which they are an artificially selected group—namely, parasitism, commensalism, and symbiosis, the class of biological phenomena in which one organism lives within, and derives its sustenance from, the tissues of another living organism. In each of the four main groups of parasitic organisms—the bacteria, the protozoa, the worms, and the fungi—a series of intergradations are to be observed in the physiological interrelations of host and parasite, from the increasing and violent struggle that continues until the destruction of one or other of the principals, to a relation of a more benign type characterized by great subordination and even tendency to usefulness on the part of the parasitic organism, and by the utmost tolerance on the part of the host[1]. In many instances the nature of the reaction is not constant, but varies with the progress of the host-parasite relation. In this intergrading series of possible host-parasite relations, the inverse relation between host vigour and parasite virulence obtains only in the instances and phases where the reaction of the host to the parasite is one of active antagonism ; here a more vigorous host means a host of greater physiological capacity to combat the progress of the invader. But when the relation between host and parasite is of a symbiotic type, a more vigorous host means a host in which more food is available for the development of the parasite. . In the group of the fungi the transition from violent and destructive parasitism to parasitism of the symbiotic type is accompanied by a transition from facultative to obligate parasitism, as if the physiological corollary of parasitism of the latter type is extreme specialization in food preferences. The series in the fungi grades from violent and destructive parasites like Botrytis, on the one hand, to, on the other hand, so benign an infestation as the seed fungus of *Lolium temulentum* (described by Freeman, 1903) in which the relation is so intimate and devoid of any untoward effect on the host, and the life history of the cohabiting organism is so parallel with that of the grass that its distinct individuality is almost open to question ".

Raines' remarks are very suggestive also in connection with the much-discussed question of the phylogenetic

[1] These various unstable relationships are beautifully shown in a recent detailed study of Mycorrhiza by Melin (1925).

relations of the three types of behaviour exhibited by the dulotic, temporary and permanent social parasites among the ants. They obviously form a series comparable with the predatory (synechthran), synœketic and symphilic series among the myrmecophiles and the ants which form compound nests with other Formicidæ. All investigators agree that the workerless, permanent parasites represent the final, degenerate and evanescent evolutionary stage in the series of social parasites, but opinions are divided in regard to the initial, or primitive stage. Wasmann and Piéron (1910) contend that it is represented by the temporary parasites and would derive their behaviour from that of queen ants which, after fecundation, seek and secure adoption in colonies of their own species (secondary pleometrosis). From temporary parasitism, according to these authors, both the dulotic and permanently parasitic behaviour are to be derived. Viehmeyer, Emery, Brun and I contend, however, that predatory behaviour of the type exhibited by *Formica sanguinea* more probably represents the initial stage and that the conciliatory, temporary and finally abject, permanent parasitism represent natural ulterior developments of this violent, or aggressive relation between host and parasite. This contention is supported, first, by the general considerations that parasitoidism and parasitism are evidently derived from predatism among a great many solitary insects and that this sequence obtains also among the myrmecophiles, the social parasites among bees, and degenerate slave-makers (e.g. Strongylognathus), the ants which form compound nests, etc.; and second by the more special consideration that the queens of certain temporary parasites, with large queen, e.g., *Formica rufa*, sometimes behave like *F. sanguinea* when establishing their colonies and that the species with small queens, red or yellow colouration and trichomes, are obviously derived forms.

The host and parasite relations above considered also have an important bearing on Wasmann's "amical selection" and "symphilic instincts". It must be admitted that the temporary and permanent parasites and at least most of the dulotic species (Strongylognathus, Polyergus, Harpagoxenus) behave like symphilic myrmecophiles, since they likewise exploit the philoprogenitive proclivities of their hosts and enter into trophallactic relations with them. We have seen that some of the temporary parasites have actually acquired symphilic characters (small stature, red or yellow colouration, trichomes, supplicatory or conciliatory behaviour). Moreover, the permanent parasites, which have lost their worker caste are therefore really solitary insects and may be regarded as symphiles. The enormous disparity between the abundance of these parasites and that of their hosts, shows clearly that the former cannot owe their peculiarities to amical selection on the part of the latter, and the fact that the parasites actually castrate or lead to castration of the colonies which they infest and never tolerate the rearing of male offspring of the host workers, proves that there can be no hereditary basis for the development of symphilic instincts on the part of the host species. As in the case of the myrmecophiles, the peculiar adaptations, both structural and behaviouristic, to particular hosts, are therefore initiated and developed entirely by the social parasites themselves and there is nothing to indicate that these adaptations require any fundamentally different biological explanation from those which have been advanced for the countless cases of parasitism among solitary insects, other animals, or plants.

XII

CONCLUSION

If the conclusions reached in the preceding lectures are correct, it is possible to sketch in rough outlines the probable course of evolution of the truly social insects belonging to the order Isoptera and the suborder Aculeata among the Hymenoptera. We have seen that the former arose among the Protoblattoids, possibly as early as Permian times and, after passing through primitive stages like those of existing Mastotermitidæ and Calotermitidæ, culminate in the specialized and exuberant Ethiopian and Oriental Termitidæ. That the character and direction of Isopteran evolution as a whole has been in the main determined by their peculiar food is obvious. A diet of cellulose or humus is responsible for their most striking peculiarities—their defenceless integument, their microphthalmy and photophobia, their architectural employment of their own fæces, their trophallactic relations with one another, the employment of symbiotic protozoa in the digestion of cellulose, etc. Only the highest termites have acquired a new habit, that of cultivating and eating fungi, a habit also intimately connected with their peculiar cellulose environment and diet. The choice of an abundant but rather indigestible food and confinement in small hard-walled cavities led to a prolongation of life as in many other wood-boring insects and a tendency to congregation of the adults and an affiliation of the parents and offspring. We observe similar, though feebler and subsocial developments among many Coleoptera of the families Ipidæ (Scolytidæ), Platypodidæ and Passalidæ, in Phrenapates among the Tenebrionidæ and Parandra among the Cerambycidæ.

The evolution of the social Aculeata exhibits a much greater diversification, intensity and acceleration of adaptive radiation. We have traced them back to Bethyloid ancestors, which were derived from primitive Terebrants, in turn descended from unknown Phytophaga. The early Vespoids and Sphecoids presumably arose among the Bethyloids, now a highly diversified complex of small archaic forms. The bees, culminating in the social Apinæ certainly had Sphecoid ancestors and the social Vespidæ and Formicidæ may be safely derived from primitive Vespoids. Here, too, as among the Isoptera, the character of the food has been a determining factor in the evolution, together with the initiative and independence of the female sex. This independence in turn naturally resulted from the possession of a highly specialized spermatheca, requiring only a single act of fecundation and permitting long retention of the sperm in a living condition. The evolution of the Hymenoptera begins with forms like the Permian Protohymenoptera, discovered by Tillyard and related to our modern Mecoptera, a group which had much greater evolutionary potentialities than the Protoblattoidea, and led to the leaf-and wood-eating Phytophaga. Some of these, like the Oryssidæ, became parasitoidal and thus opened the way for the extraordinary efflorescence of the Terebrantia, all except a few families of which (Cynipidæ, certain Chalcidoidea) have retained entomophagous habits. The Bethyloidea have preserved essentially the same habits, but some of the genera (Bethylus, Scleroderma) became more plastic and established more intimate relations between the mother and her offspring. The primitive Vespoids and Sphecoids exhibit practically the same behaviour as the Bethyloids, but with the increase in the number of flowering plants during the Cretaceous, some of the groups (Apidæ, Masarinæ) became exclusively, others (during the adult stage) partially anthophilous. Most of the ants, arising also from Vespoid (Tiphiid or Scoliid) ancestors, are still entomophagous, but many of the genera have become increasingly vegetarian.

It is interesting to note that the changes of regimen which have occurred during the course of Aculeate evolution, appear first in the adult female and are later transferred to the larva. Thus among the Phytophaga, both adults and young live on relatively coarse vegetable food (leaves and wood), but the adults of some Tenthredinidæ have become entomophagous. In the Oryssidæ and Terebrantia this habit has also been acquired by the larva. Among the Vespoids and Sphecoids the adults have become more or less nectarivorous, whereas in bees and most Masarinæ the larvæ as well as the adults feed exclusively on nectar and pollen. The ants, as previously mentioned, show a similar tendency. Without doubt the eating of nectar, seeds, fungi and honeydew, which is, of course, merely plant juices slightly modified by passing through the intestines of Aphids, Coccids, etc., was originally confined to the adult ants. But some granivorous species also feed their larvæ on seeds, and the larvæ of all the mycetophagous species have the same diet as the adults. An interesting observation communicated to me by Mr. W. F. Fiske indicates that even the habit of feeding on the saccharine excrement of Homoptera may be in process of being transferred to the larva. Some years ago in British East Africa he observed ants carrying their larvæ out of the nest and up a tree trunk to the foliage where they held them with their mouths to the anal orifices of plant lice in order that they might imbibe the honeydew! As Mr. Fiske is a very competent entomologist, I have no doubt of the accuracy of his observation, though I regret that he failed to preserve specimens of the ants, which would have enabled us to identify the species and to reinvestigate their astonishing behaviour more readily.

Not the least interesting result of our evolutionary study is the conclusion that essentially the same type of social organization and behaviour has been independently attained by at least a dozen different groups of insects. Perhaps there is no more striking example in the animal kingdom of what is called "convergent", or "parallel" development. Moreover, the various manifestations of the societal type have been arrested in their phylogeny at very different stages and there is every reason to suppose that the arrests are partly due to a stabilization or standardization of the enviromental conditions and partly to a concomitant constitutional standardization or incapacity for further development on the part of the insects themselves. Even the highest and most specialized forms (ants, termites, honeybees) present the appearance of having long ago completed their evolution. The formation of new varieties and subspecies is probably still slowly proceeding, but these are feeble oscillations which do not involve significant deviations in the essential features of the social organization.

In former lectures I occasionally indulged in comparisons of the social insects with man and I should like to devote a considerable portion of this lecture to a more serious effort in the same direction. This is not, in my opinion, a futile undertaking, although it can have only a theoretical interest. Of course, innumerable comparisons have been made from all sorts of motives, both serious and satirical, and often on the basis of inadequate or erroneous conceptions of the peculiarities of both insect and human societies. On the whole, the results have been so unsatisfactory that the sociologists have turned from them with disdain. But although it must be granted that, owing to his very peculiar structure, endowment and behaviour, man can from the social insects learn nothing that might be applied with advantage to the solution of his intricate social problems, it must be remembered that he will never cease to be an animal and that his activities are broadly and irrevocably grounded in the nutritional, reproductive, relational and appetitive functions common to all other living organisms. Even if the sociologist, therefore, prefers to minimize the importance of animal societies in his own studies, the biologist, who must always regard man as a Primate, is perfectly justified in considering his societies as animal societies. And the sociologist has the less reason to object to this procedure because, according to his own admission, sociology is still a rudimental and speculative science.

(1) The question as to whether animal and human societies represent entities or wholes comparable with but of a higher order than the individual organism, which is known to be a colony of cells, has often been discussed, especially by Herbert Spencer, Espinas, Haeckel, Ribot, de Lilienfeld, De Greef, Fouillée, Novicow, Waxweiler, René Worms, Barth, and others. Recently the whole subject has been submitted to a searching examination by Ferrière (1915). Interest has, of course, centered in human society. The differences between the latter and animal societies are so great that many sociologists have abandoned the problem as merely verbal or academic. Humanity is really a very intricate association of groups held together in their coöperations and antagonisms by activities partly conscious and partly unconscious, without definite spatial boundaries but acting in time and of such a complicated and mobile character that each individual is simultaneously a functional member of several groups. Ferrière therefore declines to regard human society as an organism in the static sense, though he admits that it is organismal in a dynamic sense. He seems to obscure the problem, however, by neglecting to give a correct estimate of such societies as those of the insects. These are certainly discrete entities, like the single Metazoon or Metaphyte in possessing a definite boundary, stature, structure and ontogeny and consisting of polymorphic, mutually dependent elements. They may therefore be called superorganisms and constitute a very interesting intermediate stage between the solitary Metazoon and human society. Each insect society is a family, like a human family or the horde which probably constituted human society in its early postanthropoid stage. It might be possible therefore to regard the human society of the present as a fusion of societies—a hyper-hyper-organism, or hyperorganism of the second degree. The distinction emphasized by Fouillée and Ferrière, that the individuals in animal and human society retain their

mobility and are spatially isolated, whereas the histological elements of the organism are fixed and contiguous, is perhaps not as significant as they imagine. The blood-cells are also a part of the organism and the distances separating the tissue elements are relative. Even the cells, serum, lymph, etc., are masses of electrons separated by distances so enormous that if the human body could be compressed till all its electrons were in contact with one another, it would have a total bulk of only a few cubic millimeters. The interrelationships, or what happens between the electrons, atoms, molecules, micellæ, cells and individual organisms is the significant matter and it is on these fields between the elements that scientists are increasingly concentrating their attention.

(2) The distinctive characteristic of social life has been differently conceived by different sociologists: to Herbert Spencer it is "coöperation" among the members, to Tarde (1921) their "imitation" of one another, to Durkheim (1922) their "constraint", to De Grange (1923) their "accumulative activities". Undoubtedly all these characteristics, so conspicuous in human societies, can also be detected among wasps, bees, ants, and termites. Imitation and restraint, unless they be taken in the sense of biological dependence, are less obvious, but coöperation and especially accumulation, as shown in the hoarding of foods, the preëmpting of territory, i.e., the trophoporic field with its droves of Aphids, Coccids, etc., the increase of the brood and acquisition of myrmecophiles, termitophiles, and myrmecophytes, the construction of the nest, etc., are unmistakable. This very accumulation, following naturally as it does from the activities of numerous consociated individuals and leading to overproduction or overaccumulation, is of basic importance in all social life, as suggested by De Grange and also by Le Dantec (1918). According to the latter: "The only general formula applicable to all associations, no matter what they are, is the following: in order that the association may persist, every associate must derive from the association compensating advantages and even such as more than

make up for the inconveniences that result from the competition of specific appetites. At any rate, so far as associations of free individuals are concerned, it will be necessary to estimate the excess of individual *production* over individual consumption and not, as in the case of pluricellular individuals, the collaboration of each cell in the labour of the whole, which alone is capable of assuring the life of the agglomeration. It is mainly in this very precise peculiarity that associations of free individuals differ from agglomerations of fixed cells that have been reduced to the status of simple histological elements ".

Not only is it true, as De Grange says, that " the first condition of the operation of the societal process is the accumulation of material products ", but also that the overaccumulation, which naturally ensues, especially in human societies has consequences of enormous import in its reactions on the individual. The overproduction and overaccumulation of our civilized societies are so enormous, the products of our activities, the excreta of our social metabolism so abundant, not only in the form of material objects but also in the more intangible form of mores, rites, superstitions, fashions, laws, institutions, etc., that the generations as they succeed one another find living, to say nothing of progressing, increasingly difficult. Habituation to these conditions, moreover, has created in us such a reverence for the past and the old that though we call our governments monarchies, republics, aristocracies, or democracies, they might all be more properly called gerontocracies. There is, to be sure, a constant sloughing of the outworn, but it is a very slow process. From time to time, when society becomes thoroughly auto-intoxicated with its accumulations, only war or revolution can break the crust of social deposits and furnish an opportunity for further advance. Such unfortunate methods will probably remain our sole resource till we acquire sufficient knowledge and courage ruthlessly to " scrap ", as we say in America, a lot of our social inheritances, traditions and habits—till we decline to accord to the aged, whose only desire is to "sit tight", an undue preponderance in our affairs and turn them over to the young, who have at least a love of novelty and experimentation. To put it "crue, nue, verte et sans phrase", our political bodies, universities, academies, churches and other institutions contain far too many old incompetents, and as Goethe said to Eckermann, our state funerals do not succeed one another with sufficient frequency. And the conditions become the more intolerable, the more advancing medical science prolongs human life and increases the population in general and the number of old fools in particular.[1] How differently the problem is solved by the social insects! When the colony has reached the acme of its accumulations, it sends forth fresh young individuals to found new colonies, just as the mature Metazoan body gives off young germ cells, and is left behind to succumb with modest resignation and good cosmic manners. This method, of course, does not make for rapid and excessive phylogenetic accumulation, but it makes for greater stability over long periods of time, as we have seen in the evolution of the social insects from pretertiary to the present time, and also presents more frequently recurring opportunities for a certain number of new adaptations as they are required by the very gradually changing environment.

(3) Espinas long ago divided animal societies into nutritional and reproductive, the former represented by such forms as the Siphonophores, the latter by the social insects. The distinction remains valid to-day. In human societies the most distinctive activities are evidently psychical, but in each type of society, the higher embraces also the salient characteristics of the next lower, as in the old Aristotelian schema in which nutrition is given as characteristic of plants, nutrition and sensation of animals and nutrition, sensation and reason of man.

[1] This language is very shocking, but it would be easy to draw up a list of fifty well-meaning superannuated reactionaries, headed by William Jennings Bryan, who within recent years have done more harm to civilization than any equal number of criminals one might select.

There can be no doubt that the *raison d'être* of insect societies is reproduction. If they have any purpose in nature it is to produce as many young and as many colonies as possible. But what is so obviously a fact, need not be described in teleological language. It follows, of course, that the female sex is preëminent, and this is true even in termite societies as shown by the enormous somatic and gonadic development of the queens in the higher species. The Aculeate societies are frankly female, the male being reduced to a merely temporary, fecundative agency. Thus the insect colony, both in its ontogenetic and phylogenetic development depends on an élite of fertile females, a very different condition from what we find in human societies whose progress depends on an élite of intellectual individuals of both sexes. It is well known to the biologist that there is a kind of antagonism between sexual and intellectual activities and this also has been emphasized by certain historians (e.g., Towner, 1923) in human civilizations. Societies decay intellectually or rather approach the social insect type when they incline too much towards reproduction, a conclusion which should please the advocates of birth-control.

(4) The extraordinary deployment of reproduction in the insect society has led directly to the development of castes and polymorphism, through a few fertile individuals, the queens, specializing in the production of eggs, while the sexually poorly endowed majority, the workers, specialized in providing the necessary food and accommodations for the young. In prolific species these latter functions are so exacting that a further caste, the soldiers, may be developed to take over the defence of the society as a whole or to perform other duties which cannot be adequately performed by the foraging and nursing workers. These various differentiations are so ancient and have become so rigidly established, at least in the higher social insects (ants and termites), that the specific structure and behaviour of the individuals belonging to the different castes are definitively fixed on their emergence from the pupæ. Human caste development, though

likewise depending on a division of labour, is, of course, very different and of a purely professional origin. This whole subject has been so thoroughly studied by Durkheim (1922) that it need not be considered here. It will suffice to note that the normal human individual, though born into his society with peculiar hereditary endowments, is nevertheless so generalized and plastic that he can be converted through education into a more or less efficient member of any caste or even of several castes (professions). And, as we all know, after exercising his profession for many years, he may acquire the behaviouristic idiosyncracies or even the physical stigmata of his particular calling. Even the actor, whose profession consists in a neuromuscular mimicry of individuals belonging to a number of castes, eventually acquires a characteristic facial and postural expression.

(5) Since the castes are the result of a pronounced specialization of structure and behaviour, their members necessarily become partial or more or less abortive representatives of their species and therefore exhibit traits which have often been called "degenerative". This term, however, always has a pathological connotation. But in as much as the peculiarities under discussion have persisted for hundreds of thousands of years, not only without injury to the species but as necessary adaptations to a peculiar social medium, they are to be more properly regarded as specializations. That they may endanger the life of the organism when the social or environmental medium changes, is obvious, but this is true of all precise adaptations even when they are of a kind to which we never think of applying the terms "degenerative" or "pathological". The confusion arises from our ineradicable anthropomorphism and our constant occupation with "values", especially in everything that relates to life and death. Probably the best way out of the difficulty is to avoid such terms as "degenerative" and to employ instead such neutral descriptive expressions as "atrophic", "hypertrophic" "involutionary", "regressive", etc.

The facts relating to the specialization of characters in the insect castes may be profitably considered in greater detail. There are certainly unmistakeable signs of morphological involution, or regression in the individual insect as a result of adaptation to the social medium and they increase in number and degree as we pass from the more primitive to the more advanced societies. They include not only such traits as the greater simplification of the wing-venation and mouthparts in the social as compared with the solitary bees, the aptery, microphthalmy, simplification of the thorax, genitalia, etc. of the workers and soldiers as compared with the reproductive forms in ants and termites, as we pass from the primitive species with small to the highly specialized species with very populous colonies.[1] The internal organs also show increasingly regressive alterations during the adult life of the sexual forms, and especially of the queen, which represents the complete or perfect type of the species in nearly all social insects. The observations of Holmgren (1909) and von Rosen (1913*b*) on termites are particularly instructive in this connection. The former found that in old queens the brain shrinks to two-thirds of its original size and that this shrinkage, which occurs within a few days after the colony is established, is due largely to an atrophy of the ommatidia and optic nerves and lobes. But there are also concomitant hypertrophic changes in many organs. Thus the ganglia of the sympathetic nervous system increase in volume even in the king. The cross-section of the postcephalic ganglion of the young king and queen alike measure only 67μ, in the old king the dimensions increase to 135-162μ, in the old queen to 175-190μ. The heart, fat-body and

[1] According to Langhofer (1897), the mouthparts of the honeybee and humblebee are less developed than in such solitary bees as Trachusa and Anthidium, and Tosi found that the proventriculus of the honeybee is less developed than in Bombus. Friese (1923) points to the decided reduction of the wing-venation in the social bees. Among the Formicidæ the venation shows increasing simplification as we pass from the primitive Ponerinæ and Dorylinæ to the highly specialized Formicinæ. The same is true in the reduction of the sting, which is highly developed in the lower groups and vestigial in the Dolichoderinæ and Formicinæ, as it is also in the Meliponinæ among the social bees.

alimentary canal also become enlarged, especially in the old queen, the increase in the gut being due to the purely salivary diet. The respiratory organs, too, are changed, the spiracles becoming enlarged and the number of tracheal ramifications increased, probably in direct adaptation to confinement in the narrow royal chamber and the hypertrophy of the fat-body and other viscera. Von Rosen found a very similar atrophy of the adult visual apparatus in the neotenic royalty of *Calotermes flavicollis*. Probably changes of the same kind will be observed to occur in queen ants. The extraordinary involution of the voluminous thoracic musculature detected by Janet (1907) and C. Pérez (1912, 1920) in queen ants after their nuptial flight, has been noticed in a former lecture. Besides these and similar ontogenetic changes which occur in social insects that become increasingly adapted to particular functions during their later life, there are others that have become fixed in the phylogeny or during the larval stage in correlation with other traits. Thus in the queens of the honeybee and parasitic ants the brain, even during the pupal stage is less developed than that of the worker and in the honeybee there is also an atrophy of the mouth-parts, sting, spurs of the hind legs, salivary glands, etc. In some castes of ants, like the plerergates of the honey ants, the crop hypertrophies in the adult stage. But the worker honeybee, though structurally very near the original female type of the species and exhibiting little atrophy, except in the genitalia, nevertheless shows regressive behaviour in its dependence on the presence of other workers. According to von Buttel-Reepen (1905), she dies in a few hours, when isolated from the colony, although a worker ant may be kept alive in isolation for six months (Miss Fielde, 1902, p. 599). When the temperature sinks below a certain point the isolated honeybee cannot keep warm enough to remain alive. But even the ants differ from the honeybee only in the degree of their dependence on their sisters.

Holmgren finds that the termites as a whole have a

rather uniform bodily structure and that notwithstanding the evidence of the acquisition of morphological traits during phylogeny, the evidence of losses is more apparent. The lower termites contain everything *in nuce*, and although specializations have arisen at all stages of phylogenetic development, nothing really new has appeared but only modification or reductions of what was originally present. He therefore expresses his conclusions as follows: "From a consideration of communal life, we should expect theoretically that it would gradually lead to a degeneration of its participants, since natural selection should bring about not the higher differentiation of the individuals but only that of the colonies. Since, in all probability, the differentiation of the castes is a result of social life and the division of labour, in all likelihood, the immediate cause of caste differentiation, the degenerative effects of social life might be attributed to the ever increasing division of labour and caste differentiation. On the whole, therefore, we might postulate a regressive development in the termites". A comprehensive view of the Formicidæ leads to the same conclusion. The potentialities of the various caste developments in the highest subfamilies are all present in the primitive Ponerinæ, and if we compare e.g., a Pheidole, a Dorylus or a Camponotus worker with a worker of the Australian genus Myrmecia, we are compelled to admit that as an individual, the latter is a far more highly organized and efficient insect.

It is interesting to observe that effects analogous to those we have been considering in insects may also be detected in civilized man compared with the savage or barbarian. Compare for example, the independence and individual resourcefulness of a great general of the Homeric age, Achilles or Hector, who indulged in single combats, with the dependence of Foche and Pershing on their armies and the remainder of society during the recent war. No future Homer will describe these highly civilized gentlemen as challenging Hindenburg or Luedendorf to single combat in order to decide the outcome of a battle. The single large-eyed Myrmecia worker goes forth alone to attack other insects, whereas the blind Dorylus worker has to maintain constant olfactory and tactile communication with the marauding column or it is hopelessly impotent. Even between the various groups in civilized societies similar contrasts occur, e.g., between bank robbers and bank clerks. The amount of space given to the former as compared with the latter in the newspapers, especially in America, shows where our interests lie. We are not so intensely interested in bank robbers because they are bad, but because we cannot help admiring their bold resourcefulness and perhaps because we have a sneaking realization of our own abject dependence on the laws and public opinion.

Involutionary effects of social life on the human individual have long been noticed and at the present time are the subjects of much scientific and historical literature just as they were during the latter portion of the eighteenth century. The study of Le Bon (1925), Martin (1920), McDougall (1920), Freud (1921), and others of the behaviour of those temporary associations, known as crowds and mobs, has brought to light distinctly regressive manifestations in the psychological field. As Freud says: "An individual in a group is subjected, through its influence, to what is often a profound alteration of his mental activity. His emotions become extraordinarily intensified, while his intellectual ability becomes markedly reduced, both processes being evidently in the direction of an approximation to the other individuals in the group; and this result can only be reached by the removal of those inhibitions upon his instincts which are peculiar to each individual and by his resigning those expressions of his inclinations which are especially his own"[1]. Of course, the crowd displays in an acute and

[1] It is interesting to note that the higher social insects, especially wasps, honeybees and ants display phenomena similar to those of human mobs—high emotivity, suggestibility, rage, etc. These manifestations, moreover, occur only in populous colonies, since a mob implies a number of coöperating individuals; feeble or incipient colonies of bees, wasps and ants are always very timid.

exaggerated form the tendencies that are present in a chronic but less obvious form in the masses and groups of society as a whole. But there are more startling evidences of the involutionary effects of social life. The psychologists find that few of us attain an intellectual age of eighteen years and that most of us retain throughout life the mentality of children of fourteen years or less. Even this is too flattering, for the psychoanalysts call attention to the great difficulty which the best of us experience in preserving an adult or even an adolescent attitude towards reality and the extraordinary facility with which we lapse into infantilism. They also point to the appalling prevalence in our civilization of the neuroses and psychoses in which this babyish behaviour is chronic or habitual. And the sociologist proves statistically the great prevalence of suicide and its intimate dependence on civilization (Durkheim, 1922). Innumerable individuals destroy themselves because they are simply bored to death by society. These and many other considerations have led such different investigators as Dide and Juppont, Boas and Stärcke, not to mention historians like Spengler and Towner, to express equally pessimistic views of human capacities and civilizations. According to Dide and Juppont (1924) " a consideration calculated to excite human humility forces itself upon us. The subordinate forms of consciousness, closely connected with instinct, still remain vibrant with formative energy, while the cognitive functions, though of more recent origin, not only seem to have completed their evolution, but even to manifest a certain regression in man. And this agrees with the facts revealed by comparative cytogeny. Poorly differentiated cells are endowed with prodigious reproductive activity whereas highly differentiated cells, especially those of the cerebral cortex, have lost all their reproductive powers. One of us has long been making a special study of these phenomena and though he has detected a few lingering traces of direct division among the pyramidal nerve cells of rodents, has vainly sought for any similar tendency in man. Instead he has been able to note, in works still unpublished, the constancy of amitotic divisions in the cells of the sympathetic, which is the organ of the visceral functions. These facts, which are no longer open to discussion, leave little doubt concerning the future destinies of the superior biological races ".

According to Boas (1924), " when forming our judgment of the significance of racial differences, we must remember that the races of man may not be compared with wild forms of animal life, but that man is the oldest domesticated form. The use of artificially prepared food, which set in with the invention of fire during the glacial period, marks the beginning of the period of domestication ; it may indeed go back still further, to the beginning of the use of tools. This view was first expressed by Le Salles in 1849. Later Hahn called attention to it ; I dwelt on it in 1911, and in 1913 Eugen Fischer, and later on Klatt, have taken up the question from the standpoint of anatomical evidence. Blondness, blue eyes and a fair skin, as well as blackness and curly hair, are traits of domestication. So are the reduction of the size of the face and the increase in its length. Permanence of the female breast, anomalies in sexual behaviour and the lack of a mating season belong in the same category ". The psychoanalyst Stärcke (1921), impressed like so many others with the prevailing neurotic behaviour of our civilization, goes so far as to regard it as a peculiar disease which he calls " metaphrenia ", which is " imposed on a certain portion of society in order to obtain a certain extra gain whereby all profit ". And he concludes his chapter on metaphrenia with the following statement : " The sublimations and reaction-formations of the social human being follow the mechanisms of the obsessional neuroses (there is here an agreement of the moral, logical, and æsthetic compulsion with that of the neuroses). They also tend to the return of repressed material. We see the civilization of a people or a race built up in cycles according to the mechanisms of the obsessional neurosis, until it becomes no longer bearable ;

then there comes about a limitation of the useful effect through the return of the repressed material in disguised form, and a breaking through of forbidden things in war and revolution, according to the principles of manic psychoses, while various "isms" analogous to the paranoid fields are not lacking".

The notion of "domestication", which in another passage Stärcke would substitute for "sublimation", is also applicable to the social insects. But domestication is closely akin to parasitism. The difference between the enormous advantages, merely in food and protection, accruing to the individual as a member of a society, compared with its own feeble contributions, is so enormous as to be in many respects comparable with the difference between host and parasite. Consequently the regressive mental and physical traits of the individual resemble those of parasites, as we see in our domesticated animals. But domestication produces opposite effects in different organisms. In some it produces sterility, whereas in others it entails atrophy of the organs of relation (sensory and neuromuscular system) and hypertrophy of the gonads, alimentary canal and other viscera. In the social insects both effects may be observed, the former in the workers and to some degree in the soldiers, the latter in the queens. There are obvious analogies in human society. This ambivalent effect of domestication has been obscured, so far as domestic animals are concerned, by the fact that man could select and retain only fertile species, and so far as plants are concerned, by the fact that they can be propagated asexually.

(6) But notwithstanding all the involutionary traits that have been cited as characterizing the social individual, it must be admitted that societies as wholes are very powerful agents as compared with any solitary, non-social, organisms. The only dangerous rivals of social species, therefore, are other social species. They constitute, in fact, the most virulent forces in the fauna, since they are so destructive either as predators or as competitors, not only to the solitary species in their environment but to

one another. Hence we should expect the activities of social organisms, when unrestrained by climatic or other environmental factors, to lead eventually to the extermination of the solitary and the survival of a few very powerful and prolific social species. It is scarcely necessary to call attention to the behaviour of civilized man in this connection. Having exterminated nearly all the primitive human societies and nearly all the Mammalia, except the few he has been able to domesticate, he is now engaged in destroying much of the remainder of the life of the planet. And the social insects assume a similar rôle among the terrestrial Arthropoda, especially in those portions of the tropics where they have full sway and have not yet come in contact with civilized man. But before continuing this topic something must be said about the death of societies.

Even the most pessimistic thinkers (Spengler, etc.) do not believe that human societies die from the effects of their culture and internecine wars. They merely return after each great efflorescence to dormant, Fellahin conditions, like those of the Egyptians, Chinese and the peoples of the Dark Ages in Europe, for nothing short of a cosmic catastrophe can exterminate the human race. Insect societies are more like single organisms in that they die. Unfortunately our knowledge of the duration, decline and extinction of their perennial societies is not as precise as we could wish. We know, of course, the conditions in the wasps, humblebees and Halicti of temperate regions, where the colonies exist for only six to eight months and then disappear, leaving the young fertilized queens to create new colonies during the ensuing year. The life of an ant colony containing only a single fertilized queen, must end soon after her death. In some of our species she is known to live at least fifteen years and this, since the workers are short-lived (three to four years), would be about the length of the colony's life. But when fertilized daughters of the queen are adopted from year to year by the colony, it may live much longer—to an age of forty or fifty years or perhaps more, but in regard to this matter and the longevity of termite colonies, our knowledge is meagre. Since, however, the life of the species does not depend on the life of the single colony, another and more obscure problem is suggested by the question as to how the species of social insects become extinct. Their colonies, as we have seen, are so much more powerful and persistent than individual organisms and through their wide dispersal so difficult to destroy even by larger animals (birds, toads, ant-eaters, etc.) or by merely local or temporary climatic conditions, that the extermination of the species of social insects must be due to other causes. That it has occurred repeatedly during geological time is certain, and that it is still taking place is highly probable. In boreal and temperate regions, no doubt, the severest effects are produced on wasp and humblebee colonies by adverse climatic conditions. Concerning the causes of extinction among the Vespinæ, Apinæ and Meliponinæ of the tropics we know nothing, though their colonies are much fewer in number than those of the ants and termites. Their growth and decay are probably regulated by the available insect and plant foods. The Doryline ants are certainly formidable enemies of the wasps. The greatest enemies of the termites are the ants and of the ants other ants, and undoubtedly civilized man is the most implacable enemy of both of these groups of social insects. The abundance of species and survival of many primitive types in the portions of the East Indian, Australian and Neotropical regions that have remained untouched by civilized man shows that in these places a nearly perfect biocœnotic balance has been preserved for ages, but wherever he is settling the balance is being rapidly upset. The clearing of the land of indigenous vegetation and the introduction of foreign cultivated plants and domesticated animals is proving to be the most serious factor in the extermination of ants and termites. The effects of civilization are therefore in great measure indirect. This is apparent in North America, where all the more primitive ants, like the Ponerinæ, which for ages have been adjusted to very precise edaphic and other conditions, are rapidly disappearing. Only the more plastic and adaptable native Myrmicinæ and Formicinæ manage to survive in the settled areas and with the increase in the density of human population they, too, are tending to disappear. In the tropics an even more rapid extermination is due to certain extremely prolific and adaptable species of ants, which are able to live even on ships and have therefore been transported to all parts of the world where they quickly take possession of such areas as have a mean annual temperature like that of their original habitat. Two of these species, the Myrmicine *Pheidole megacephala* and the Dolichoderine *Iridomyrmex humilis*, may be briefly considered.

Pheidole megacephala is supposed to have originated in Madagascar, because it has the greatest number of varieties and allied species on that island. Within the past century it has been carried to all parts of the tropics and exterminates the native ant fauna wherever man has prepared the way for its conquests by destroying the native vegetation and introducing extensive cultures of foreign plants. The latter soon become infested with insects and especially Coccids, which are favourite wards of *megacephala*. Heer showed that by 1852 this ant had taken complete possession of the island of Madeira. It has also exterminated all or nearly all the ants on some of the West Indian Islands, notably Bermuda, the Bahamas, St. Thomas and Culebrita, near Porto Rico, and has made considerable progress in Cuba and Hawaii. In larger areas, especially on the continents, its progress is slower, because the same is true of its advance agent, civilized man. Several years ago I was able to witness its gradual invasion along the eastern coast of Australia where at widely separated points the British colonists have founded a number of towns. *Ph. megacephala*, brought by commerce to these settlements, is at first confined to their streets and gardens, completely replaces the native ant fauna, establishes its nests everywhere about the roots of the trees and pastures Coccids on their bark and foliage. On proceeding inland a short distance to a zone which is being cleared and brought under cultivation, one observes *megacephala* in active conflict with the native ants, entering their nests, carrying away their brood and assassinating the adults. I actually witnessed the destruction of a colony of the superb *Podomyrma femorata*, a large, rather stolid ant, which lives in burrows in the hard wood of standing trees. These ants were quite unable to defend themselves against the hordes of minute Pheidole workers. The colony was completely destroyed, the larvæ and adults killed and slowly dragged away. Proceeding still farther inland one comes upon the untouched, original forest, which, owing to the exquisitely balanced interrelationships of its faunal and floral components, *megacephala* is unable to penetrate, till man intervenes with his devastations.

The history of *Iridomyrmex humilis* (Fig. 79) is similar. It seems to have originated in Argentina and has therefore been called the Argentine ant. After entering New Orleans, it increased to enormous numbers and spread rapidly over the Southern States and California, everywhere exterminating the native ants. It crossed the Atlantic and became a pest in such widely separated countries as Cape Colony and Portugal. According to Stoll (1898), it landed also in Madeira and succeeded in supplanting *Pheidole megacephala*, which had held possession of the island since the days of Heer. More recently *humilis* has appeared in France and the Canary Islands (see Wheeler, 1927).[1] The fact that wherever this ant comes in contact with *megacephala*, the latter has to give way, is very interesting in connection with my remarks on the contrast between the vigour of the colony as contrasted with the feebleness of its members. *I. humilis* is a soft-bodied species, without a sting, whereas *megacephala* has a hard integuemnt, a sting and a large-headed soldier caste with powerful mandibles.

[1] In the second edition of his "British Ants" (1927), Donisthorpe records *I. humilis* as having been recently introduced into houses in various parts of Great Britain, as far north as Edinburgh and into Guernsey and Germany (Hamburg). He observed it at Palermo in Sicily and Kutter (1927) found it established in the vicinity of Naples.

Although the Argentine ant is therefore individually the weaker, it surpasses the Pheidole in the fecundity of its queens, the size of its colonies, its activity, enterprise and resourcefulness—all traits which make it as much superior to the Pheidole as the latter is superior to the other ants when introduced into a fauna that has been more or less weakened by the agency of man. We can foresee that as civilization advances man will exterminate the terrestrial fauna and flora, except the portions from which he can derive some benefit. But the most prolific and resourceful ants will remain as his annoyers and competitors and will probably be among the last insects to disappear. In his struggle with them he will probably succeed best through indirect methods, i.e., by modifying or abolishing the environmental conditions essential to their survival.

BIBLIOGRAPHY

ADLERZ (G.): **1884.** Myrmekologiska Studier. I. *Formicoxenus nitidulus* Nyl. Ofv. Kgl. Vet. Akad. Förh., 8, pp. 43-64, 2 pls. **1896.** Myrmekologiska Studier. III. *Tomognathus sublævis* Mayr. Bih. till K. Svensk. Vet. Akad. Handl. 21, pp. 3-76, 1 pl.

ALEXANDER (S.): **1920.** Space, Time and Deity, 2 vols. London, Macmillan & Co.

VON ALTEN (F.): **1910.** Zur Phylogenie des Hymenopterengehirns. Jen. Zeitschr. Naturw. 46, pp. 511-590, 4 pls., 28 figs.

ALTSON, (A. M.): **1920.** The Life-history and Habits of two Parasites of Blowflies. Proc. Zool. Soc. London, pp. 195-243, 20 figs.

ALVERDES (F.): **1925.** Tiersoziologie. Vol. I of R. Thurnwald's "Forschungen zur Völkerpsychologie und Soziologie." Leipzig, Hirschfeld. (English translation, *sub. tit.* Social Life in the Animal World. London: Kegan Paul & Co.; New York: Harcourt, Brace & Co., Inc.)

ANDRÉ (Ed.): **1881.** Spécies des Hyménoptères d'Europe et d'Algérie. Tome 2, Beaun.

ANDRÉ (Ern.): **1892.** Voyage de M. Chaper a Bornéo. Catalogue des Fourmis et Description des Espèces Nouvelles. Mém. Soc. Zool. France 5, pp. 46-55, 5 figs. **1895.** Notice sur les fossiles de l'ambre de la Baltique et déscription de deux espèces nouvelles. Bull. Soc. Ent. France 20, pp. 80-84. **1903.** Famille Mutillidæ, in Wytsman's "Genera Insectorum," pp. 1-77, 3 pls.

ARMBRUSTER (L.): **1914.** Probleme des Hummelstaates. Biol. Centralbl. 34, pp. 685-707, 1 pl., 1 fig. **1916.** Zur Phylogenie der Geschlechtsbestimmungsweise bei Bienen. Zool. Jahrb. Abt. Syst. 40, pp. 328-388, 1 pl., 6 figs.

ARNOLD (G.): **1916.** A Monograph of the Formicidæ of South Africa. Myrmicinæ. Ann. South Afr. Mus. 14, p. 252.

ASHMEAD (W. H.): **1890.** *Neolarra pruinosa* n. gen. n. sp. Colorado Biol.Ass., p. 8. **1893.** Monograph of the North American Proctotrypidæ. Bull. No. 45, U.S. Natl. Mus., Washington, 472 pp., 18 pls. **1896.** The Phylogeny of the Hymenoptera. Proc. Ent. Soc. Washington 3, pp. 323-336, 2 figs. **1902-1903.** Classification of the Fossorial, Predaceous and Parasitic Wasps, or the Sub-family Vespoidea. Paper No. 10, Canad. Ent., 1902, pp. 287-291; Paper No. 12, *ibid*, 1903, pp. 39-44.

ASSMUSS (E.): **1865.** Die Parasiten der Honigbiene und die durch dieselben bedingten Krankheiten dieses Insects. Berlin, E. Schotte & Co., 56 pp., 3 pls.

ASSMUTH (J.): **1913.** *Termitoxenia assmuthi* Wasm. Anatomisch-histologische Untersuchung. Nova Acta Leop. Carol. Akad. Naturf. Halle, 98, pp. 191-316, 11 pls.

AUDOUIN (J. V.): **1842.** Histoire Naturelle des Insectes Nuisibles. p. 189.

AURIVILLIUS (C.): **1896.** Ueber Zwischenformen zwischen socialen und solitären Bienen. Festschr. f. Wilh. Lilljeborg, Upsala, p. 69-77.

BAKER (C. F.): **1896.** On the Affinities of Neolarra. Proc. Ent. Soc. Washington, 4, p. 20-23, 1 fig.

BANKS (N.): **1904.** (On Social Spiders.) Proc. Ent. Soc., Washington 6, p. 148.

BANKS (N.) and SNYDER (T. E.): **1920.** A Revision of Nearctic Termites with Notes on Biology and Geographic Distribution. Bull. No. 108, U.S. Natl. Mus., 228 pp., 35 pls., 70 figs.

BARBOUR (T.): **1926.** Reptiles and Amphibians, their Habits and Adaptations. Boston, Houghton, Mifflin Co.

BECHER (E.): **1917.** Die fremddienliche Zweckmässigkeit der Pflanzengallen und die Hypothese eines überindividuellen Seelischen. Leipzig, 149 pp.

BELT (T.): **1874.** The Naturalist in Nicaragua. London, Bumpus.

BENNETT (C. B.): **1904.** Earwigs (*Anisolabis maritima*). Psyche 11, p. 47-53, 2 figs.

BEQUAERT (J.): **1916.** On the Occurrence of *Vespa austriaca* Panzer in the Northeastern United States. Bull. Brooklyn Ent. Soc. 11, p. 101-107. **1918.** A Revision of the Vespidæ of the Belgian Congo, based on the Collection of the American Museum Congo Expedition, with a List of the Ethiopian Diplopterous Wasps. Bull. Amer. Mus. Nat. Hist. 39, p. 1-384, 6 pls. 267 figs. **1922.** (On Bengalia) In Wheeler, "Ants of the Belgian Congo", *Ibid.* 45, p. 281. **1924.** Galls that Secrete Honeydew. A contribution to the Problem as to whether Galls are Altruistic Adaptations. Bull. Brooklyn Ent. Soc. 19, p. 101-124.

BERGER: **1878.** Untersuchungen uber den Bau des Gehirns und der Retina der Arthropoden. Arb. Zool. Inst. Wien, 1.

BERLESE (A.): **1909.** Gli Insetti, loro organizazione, sviluppo abitudini e rapporti coll'uomo. Vol. I., Milano, Soc. Edit. Libr.

BERNARD (L. L.): **1924.** Instinct, a Study in Social Psychology. N.Y., Henry Holt.

BERRY (E. W.): **1927.** The Baltic Amber Deposits. Scient. Month., p. 268-278, 3 figs.

BETHE (A.).: **1898.** Dürfen wir den Ameisen und Bienen psychische Qualitäten zuschreiben. Arch. gesamt. Physiol. 70, p. 15-100, 2 pls., 5 figs. **1900.** Noch einmal über die psychischen Qualitäten der Ameisen. *Ibid.*, 79, p. 39-52. **1902.** Die Heimkehrfähigkeiten der Ameisen und Bienen, zum Teil nach neuen Versuchen. Biol. Centralbl. 22, p. 193-215, 234-238.

BIEDERMANN (W.): **1911.** Die Aufnahme, Verarbeitung und Assimilation der Nahrung. In Winterstein's "Handbuch der vergleichenden Physiologie", 2, Jena, G. Fischer.

BISCHOFF (H.): **1909.** Neue Beiträge zur Lebensweise der Trigonaliden. Berlin. Ent. Zeitschr. 54, p. 76-80, 1 fig.

BLANCHARD (R.): **1889.** Pseudoparasites. Art. Extr. Diction. Encycl. des Sciences Médicales (2) 27, p. 704-709.

Bönner (W.): **1915.** Der temporäre soziale Hyperparasitismus von *Lasius fuliginosus* und seine Beziehungen zu *Claviger longicornis* Müll. Zeitschr. wiss. Insektenbiol. 11. p. 14-20.

Borgmeier (T.). **1923.** Beitrag zur Biologie der Feuerameise und ihrer Gäste (*Solenopsis geminata sævissima* Sm.) Zeitschr. deutsch. Ver. Wiss. Kunst São Paulo, 3, p. 1-19, 1 fig.

Börner (C.): **1919.** Stammesgeschichte der Hautflügler. Biol. Zentralbl. 39, p. 145-186, 6 figs.

Böttger (O.): **1910.** Das Gehirn eines niederen Insektes *Lepisma saccharina.* Jen. Zeitschr. Naturw. 46, pp. 801-844, 2 pls., 6 figs.

Bouvier (E. L.): **1918.** La Vie Psychique des Insectes. Paris, Flammarion. **1926.** Le Communisme chez les Insects. Paris, Flammarion.

Boveri (T.): **1915.** Ueber die Entstehung der Eugsterschen Zwitterbienen. Arch. Entwick. Mech. Org. 41, p. 264-311, 2 pls.

Bradley (J. C.): **1912.** Some Features of the Hymenopterous Fauna of South America. Actes Soc. Scient. Chili 30 (1920), p. 51-74. **1922.** The Taxonomy of the Masarid Wasps, including a Monograph on the North American Species. Univ. California Publ. Tech. Bull. Vol. 1, p. 369-464, 15 pls.

Branner (J. C.): **1900.** Ants as Geologic Agents in the Tropics. Journ. Geol. Chicago 8, p. 151-153, 3 figs.

Brauer (F.): **1869.** Beschreibung der Verwandlungsgeschichte der *Mantispa styriaca* Poda, und Betrachtungen über die sogenannte Hypermetamorphose Fabre's. Verh. zool. bot. Ges. Wien 19 p. 831-840, 1 pl. **1883a.** Zwei Parasiten des *Rhizotrogus solstitialis* aus der Ordnung der Dipteren. Sitzb. k. Akad. Wiss. Wien, 88, p. 865-877, 2 pls. **1883b.** Ergänzende Bemerkungen zu A. Handlirsch's Mitteilungen uber *Hirmoneura obscura* Meig. Wien. Ent. Zeitg. 2, p. 26.

Brauns (H.): **1902.** Eucondylops n.g. Apidarum. Zeitschr. System. Hym. Dipt. 2, p. 377-380. **1910-1911.** Biologisches über südafrikanische Hymenopteren. Zeitschr. wiss. Insektenbiol. 6, 1910, p. 384-387, 445-447; 7, 1911, p. 16-19, 90-92, 117-120, 238-240. **1926.** A Contribution to the Knowledge of the Genus Allodape, St. Farg. et Serv. Ann. South Afric., Mus. 23, p. 417-434, 2 pls., 2 figs.

Breitenbach (W.): **1878.** Ueber *Halictus 4-cinctus* F. und *Sphecodes gibbus* L. Stettin. Ent. Zeitg., p. 241-243.

Bretschneider (F.): **1915.** Neuere Untersuchungen über das Gehirn der Insekten. Naturw. Wochenschr. n. F. 14, p. 17-24, 17 figs.

Bridwell (J. C.): **1917a.** Notes on a Peregrine Bethylid. Proc. Hawaiian Ent. Soc. (1916), 3, p. 276-279. **1917b.** A Note on an Epyris and its Prey. *Ibid.* (1916) 3, p. 262-263. **1917c.** Notes on Bruchidæ and their Parasites in the Hawaiian Islands. *Ibid.* 3, p. 465-500. **1919.** Some Notes on Hawaiian and other Bethylidæ, with Descriptions of New Species. *Ibid.* (1918) 4, p. 21-38. **1920.** Some Notes on Hawaiian and other Bethylidæ, with the Description of a New Genus (Second paper). *Ibid.* (1919) 4, p. 291-314.

Browne (F. B.): **1922.** On the Life History of *Melitobia acasta* Walk., a Chalcid Parasite of Bees and Wasps. Parasitology, 14.

Bruch (C.): **1917a.** Observaciones sobre *Hirmoneura exotica* Wiedem. Physis 3, p. 427-430, 2 figs. **1917b.** Nuevas capturas de Insectos Mirmecofilos. *Ibid.* 3, p. 458-465, 3 figs. **1926.** Orugas mirmecofilos de *Hamearis epulus signatus* Stichel. Revist. Soc. Ent. Argentina 1, p. 1-8, 3 pls. 2 figs.

Brues (C. T.): **1906.** Fossil Parasitic and Phytophagous Hymenoptera from Florissant. Bull. Amer. Mus. Nat. Hist. 22, p. 491-498, 7 figs. **1908a.** New Phytophagous Hymenoptera from the Tertiary of Florissant, Colorado. Bull. Mus. Comp. Zool. 51, p. 259-276, 10 figs. **1908b.** Some Stages in the Embryology of Certain Degenerate Phoridæ and the Supposed hermaphroditic genus Termitoxenia. Science n.s. 27, p.942. **1908c.** The Correlation between Habits and Structural Characters among Parasitic Hymenoptera. Journ. Econom. Ent. 1, p. 123-128. **1910a.** The Parasitic Hymenoptera of the Tertiary of Florissant, Colorado. Bull. Mus. Comp. Zool. 54, pp. 3-125, 1 pl., 88 figs. **1910b.** Some Notes on the Geological History of the Parasitic Hymenoptera. Journ. N. Y. Ent. Soc. 18, pp. 1-22. **1919a.** A New Chalcid Fly Parasitic on the Australian Bull-Dog Ant. Ann. Ent. Soc. Amer. 12, pp. 13-21, 2 pls. **1919b.** The Classification of Insects on the Characters of the Larva and Pupa. Biol. Bull. 37, pp. 1-21. **1921.** Correlation of Taxonomic Affinities with Food Habits in Hymenoptera, with Special Reference to Parasitism. Amer. Natural. 55, pp. 134-164. **1922.** On the Hymenopterous Genus Harpagocryptus and its Allies. Psyche 29, pp. 101-109. **123a.** Termitobracon, a Termitophilous Braconid from British Guiana. Zoologica 3, p. 427-432, 2 figs. **1923b.** Some New Fossil Parasitic Hymenoptera from Baltic Amber. Proc. Amer. Acad. Arts. Sci. 58, pp. 327-346. **1923c.** A Fossil Genus of Dinapsinæ from Baltic Amber. Psyche 30, pp. 31-35, 1 fig. **1924a.** Triungulins of a Meloid Beetle borne by Xylocopa, with Remarks on this Type of Larva in the Coleoptera and Strepsiptera. Zoologica 5, pp. 125-136, 5 figs. **1924b.** Some South African Parasitic Hymenoptera of the Families Evaniidæ, Braconidæ, Alysiidæ and Plumariidæ, in the South African Museum, with a Catalogue of the Known Species. Ann. South Afric. Mus. 19, pp. 1-150.

Brues (C. T.) and Melander (A. L.): **1915.** Key to the Families of North American Insects. Boston and Pullman, Wash.

Brun (R.): **1910.** Zur Biologie und Psychologie von *Formica rufa* und anderen Ameisen. Biol. Centralbl. 30, pp. 524-545. **1912a.** Zur Psychologie der künstlichen Allianzkolonien bei den Ameisen *Ibid.* 32, pp. 308-322. **1912b.** Weitere Beiträge zur Frage der Koloniegründung bei den Ameisen mit besonderer Berücksichtigung der Phylogenese des socialen Parasitismus und der Dulosis bei Formica. *Ibid.* 32, pp. 154-180, 216-226. **1913a.** Zur Biologie von *Formica rufa* und *Camponotus herculeanus* i. sp. Zeitschr. wiss. Insektenbiol. 9, pp. 15-19. **1913b.** Ueber die Ursachen der künstlichen Allianzen bei den Ameisen. Journ. Psych. Neurol. 20, pp. 171-181. **1917.** Die moderne Ameisenpsychologie—ein anthromorphischer Irrtum. Biol. Centralbl. 37, pp. 357-372. **1923a.** Vergleichende Untersuchungen über Insektengehirne, mit besonderer Berücksichtigung der pilzförmigen Körper (Corpora pedunculata Dujardini). Schweiz.

Arch. Neurol. Psychiatr. 13, pp. 144-172, 5 figs. **1923b.** Selektionstheorie und Lustprinzip. Internat. Zeitschr. Psychoanal. 9, pp. 183-200. **1924.** Das Leben der Ameisen. Teubner's Naturw. Biblioth. Bd. 31. Leipzig-Berlin, Teubner.

BRUNELLI (G.) : **1904.** Collezionismo e Ibernazione nell' Origine degli Istinti delle Api Solitarie e Sociali. Rivist. Ital. Sc. Nat. 24, pp. 3-7.

BUCKINGHAM (E. N.) : **1911.** Division of Labour among Ants. Proc. Amer. Acad. Arts Sci. 46, pp. 425-507, 1 pl., 19 figs., 8 tabl.

BUGNION (E.) : **1909.** Le Termite noir de Ceylon (*Eutermes monoceros* Koen.) Ann. Soc. Ent. France 78, pp. 271-281, 3 pls.; Arch. Sc. Phys. Nat. Genève, 1909. **1910.** La Structure Anatomique du Trigonalys Hahni Spin. Mitt. Schweiz. Ent. Ges. 12, pp. 14-20, 4 pls. **1912.** Observations sur les Termites. Différentiation des Castes. C. R. Séances Soc. Biol. 72, pp. 1091-1094, 2 figs. **1913a.** Termitoxenia, Etude Anatomo-histologique. Ann. Soc. Ent. Belg. 57, pp. 23-44, 3 pls. **1913b.** La Différentiation des Castes chez les Termites. Bull. Soc. Ent. France, pp. 213-218. **1914.** La Biologie des Termites de Ceylan. Bull. Mus. d'Hist. Nat., pp. 170-204, 8 pls. **1923.** La Guerre des Fourmis et des Termites, la Genèse des Instincts expliquée par cette Guerre. In Forel's "Le Monde Social des Fourmis", III., Appendix, pp. 173-225, 8 pls.

BUGNION (E.) and POPOFF (N.) : **1910.** Le Termite à Latex de Ceylan, *Coptotermes travians* Haviland. Mém. Soc. Zool., France, pp. 107-123, 2 pls., 1 fig.

BURKE (H. E.) : **1917.** Oryssus is Parasitic. Proc. Ent. Soc., Washington 19, pp. 87-89.

BUSCALIONI (L.) and COMES (S.) : **1910.** La digestione delle membrane vegetali per opera dei Flagellati contenuti nello intestino dei Termitide e il problema della Simbiosi. Att. Accad. Gioenia Sc. Nat. (5) 3, 16 pp., 4 figs.

VON BUTTEL-REEPEN (H.) : **1903.** Die stammesgeschichtliche Entstehung des Bienenstaates, sowie Beiträge zur Lebensweise der solitären und sozialen Bienen (Hummeln, Meliponen, etc.) Leipzig, G. Theime, xii. + 138 pp., 20 figs. **1905.** Soziologisches und Biologisches vom Ameisen–und Bienenstaat. Wie entsteht eine Ameisenkolonie. Arch. Rassen Ges. Biol. 2, pp. 1-16, 1 fig. **1906.** Apistica, Beiträge zur Systematik, Biologie, etc. der Honigbiene (*Apis mellifica* L.), ihrer Varietäten und der übrigen Apis Arten. Mitteil. Zool. Mus., Berlin 3, pp. 121-201, 8 figs. **1911.** Atavistische Erscheinungen im Bienenstaat (*Apis mellifica* L.). Müssen wir dem Bienenei zwei oder drei Keimesanlagen zuschreiben ? Entdeckung der "Sporen" (Calcaria) bei der Honigbiene, 1er Congr. Internat. d'Ent. Bruxelles, pp. 113-132, 3 figs., 1 pl. **1915.** Leben und Wesen der Bienen. Braunschweig, Viehweg u. Sohn.

DU BUYSSON (R.) : **1903-1905.** Monographie des Guêpes ou Vespa. Ann. Soc. Ent. France, 72, pp. 260; 74, pp. 485.

CAFFREY (D. J.) : **1921.** Biology and Economic Importance of *Anastatus semiflavidus*, a recently described Egg Parasite of *Hemileuca oliviæ*. Journ. Agric. Research 21, pp. 373-384, 1 pl., 3 figs.

CARPENTER (G. D. H.): 1919. The Fly *Bengalia depressa* Walk., attacking a wingless Termite. Proc. Ent. Soc., London, p. xlviii.

CARPENTER (G. H.) and PACK-BERESFORD (D. R.): 1903. The Relationship of *Vespa austriaca* to *Vespa rufa*. Ent. Month. Mag. (2) 14, pp. 230-244, 1 pl.

CAUDELL (A. N.): 1923. Grylloblatta in California. Canad. Ent. 55, pp. 148-150. 1924. Notes on Grylloblatta with Description of a New Species. Proc. Washington Acad. Sci. 14, pp. 369-371.

CAUDELL (A. N.) and KING (J. L.): 1924. A New Genus and species of the Notopterous Family Grylloblattidæ from Japan. Proc. Ent. Soc. Washington 26, pp. 53-60, 1 pl.

CAULLERY (M.): 1922. Le Parasitisme et la Symbiose. Paris, Doin.

CHAMPY (C.): 1924. Les Caractères Sexuels considérés comme phénomènes de développment et dans leurs rapports avec l'Hormone sexuelle. Paris, Doin.

CHAPMAN (T. A.): 1890. Some Facts towards a Life-history of *Rhipiphorus paradoxus*. Ann. Mag. Nat. Hist. 6, pp. 314-326, 1 pl. 1878. On the Economy, etc., of Bombylius. Ent. Month. Mag. 14, pp. 196. 1902. On the Larva of *Liphyra brassolis* Westw. Entomologist 35, pp. 225-228, 252-255, 4 pls. 1915a. What the Larva of *Lycæna arion* does during its Last Instar. Trans. Ent. Soc. London, pp. 291-297, 9 pls. 1915b. Observations Completing an Outline of the Life History of *Lycæna arion* L. *Ibid.*, pp. 298-312, 3 pls.

CHILD (C. M.): 1924. Physiological Foundations of Behaviour. New York, Henry Holt & Co.

CHOBAUT (A.): 1891. Mœurs et Métamorphoses de *Emenadia flabellata* F., Insecte Coleoptère de la Famille des Rhipiphorides. Ann. Soc. Ent. France 60, 1891. pp. 647-456, 6 figs.

CLAUSEN (C. P.): 1923. The Biology of *Schizaspidia tenuicornis* Ashm., a Eucharid Parasite of Camponotus. Ann. Ent. Soc. Amer. 16, pp. 195-217, 2 pls.

CLAUSEN (C. P.), KING (J. L.) and TERANISHI (C.): 1927. The Parasites of *Popillia japonica* in Japan and Chosen (Korea) and their Introduction into the United States. Bull. No. 1429, U.S. Dept. Agric. 55 pp., 1 pl., 35 figs.

CLEVELAND (L. R.): 1923a. Symbiosis between Termites and their Intestinal Protozoa. Proc. Natl. Acad. Sci. 9, pp. 424-428. 1923b. Correlation between the Food and Morphology of Termites and the Presence of Intestinal Protozoa. Amer. Journ. Hygiene 3, pp. 444-461. 1924. The Physiological and Symbiotic Relationship between the Intestinal Protozoa of Termites and their Host, with special reference to *Reticulitermes flavipes* Kollar. Biol. Bull. 46, pp. 177-225. 1925a. The Method by which *Trichonympha campanula*, a Protozoan in the Intestine of Termites, ingests solid particles of wood for food. *Ibid.* 48, pp. 282-288, 1 pl., 1 fig. 1925b. The Ability of Termites to live perhaps indefinitely on a Diet of pure Cellulose. *Ibid.* 48, pp. 289-293. 1925c. The Feeding Habit of Termite Castes and its Relation to their Intestinal Flagellates. *Ibid.* 48, pp. 295-308, 1 pl. 1925d. The Effects of Oxygenation and Starvation on the Symbiosis between the Termite, Termopsis, and its

Intestinal Flagellates. *Ibid.* 48, pp. 309-326. **1925e.** Toxicity of Oxygen for Protozoa in Vivo and in Vitro : Animals defaunated without Injury. *Ibid.* 48, pp. 455-468.

COCKERELL (T. D. A.) : **1906a.** A New Fossil Ant. Ent. News, pp. 27-28. **1906b.** Fossil Hymenoptera from Florissant, Colorado. Bull. Mus. Comp. Zool. 50, pp. 33-58 **1909a.** Descriptions of Hymenoptera from Baltic Amber. Schrift. Physik. ökonom. Ges. Königsberg i. Pr. 50, pp. 1-25, 14 figs. **1909b.** Some European Fossil Bees. Entomologist, pp. 313-317. **1913.** The Genera Paratermes and Hodotermes. Ent. News, 24, pp. 6-8. **1915.** British Fossil Insects. Proc. U.S. Natl. Mus. 49, pp. 469-499, 6 pls. **1916a.** Some American Fossil Insects. Proc. U.S. Natl. Mus. 51, pp. 89-106, 1 pl. **1916b.** Insects in Burmese Amber. Amer. Journ. Sci. 42, pp. 135-138, 4 figs. **1917.** Descriptions of Fossil Insects. Proc. Biol. Soc. Washington 30, pp. 79-81, 4 figs. **1920a.** Fossil Arthropods in the British Museum I., Ann. Mag. Nat. Hist. (9) 5, pp. 273-279. **1920b.** Eocene Insects from the Rocky Mountains. Proc. U.S. Natl. Mus. 57, pp. 233-260, 5 pls. **1921a.** Some Eocene Insects from Colorado and Wyoming. *Ibid.* 59, pp. 29-39, 1 pl., 9 figs. **1921b.** Fossil Arthropods in the British Museum V. Oligocene Hymenoptera from the Isle of Wight. Ann. Mag. Nat. Hist. (9) 7, pp. 1-25, 32 figs. **1923a.** Fossil Insects from the Eocene of Texas. Amer. Journ. Sci. 5, 1923, pp. 397-400, 1 fig. **1923b.** The Earliest Known Ponerine Ant. Entomologist 56, pp. 51-52, 1 fig.

COLLIN (J. E.) : **1922.** Description of a New Genus and two new species of Cecidomyidæ, and six new species of Acalyptrate Muscidæ (Ephydridæ and Milichiidæ). Trans. Ent. Soc. London, pp. 504-517, 4 pls.

COLLINS (R. L.) : **1925.** A Lower Eocene Termite from Tennessee. Amer. Journ. Sci. 9, pp. 406-410, 7 figs.

COMSTOCK (J. H.) : **1912.** The Spider Book. Garden City, N.Y. Doubleday Page & Co. **1924.** An Introduction to Entomology Ithaca, N.Y., Comstock Publ. Co., xix.+1044 pp., 1228 figs.

CRAGG (F. W.) : **1917-18.** The Mouthparts of *Ochromyia jejuna*, a Predacious Muscid. Ind. Journ. Med. Research 5, pp. 516, 2 pl.

CRAIG (W.). : **1918.** Appetites and Aversions as Constituents of Instincts. Biol. Bull. 34, pp. 91-107.

CRAMPTON (G. C.) : **1915.** The Thoracic Sclerites and the Systematic Position of *Grylloblatta campodeiformis* Walker, a Remarkable Annectant " Orthopteroid " Insect. Ent. News 26, pp. 337-350, 1 pl. **1920.** The Terminal Abdominal Structures of the Primitive Australian Termite *Mastotermes darwiniensis* Froggatt. Trans. Ent. Soc. London, pp. 137-145.

CRAWLEY (W. C.) : **1909.** Queens of *Lasius umbratus* Nyl. accepted by Colonies of *Lasius niger* L. Ent. Month. Mag .(2) 20, pp. 94-99. **1910a.** Summary of Experiments with Fertile Females of Several Species of Ants. Ent. Rec. 22, pp. 152-156. **1910b.** How Ants greet Members of the same Colony. Ent. Rec. 22, pp. 43-44. **1912.** Parthenogenesis in Worker Ants, with Special Reference to two Colonies of *Lasius niger*, Linn. Trans. Ent. Soc. London, pp. 657-663.

CRAWLEY (W. C.): and BAYLIS (H.A.): 1921. Mermis parasitic on Ants of the Genus Lasius. Journ. Roy. Micr. Soc., pp. 353-372, 12 figs.

CRAWLEY (W. C.) and DONISTHORPE (H.): 1913. The Founding of Colonies by Queen Ants. Internat. Ent. Congr. Oxford (1912) 2, pp. 5-55.

CREIGHTON (W. S.): 1927. The Slave Raids of *Harpagoxenus americanus*. Psyche 34, pp. 11-29, 2 figs.

CUÉNOT (L.): 1911. La Genèse des Espèces Animales. Paris, F. Alcan.

CURTIS (J.): 1845. Description of the Nests of two Hymenopterous Insects inhabiting Brazil, and of the Species by which they were constructed. Trans. Linn. Soc. 19, pp. 249-259, 1 pl.

CUSHMAN (R. A.): 1913. The Calliephialtes Parasite of the Codling Moth. Journ. Agric. Research I., pp. 211-237, 1 pl., 15 figs. 1917. Notes on the Biology of *Schizonotus sieboldii* Ratz. Proc. Ent. Soc. Washington 19, pp. 128-129, 2 pl. 1923. A New Subfamily of Braconidæ from Termite Nests. *Ibid.* 25, pp. 54-55, 1 pl.

CUTLER (W.): 1921. Observations on the Protozoa parasitic in *Archotermopsis Wroughtoni* D. Part III., *Pseudotrichonympha pristina*. Quart. Journ. Micr. Sci. 65, pp. 247-264, 1 pl., 8 figs.

DALY (R. A.): 1926. Our Mobile Earth. N.Y. and London, Chas. Scribner.

DAVIDSON (A.): 1913. " Masaria Vespoides ". Bull. South. Cala. Acad. Sci. 12, pp. 17.

DAVIS (J. J.): 1919. Contributions to a Knowledge of the Natural Enemies of Phyllophaga. Bull. No. 13. Nat. Hist. Surv. Illinois, pp. 53-138, 13 pls., 46 figs.

DEEGENER (P.): 1912. Sinnesorgane, in Schröder's " Handbuch der Entomologie ", Bd. I., pp. 142-158. 1918. Die Formen der Vergesellschaftung im Tierreiche. Leipzig. Veit & Co.

DE GRANGE (McQ): 1923. La Courbe du Mouvement Sociétal Paris, Libr. Bibl. Aug. Comte.

DEMOLL (R.): 1917. Die Sinnesorgane der Arthropoden, ihr Bau und ihre Funktion. Braunschweig.

DESCY (A.): 1924. Recherches sur la Sexualité et l'Instinct chez les Hyménoptères. Bull. Biol. France Belg. 58, pp. 1-37.

DESNEUX (J.): 1904. A propos de la Phylogénie des Termitides. Ann. Soc. Ent. Belg. 48, pp. 278-286, 2 figs.

DIDE (M.) and JUPPONT (P.): 1924. La Métaphysique Scientifique. Paris, F. Alcan.

DIETL (M. J.): 1876. Die Organisation des Arthropodengehirns Zeitschr. wiss. Zool. 27, pp. 489-517, 3 pls.

DIGUET (L.): 1909a. Sur l'Araignée Mosquero, C. R. Acad. Sci., Paris 148, pp. 735-736. 1909b. Le Mosquero. Bull. Soc. Nat. Acclim. Paris 56. pp. 368-375.

DIMMOCK (G.): **1886.** Sphærularia in America. Amer. Natural. 20, pp. 73-75.

DISTANT (W. L.): **1898.** Zoological Rambles in the Transvaal. Zoologist (4) 2, pp. 249-260, 1 pl.

DODD (F. P.): **1902a.** Notes on the Queensland Green Tree Ants *Oecophylla smaragdina* Fabr. ?) Victorian Naturalist 18, pp. 136-140. **1902b.** Contribution to the Life History of *Liphyra brassolis* Westw. Entomologist 35, pp. 153-156. **1912.** Some Remarkable Ant-friend Lepidoptera of Queensland. Trans. Ent. Soc. London 10, pp. 577-590, 1 pl.

DOFLEIN (F.): **1920.** Mazedonische Ameisen, Jena.

DÖNHOFF (E.): **1882.** Beiträge zur Physiologie. Arch. Anat. Physiol. Abt. Physiol., pp. 157-163.

DONISTHORPE (H. S. J. K.): **1897.** Myrmecophilous Coleoptera in 1897. Ent. Rec. 9, p. 246. **1910.** Some Experiments with Ants' Nests. Trans. Ent. Soc. London, pp. 142-150. **1911.** Further Observations on Temporary Social Parasitism and Slavery in Ants. *Ibid.*, pp. 175-183. **1912.** Myrmecophilous Notes for 1911. Ent. Rec. 24, pp. 4-10, 34-40. **1915-1927.** British Ants, their Life History and Classification. Second Edition. London, George Routledge & Sons, Ltd., 1927. **1920.** British Oligocene Ants. Ann. Mag. Nat. Hist. (9) 6, pp. 81-94, 1 pl. **1921a.** Myrmecophilous Notes for 1920. Ent. Rec. 33, pp. 21-25. **1921b.** The Subfamilies of Formicidæ. Proc. Ent. Soc. London, pp. xl-xlvi., 5 figs. **1922.** The Colony Founding of *Acanthomyops* (*Dendrolasius*) *fuliginosus* Latr. Biol. Bull. 42, pp. 173-184. **1927.** Guests of British Ants, their Habits and Life Histories. London, George Routledge & Sons, Ltd., 1927.

DOTEN (S.B.): **1911.** Concerning the Relation of Food and Reproductive Activity and Longevity in Certain Hymenopterous Parasites. Techn. Bull. No. 78, Agric. Exper. Stat. Univ. Nevada, pp. 7-30, 10 pls.

DREVER (J.): **1917.** Instinct in Man, a Contribution to the Psychology of Education. Cambridge Univ. Press.

DREYLING (L.): **1905.** Die wachsbereitenden Organe bei den geselliglebenden Bienen. Zool. Jahrb. Abt. Anat. 22, pp. 289-330, 2 pls., 1 fig.

DUCKE (A.): **1902.** As Especias Paraenses do Genero Euglossa. Bol. Mus. Parà. 3, pp. 561-577, 1 pl. **1903.** Biologische Notizen über einige südamerikanische Hymenoptera. Allgem. Zeitschr. Ent. 8, pp. 367-372, 5 figs. **1905.** Biologische Notizen über einige südamerikanische Hymenoptera. Zeitschr. wiss. Insektenbiol. 1, pp. 175-177. **1910.** Révision des Guêpes Sociales Polygames d'Amérique. Ann. Mus. National. Hungarici 8, pp. 449-544, 17 figs. **1914.** Ueber Phylogenie und Klassifikation der sozialen Vespiden. Zool. Jahrb. Abt. Syst. 36, pp. 303-330, 17 figs., 1 table. **1916.** Enumeraçao do Hymenopteros Collegidas pela Commissão. Revisão das Especias de Abelhas do Brasil. Com. Lin. Telegraph. Estrateg. Matto Grosso ao Amazonas., No. 35, 175 pp., 7 pls.

DUFOUR (L.): 1837. Recherches sur quelques Entozoaires et larves parasites des Insectes Orthoptères et Hyménoptères. Ann. Sc. Nat. Zool. (2) 7, pp. 5-20, 1 pl.

DUNBAR (C. O.): **1924.** Kansas Permian Insects. Part I. The Geologic Occurrence and the Environment of the Insects. Amer. Journ. Sci. (5) 7, pp. 171-209, 1 pl., 4 figs.

DUNN (E. R.): **1926.** The Frogs of Jamaica. Proc. Boston Soc. Nat. Hist. 38, pp. 111-130, 2 pls.

DÜRKEN (B.) and SALFELD (H.): **1921.** Die Phylogenese, Fragestellungen zu ihrer exakten Erforschung. Berlin, Gebr. Bornträger.

DURKHEIM (E.): **1922.** De la Division du Travail Social. 4me ed. Paris, F. Alcan.

DUTT (G. R.): **1912.** Life Histories of Indian Insects. Mem. Dept. Agric. India. Ent. Ser. iv., No. 4, p. 251.

EDWARDS (F. W.): **1922.** A Revision of the Genus Harpagomyia de Meij. Trans. Ent. Soc. London, pp. 496-503, pl. 16, figs. 5-12.

EIDMANN (H.): **1922.** Die Durchlässigkeit des Chitins bei osmotischen Vorgängen. Biol. Zentralbl. 42, pp. 429-435, 1 fig.

EMELJANOFF (N.): **1924.** Intersexualität bei *Lymantria dispar* L. unter Einwirkung der Temperatur. Biol. Zentralbl. 44, 1924, pp. 106-110, 4 figs.

EMERSON (A. E.): **1926.** Development of Soldier Termites. Zoologica 7, pp. 69-100, 10 figs.

EMERY (C.): **1890.** Studii sulle Formiche della Fauna Neotropica. Bull. Soc. Ent. Ital. 22, pp. 38-40, 5 pls. **1891.** Le Formiche dell'Ambra Siciliana nel Museo Mineralogico dell 'Università di Bolgona. Mem. R. Accad. Sc. Ist. Bologna (5) 1, pp. 141-165, 3 pls. **1894.** Die Entstehung und Ausbildung des Arbeiterstandes bei den Ameisen. Biol. Centralbl. 14, pp. 53-59. **1895.** Die Gattung Dorylus Fab. und die systematische Eintheilung der Formiciden. Zool. Jahrb. Abt. Syst. 8, pp. 685-778, 4 pls., 41 figs. **1896.** Le Polymorphisme des Fourmis et la Castration Alimentaire. C. R. 3me Congr. Internat. Zool. Leyden, pp. 395-410. **1898.** Beiträge zur Kenntniss der paläarktischen Ameisen. Ofvers. Finsk. Vet. Soc. Förh. 20, 28 pp. **1904.** Zur. Kenntniss des Polymorphismus der Ameisen. Zool. Jahrb. Suppl. VII., p. 587-610, 6 figs. **1905.** Deux Fourmis de l'Ambre de la Baltique. Bull. Soc. Ent. France, pp. 187-189, 2 figs. **1906.** Zur Kenntniss des Polymorphismus der Ameisen. Festschr. f. J. Rosenthal. Leipzig. G. Theime pp. 35-40. **1908a.** Remarques sur les Observations de M. de Lannoy touchant l'existence de *L. mixtus* dans les fourmilières de *L. fuliginosus*. Ann. Soc. Ent. Belg. 52, pp. 182-183. **1908b.** Osservazioni ed Esperimenti sulla Formica Amazzone. Rend. R. Accad. Sc. Ist. Bologna, pp. 49-62. **1909a.** Ueber den Ursprung der dulotische, parasitischen und myrmecophilen Ameisen. Biol. Centralbl. 29, pp. 352-362; also in Rend. R. Accad. Sc. Ist. Bologna, pp. 36-50. **1909b.** Nuove Osservazioni ed Esperimenti sulla Formica Amazzone *Ibid.*, pp. 31-36. **1910.** Il Polimorfismo e la Fondazione delle Società negli Insetti Sociali. Scientia 7, pp. 336-349. **1911a.** Ulteriori Osservazioni ed Esperimenti sulla Formica Amazzone Rend. R. Accad. Sc. Ist. Bologna, pp. 60-75. **1911b.** Beobachtungen und Versuche an *Polyergus rufescens*. Biol. Centralbl. 31, pp. 625-642. **1915a.** La Vita delle Formiche. Torino, Fratelli

Bocca. vii. + 251 pp., 77 figs. 1915b. Können weisellose Ameisenvölker die fehlende Mutter aus eigenen Mitteln ersetzen? Biol. Centralbl. 35, pp. 252-254. 1916. La Fondazione delle Società di Myrmica. Considerazioni sull' Alimentazione delle Formiche. Rend. R. Accad. Sc. Ist. Bologna, pp. 45-50. 1918. Esperienze intorno alla produzione di individui sessuati nella Società delle Formiche. *Ibid.*, p. 65-72. 1920. La Distribuzione Geografica Attuale delle Formiche. Real. Acad. Lincei (5) 13, pp. 3-98. 1921. Quels sont les Facteurs du Polymorphisme du Sexe Féminin chez les Fourmis? Rev. Gén. Sciences 32, pp. 737-741. 1924. Casi di Anomalie e di Parassitismo nelle Formiche. Rend. R. Accad. Sc. Ist. Bologna, pp. 82-89, 4 figs.

ENDERLEIN (G.): 1914. Archihymenidæ, in W. Michaelsen's " Beiträge zur Kenntnis der Land-und Süsswasserfauna Deutsch-südwestafrikas ". Hamburg, pp. 195-199, 1 pl., 1 fig.

VON ENGELHARDT: 1914. Ueber die Hancock'sche Drüse von *Œcanthus pellucens* Scop. Zool. Anzeig. 44, pp. 219-227.

ERICHSON (W. F.): 1847. De Fabrica et Usu Antennarum in Insectis. Berolini, 13 pp. 1 fig.

ESCHERICH (K.): 1902. Biologische Studien über algerische Myrmekophilen, zugleich mit allgemeinen Bemerkungen über die Entwicklung und Bedeutung der Symphilie. Biol. Centralbl. 22, p. 638-663, 4 figs. 1909. Die Termiten oder weissen Ameisen. Leipzig. xii. + 198 pp., 1 pl., 51 figs. 1911. Termitenleben auf Ceylon. Jena, xxxii. + 263 pp., 3 pls. 1917. Die Ameise, Schilderung ihrer Lebensweise. *2te* Aufl. Braunschweig, Viehweg, xvi. + 348 pp., 98 figs.

ESPINAS (A.): 1877. Les Sociétés Animales, Etude de Psychologie Comparée. Paris, G. Baillière et Co.

EVERSMANN (E.): 1846. Brutzellen des *Halictus quadricinctus* und *Pelopæus destillatorius*. Bull. Naturf. Ges. Moskau 19, pp. 188-193.

EZIKOV (J.): 1926. Ueber den Character der Variabilität der Ameisen-Ovarien. Rev. Zool. Russe 3, pp. 333-356, 22 figs. (Russian).

FABRE (J. H.): 1879-80. Etudes sur les Moeurs et la Parthénogenèse des Halictes. Ann. Sc. Nat. Zool. (6) 9, pp. 1-27, 1 pl. 1890. Souvenirs Entomologiques, 3me sér. 2me ed. 1903. Souvenirs Entomologieques, 8me sér. Chapt. 7-9. 1920. Souvenirs Entomologiques, 3 me sér.

FAHRINGER (J.): 1914. Ueber den Nestbau zweier Bienen. Zeitschr. wiss. Insektenbiol. 10, pp. 16-20, 5 figs.

FARQUHARSON (C. O.): 1918. Harpagomyia and Other Diptera fed by Crematogaster Ants in Southern Nigeria. Proc. Ent. Soc. London, pp. xxix-xl. 1919. The Tachinid Fly Bengalia attacking Termites, *Ibid.*, pp. lii.-lxii. 1922a. Five Years' Observations (1914-1918) on the Bionomics of Southern Nigerian Insects (Edited by E. B. Poulton). Trans. Ent. Soc., London, pp. 319-448. 1922b. The Pursuit of Living Ants by the Ephydrid *Rhynchopsilota apicalis* Collin sp. nov. *Ibid.*, pp. 443-444.

FERRIÈRE (A.): 1915. La Loi du Progrès en Biologie et en Sociologie et la Question de l'Organisme Social. Paris, Giard et Brière.

FERTON (C.): 1890. L'Evolution de l'Instinct chez les Hyménoptères. Rev. Scient., pp. 45, 496-498. 1898. Sur les Mœurs des Sphecodes et des Halictus Latr. Bull. Soc. Ent. France, pp. 75-77. 1901-21. Notes Detachées sur l'Instinct des Hyménoptères Mellifères et Ravisseurs. 9 series. Ann. Soc. Ent. France 70-89.

FEYTAUD (J.): 1912. Contribution a l'Etude du Termite lucifuge (Anatomie, Fondation de Colonies Nouvelles). Arch. d'Anat. Micros. 13, pp. 481-609, 3 pls., 34 figs.

FIELDE (A. M.): 1901. A Study of an Ant. Proc. Acad. Nat. Sci. Phila. 53, pp. 425-449. 1902. Notes on an Ant. *Ibid.*, pp. 599-625, 2 figs. 1904. Power of Recognition among Ants. Biol. Bull. 7, pp. 227-250, 4 figs. 1905a. The Communal Life of Ants. Nature Study Review 1, pp. 239-250, 4 figs. 1905b. Observations on the Progeny of Virgin Ants. Biol. Bull. 9, pp. 355-360. 1905c. The Progressive Odor of Ants. *Ibid.*, 10, pp. 1-16.

FISKE (J.).: 1874. Cosmic Philosophy. Vol. I., p. viii.

FLETCHER (J.): 1908. *Vespa borealis* an Inquiline? Ann. Ent. Soc. America I., p. 30; also in Psyche 15, p. 16.

FORBES (H. O.): 1885. A Naturalist's Wanderings in the Eastern Archipelago. London, pp. 72-73.

FORBES (S.A.): 1908. On the Life History, Habits and Economic Relations of the White-Grubs and May-beetles (Lachnosterna). 24th Rep. State Ent. Nox. Benef. Insects Illinois, pp. 135-168, 3 pls.

FORD (N.): 1922. An Undescribed Planidium of Perilampus from Conocephalus. Canad. Ent., pp. 199-204, 1 fig. 1926. On the Behaviour of Grylloblatta. *Ibid.* 58, pp. 66-70, 1 fig.

FOREL (A.): 1874. Le Fourmis de la Suisse (2nd ed., 1920). 1890. Un Parasite de la *Myrmecia forficata* F. Ann. Soc. Ent. Belg. 20, p. 8. 1894. Ueber den Polymorphismus und Ergastomorphismus bei den Ameisen. Vers. deutsch. Naturf. Wien., pp. 142-147. 1898. La Parabiose chez les Fourmis. Bull. Soc. Vaud Sc. Nat. (4) 34, pp. 380-384. 1900. Fourmis de Japon. Nids en Toile. *Strongylognathus huberi* et Voisins. Mitteil. schweiz. Ent. Ges. 10, pp. 267-287. 1902. Beispiele phylogenetischer Wirkungen und Rückwirkungen bei den Instinkten und dem Körperbau der Ameisen als Belege für die Evolutionslehre und die psychophysiologische Identitätslehre. Journ. Psych. Neurol. 1, pp. 99-110. 1904. Ueber Polymorphismus und Variation bei den Ameisen. Zool. Jahrb. Suppl. VII., pp. 571-586. 1906. Mœurs des Fourmis Parasites des Genres Wheeleria et Bothriomyrmex. Rev. Suisse Zool. 14, p. 68. 1908. Lettre à la Société Entomologique de Belgique. Ann. Soc. Ent. Belg. 52, pp. 180-181. 1910a. Das Sinnesleben der Insekten. München. 1910b. Aperçu sur la distribution géographique et la phylogénèse des Fourmis. 1me Congr. Internat. d'Ent., pp. 81-100. 1921-23. Le Monde Social des Fourmis, 6 vols. Genève, Libr. Kundig.

FOUILLÉE (A.): 1920. L'Evolutionnisme des Idées-Forces. 6th ed. (1st ed., 1890). Paris, F. Alcan.

FOX (W. J.): **1893.** Observations on the Hymenopterous Genus Neolarra Ashmead. Ent. News, 4, pp. 292-293, 1 fig.

FREUD (S.): **1921.** Group Psychology and the Analysis of the Ego. Trans. by J. Strachey. New York, Boni & Liveright.

FRIESE (H.): **1888.** Die Schmarotzerbienen und ihre Wirte. Zool. Jahrb. Abt. Syst. 3, pp. 847-870. **1891.** Beiträge zur Biologie der solitären Blumenwespen (Apidæ). *Ibid.*, 5, pp. 751-860, 1 pl. **1903.** Neue Meliponiden II., Zeitschr. Hymen. Dipt. 3, pp. 359-361. **1905.** Ein Bienennest mit Vorratskammern (*Lithurgus dentipes* Sm.). Zeitschr. wiss. Insektenbiol. 1, pp. 118-119. **1923.** Die europäischen Bienen (Apidæ). Das Leben und Wirken unserer Blumenwespen. Berlin u. Leipzig, de Gruyter & Co.

VON FRISCH (K.): **1921.** Ueber den Sitz des Geruchsinnes bei Insecten. Zool. Jahrb. Abt. Zool. Physiol. 38, pp. 449-516. **1923.** Ueber die "Sprache" der Bienen. Jena, G. Fischer, 186 pp., 2 pls., 25 figs. (Also in Zool. Jahrb. Abt. Zool. Physiol. 40.)

FRISON (T. H.): **1916.** Note on the Habits of *Psithyrus variabilis* Cress. Bull. Brooklyn Ent. Soc. 11, pp. 46-47. **1921.** *Psithyrus laboriosus* Fabr. in the Nests of Bumble bees. Canad. Ent. 53, pp. 100-101. **1926.** Contribution to the Knowledge of the Bumble Bees of Illinois and their Animate Environment. Ann. Ent. Soc. Amer. 19, pp. 203-234, 1 pl.

FROGGATT (W. W.): **1895-'97.** Australian Termitidæ. 3 Parts; Proc. Linn. Soc. N.S. Wales. 1895, p. 420; 1896, pp. 455 and 510; 1897, p. 721. **1905.** White Ants. Misc. Publ., No. 874, Dept. Agric. N.S. Wales, pp. 1-47, 2 pls., 12 figs.

FROHAWK (F. W.): **1915.** Further Observations on the Last Stage of the Larva of *Lycæna arion*. Trans. Ent. Soc. London, pp. 313-316, 2 pls.

FULLAWAY (D. T.): **1920.** New Species of Sierola with Explanatory Notes. Occas. Papers Bernice Pauahi Bishop Mus. 7, pp. 157-59, 1 pl.

FULLER (C.): **1915.** Observations on Some South African Termites. Ann. Natal Mus. 3, p.p 329-504, 11 pls., 16 figs.

FULTON (B.B.): **1915.** The Tree Crickets of New York: Life History and Bionomics. Techn. Bull., No. 42, N.Y. Exper. Stat., 47 pps., 6 pls., 21 figs. **1924a.** Some Habits of Earwigs. Ann. Ent. Soc. Amer. 17, pp. 357-367, 1 fig. **1924b.** The European Earwig. Bull., No. 207, Oregon Agric. Coll. Exper. Stat., pp. 5-29, 6 figs.

GABRITSCHEVSKY (E.): **1924.** Farbenpolymorphismus und Vererbung mimetischer Varietäten der Fliege *Volucella bombylans* und anderer "hummelähnlicher" Zweiflügler, Zeitschr. indukt. Abstam. Vererbl. 32, pp. 321-353. **1926.** Convergence of Coloration between American Pilose Flies and Bumble Bees (Bombus). Biol. Bull. 51, pp. 269-282, 4 pls.

GENIEYS (P.): **1924.** Contribution à l'Etude des Evaniidæ; *Zeuxevania splendidula* Costa. Bull. Biol. France Belg. 58; pp. 482-494, 1 pl., 3 figs.

GIARD (A.): **1889a.** Sur une Galle produite chez le *Typhlocyba rosæ* L. par une Larve d'Hyménoptère, C. R. Acad. Sc. Paris 109, p. 79. **1889b.** Sur la Castration Parasitaire de Typhlocyba par une Larve d'Hyménoptère (*Aphelopus melaleucus* Dalm.) et par une Larve de Diptère (*Atelura spuria* Meig.) *Ibid.*, 109, pp. 708-710. **1905.** La Pœcilogonie. Bull. Sci. France Belg. 39, pp. 153-187.

GIRAUD (J.): **1871.** Notes sur les Mœurs du *Ceramius lusitanicus* Klug. Ann. Soc. Ent. France (5) 1, pp. 375-379.

DE GOURMONT (R.): **1924.** Physique de l'Amour. Essai sur l'Instinct Sexuel. Paris, Mercure de France. English translation, by Ezra Pound, *sub. tit.* The Natural Philosophy of Love, London, 1926.

GRANDI (G.): **1921.** Ricerche sul Gen. Philotrypesis Först. Boll. Lab. Zool. Gen. Agrar. Portici 15, pp. 33-190, 44 figs. **1923.** Gli Insetti dei Caprifichi. Rivist. Biol. 5, pp. 5-26, 15 figs. **1925.** Biologia, Morfologia e Adattamento negli Insetti dei Fichi, Atti Soc. Ital. Sc. Nat. 63, pp. 288-311.

GRASSI (B.) and SANDIAS (A.): **1893-'94.** Constituzione e Sviluppo della Società dei Termitidi, Atti Accad. Gioenia (4) 6-7, 151 pp., 5 pls.

GREEN (E. E.): **1906-'07.** Spolia Zelanica 3, p. 220, 1906; pp. 183-184, 1907. **1908.** Tongue of Ochromyia. Proc. Ent. Soc. London, pp. xxvi.-xxvii.

GRIFFINI (A.): **1905.** Studi sui Lucanidi. 1e Considerazioni Generali sulla grande Variazione di caratteri nei maschi dei Lucanidi. Torino, P. Gerbone, 40 pp., 4 figs.

GUTBIER (A.): **1914.** Ueber einige Hymenopterennester aus Turkestan. Zeitschr. wiss. Insektenbiol. 10, pp. 339-345, 6 figs.

HALIDAY (A. H.): **1835.** Notes on the Bethyli and on *Dryinus pedestris*, Ent. Mag 2, pp. 219-220.

HAMAKER (J. J.): **1898.** The nervous System of *Nereis virens* Sars. Bull. Mus. Comp. Zool. 32, pp. 89-124, 5 pls.

HANCOCK (J. L.): **1905.** The Habits of the Striped Meadow-Cricket (*Oecanthus fasciatus*). Amer. Natural. 39, pp. 1-11, 1 fig.

HANDLIRSCH (ADAM): **1882-83.** Die Metamorphose und Lebensweise von *Hirmoneura obscura* Meig. einem Vertreter der Dipteren-Familie Nemestrinidæ. I. Wien. Ent. Zeitg. 1, 1882, pp. 224-228; II., *Ibid.* 2, 1883, pp. 11-15, 1 pl.

HANDLIRSCH (ANTON): **1887-'95.** Monographie der mit Nysson und Bembex verwandten Grabwespen. Sitzb. K. Akad. Wiss. Wien., 8 parts. **1903.** Zur Phylogenie der Hexapoden. *Ibid.* 112, pp. 1-23, 1 pl. **1904.** Zur Systematik der Hexapoden. Zool. Anzeig. 27, pp. 773-769. **1908.** Die fossilen Insekten und die Phylogenie der rezenten Formen. 2 vols. Leipzig, W. Engelmann. **1910.** Fossile Wespennester. Ber. Senkenb. Naturf. Ges. Frankf. a. M., pp. 265-266, 1 fig. **1913a.** Beiträge zur exakten Biologie. Sitzb. Akad. Wiss. Wien. 122, pp. 361-481, 1 diagram, 5 maps. **1913b.** Ueber einige Beziehungen zwischen Paläontologie, geographische Verbreitung und Phylogenie der Insekten. Internat. Congr. Ent. Oxford 2 (1912), pp. 248-270, 3 pls. **1924.** Systematische Uebersicht. Kap. 9. in C. Schröder's "Handbuch der Entomologie". Jena, G. Fischer, pp. 712-825, 112 figs.

HARRIS (R. G.): **1923.** Occurrence, Life-Cycle and Maintenance under Artificial Conditions of Miastor. Psyche 30, pp. 95-101. **1924.** Sex of Adult Cecidomyidæ (Oligarces sp.) arising from Larvæ produced by Parthenogenesis. *Ibid.* 31, pp. 148-154.

HARRIS (T.): **1862.** Treatise on Some of the Insects Injurious to Vegetation. New ed. Boston, Crosby & Nichols.

HÄRTER: **1890.** Biologische Beobachtungen an Hummeln. 27. Ber. Oberhess. Ges. Natur. Heilk., pp. 59-75.

HASE (E.): **1926.** Ueber die Nester der Wachsmotten und der Aphomiaraupen. Arb. Biol. Reichsanst. Land Forstwiss. 14, pp. 555-565, 5 figs.

HAUPT (P.): **1922.** Manna, Nectar and Ambrosia. Proc. Amer. Philosoph. Soc. 61, pp. 227-236.

HAUSER (G.): **1880.** Physiologische und histologische Untersuchungen über das Geruchsorgan der Insekten. Zeitschr. wiss. Zool. 34, pp. 1-37, 3 pls.

HAYES (W. P.): **1927.** Another Host of *Pristocera armifera* (Say). Proc. Ent. Soc. Washington 29, pp. 20-22.

HEARD (G.): **1924.** Narcissus, an Anatomy of Clothes. London, Kegan Paul & Co., Ltd.; New York, E. P. Dutton & Co.

HEATH (H.): **1902.** The Habits of California Termites. Biol. Bull. 4, pp. 47-63, 3 figs. **1907.** The Longevity of Members of the Different Castes of *Termopsis angusticollis*. *Ibid.* 13, pp. 161-164. **1927.** Caste Formation in the Termite Genus Termopsis. Journ. Morph. 43, pp. 387-425, 3 pls., 3 figs.

HEER (O.): **1849.** Die Insektenfauna der Tertiärgebilde von Oeningen und von Radoboj in Croatien II. Neue Denkschr. allg. schweiz. Ges. gesamt. Naturw. 11, pp. 1-264, 17 pls. **1852.** Ueber die Hausameise Madeiras. Züricher Jugend, 1852, von der naturf. Gesell, 54. Stück., 24 pp., 1 pl.

HEGH (E.): **1922.** Les Termites. Partie Générale. Brussels, Imp. Industr. Financ., 756 pp., 460 figs., 1 map.

HEIKERTINGER (F.): **1919.** Die metöke Myrmekoidie, Tatsachenmaterial zur Lösung des Mimikryproblems. Biol. Zentralbl. 39, pp. 65-102, 13 figs. **1925.** Ueber die Begriffe " Mimicry " und " Mimese ", mit besonderer Berücksichtigung der Myrmekoidie. *Ibid.* 45, pp. 272-289. **1925-'27.** Die Ameisenmimese. *Ibid.* 45, 1895, pp. 705-727, 6 figs.; 46, 1926, pp. 351-382, 385-405, 593-625, 9 figs.; 47, 1927, pp. 462-501, 47 figs.

HENNING (H.): **1916.** Der Geruch. Leipzig.

HENTZ (N. M.): **1875.** Collection of Arachnological Writings, edited by E. Burgess. Occas. Papers Boston Soc. Nat. Hist. 2, 171 pp., 21 pls.

HERBST (P.): **1922.** Revision der Halictus-Arten von Chile. Ent. Mitteil. 11, pp. 180-191.

HERMS (H. B.): **1907.** An Ecological and Experimental Study of Sarcophagidæ with Relation to Lake Beach Débris. Journ. Exper. Zool. 4, pp. 45-83, 7 figs.

HERRICK (C. J.): 1903. The Organ and Sense of Taste in Fishes. Bull. U. S. Fish. Comm. (1902), pp. 237-272.

HESELHAUS (F.): 1922. Die Hautdrüsen der Apiden und verwandter Formen. Zool. Jahrb. Abt. Anat. 43, pp. 369-464, 11 pls.

HEYMONS (R.): 1907. Die verschiedenen Formen der Insektenmetamorphose und ihre Bedeutung im Vergleich zur Metamorphose anderer Arthropoden. Ergebn. u. Fortschr. Zool. 1, pp. 137-188. 1908. Süsswasser-Hymenopteren aus der Umgebung Berlins. Deutsch. Ent. Zeitschr., pp. 137-150, 4 figs.

HILL (G. F.): 1925. Notes on *Mastotermes darwiniensis* Froggatt. Proc. Roy. Soc. Victoria 37 (n.s.), pp. 119-124, 1 fig.

HINGSTON (R. W. G.): 1922. A Naturalist in Himalaya. Boston, Small, Maynard & Co.

HOCHREUTER (R.): 1912. Die Hautsinnesorgane von *Dytiscus marginalis* L., ihr Bau und ihre Verbreitung am Körper. Zeitschr. wiss. Zool. 103, pp. 1-114, 102 figs.

HOFFER (E.): 1881. Biologische Beobachtungen an Hummeln und Schmarotzerhummeln. Mitth. naturw. Ver. Steiermark 18, pp. 68-92. 1888. Die Schmarotzerhummeln Steiermarks, Lebensgeschichte und Beschreibung derselben. *Ibid.* 25, pp. 82-159, 1 pl.

HOLLIDAY (M.): 1903. A Study of Some Ergatogynic Ants. Zool. Jahrb. Abt. Syst. 19, pp. 293-327, 15 figs.

HOLMGREN (N.): 1909-13. Termitenstudien, 4 Parts. K. Svensk. Vetensk. Handl. 44, 1909, 215 pp., 3 pls., 76 figs.; 46, 1911, 86 pp., 6 pls., 6 figs.; 48, 1912, 166 pp., 4 pls. 88 figs.; 50, 1913, 276 pp., 8 pls., 14 figs.

HOOD (J. D.): 1913. Notes on the Life-history of *Rhopalosoma poeyi*. Proc. Ent. Soc. Washington 15, pp. 145-148.

HOWARD (L. O.): 1905. The Insect Book. Doubleday, Page & Co. 1908. The Importation of *Tetrastichus xanthomelænæ* Rond. Journ. Econom. Ent. 1, pp. 281-289. 1910. On the Habit with Certain Chalcidoidea of Feeding at Puncture Holes made by the Ovipositor. *Ibid.* 3, pp. 257-260.

HOWES (P. G.): 1917. Tropical Life in British Guiana (Beebe). Entomology, Part III., pp. 371-450, 2 pls., 15 figs.

HUBER (J.): 1905. Ueber die Koloniegründung bie *Atta sexdens*. Biol. Centralbl. 25, pp. 606-619, 625-635, 26 figs. Transl. in Smithson Rep. (1906) 1907, pp. 355-367, 5 pls.

HUBER (P.): 1810. Recherches sur les Mœurs des Fourmis Indigènes. Paris and Geneva.

HUNTINGTON (E.): 1914. The Climatic Factor as illustrated in Arid North America. Carnegie Inst., Washington, D.C., 341 pp., 90 figs.

HUXLEY (J. S.): 1927. Further Work on Heterogonic Growth. Biol. Zentralbl. 47, pp. 151-163, 5 figs.

HYSLOP (J. A.): 1916. *Pristocera armifera* (Say) parasitic on *Limonius agonus* (Say). Proc. Ent. Soc. Washington 18, pp. 169-170, 1 pl.

VON IHERING (H.) : **1896.** Zur Biologie der socialen Wespen Brasiliens. Zool. Anzeig., pp. 449-453. **1903.** Biologie der stachellosen Bienen Brasiliens. Zool. Jahrb. Abt. Syst. 19, pp. 179-287, 13 pls., 8 figs.

VON IHERING (R.) : **1903a.** Biologische Beobachtungen an brasilianischen Bombus-Arten. Allgem. Zeitschr. Ent. 8, pp. 447-453, 5 figs. **1903b.** Contributions à l'Etude des Vespides. Ann. Soc. Ent. France, pp. 144-145. **1903c.** Zur Frage nace dem Ursprung der Staatenbildung bei den socialen Hymenopteren Zool. Anzeig., pp. 113-118. **1904.** Biologia das Abelhas Solitarias do Brazil. Bevist. Mus. Paulista 6, pp. 461-481, 5 figs.

IMMS (A. D.) : **1913.** Contributions to a Knowledge of the Structure and Biology of Some Indian Insects. II., On *Embia major* sp. nov. from the Himalayas. Trans. Linn. Soc., London, Zool. (2) 11, pp. 167-195, 3 pls., 6 figs. **1919.** On the Structure and Biology of Archotermopsis, together with Descriptions of New Species of Intestinal Protozoa and General Observations on the "Isoptera." Phil. Trans. Roy. Soc., London (B) 209, pp. 75-180, 7 pls. **1927.** On the Affinities of the Grylloblattidæ. Psyche, 34, pp. 36-39.

JACK (R. W.) : **1917.** Parthenogenesis amongst the Workers of the Cape Honey Bee. Trans. Ent. Soc., London, pp. 396-403, 2 pls.

JACOBSON (E.) : **1909.** Ein Moskito als Gast und diebischer Schmarotzer der *Cremastogaster difformis* Smith und eine andere schmarotzende Fliege. Tijdschr. Ent. 52, pp. 158-164. **1910** *Pheidologeton diversus* Jerdon und eine myrmecophile Fliegenart. *Ibid.* 53, pp. 328-335. **1911a.** Biological Notes on the Hemipteron *Ptilocerus ochraceus. Ibid.* 54, pp. 175-179. **1911b.** Nähere Mitteilungen über die myrmecophile Culicide *Harpagomyia splendens* de Meij. *Ibid.* 54, pp. 158-161, 3 pls.

JAMBUNATHAN (N. S.) : **1908.** The Habits and Life History of a Social Spider (*Stegodyphus sarasinorum* Karsch). Madura, Victoria Press, 14 pp.

JAMES (S. P.) : **1914.** Summary of a Year's Mosquito Work in Ceylon. Indian Journ. Med. Research 2, pp. 233-234.

JANET (C.) : **1896.** Sur les Rapports des Lepismides myrmécophiles avec les Fourmis. C. R. Acad. Sc. Paris 122, pp. 799, 1 fig. **1898a.** Etudes sur les Fourmis, les Guêpes et les Abeilles. 19e Note. Anatomie du Corselet de la *Myrmica rubra* Reine. Mém. Soc. Zool. France 11, pp. 393-450, 1 pl., 25 figs. **1898b.** Sur une Cavité du Tégument servant chez les Myrmicinæ a étaler, au contact de l'air, un produit de sécretion, C. R. Acad. Sc. Paris 126, pp. 1168-1171. **1903.** Observations sur les Guêpes. Paris, Carré et Naud. **1904.** Observations sur les Fourmis. Limoges. **1907.** Anatomie du Corcelet et Histolyse des Muscles Vibrateurs, après le vol nuptial, chez la Reine de la Fourmi (*Lasius niger*). Limoges, Ducourtieux et Gout. 149 + 20 pp., 13 pls., 41 figs. **1909.** Sur la Parthénogénèse Arrhénotoque de la Fourmi Ouvrière. Mém. Soc. Acad. de l'Oise, 8 pp.

JEANNEL (R.) : **1913.** Sur des Larves de Proctotrupidæ parasites d'Homoptères. Appendix to J. J. Kieffer, Proctotrypidæ, Cynipidæ et Evaniidæ, in Voyage de Ch. Alluaud et R. Jeannel en Afrique Orientale. Hyménopt. I., pp. 36-39, 1 pl.

JONES (D. W.): **1917.** The European Earwig and its Control. Bull. No. 566, U.S. Dept. Agric., 12 pp., 8 figs.

JONESCU (C. N.): **1909.** Vergleichende Untersuchungen über das Gehirn der Honigbiene. Jen. Zeitschr. Naturw. 45, pp. 111-180, 5 pls., 13 figs.

JUCCI (C.): **1924.** Su la Differenziazione de le Caste ne la Società dei Termitidi I., I Neotenici. Mem. R. Accad. Naz. Lincei (5) 14 (1923), pp. 5-238, 4 pls., 23 figs.

KEILIN (D.) and THOMPSON (W. R.): **1915.** Le Cycle évolutif des Dryinidæ. C. R. Soc. Biol., Paris, 78, pp. 83-87, 10 figs.

KELLY (E. O. G.): **1914.** A New Sarcophagid Parasite of Grasshoppers. Journ. Agric. Research, U.S. Dept. Agric. 2, pp. 435-445, 1 pl.

KEMNER (N. A.): **1922.** Die Larve der Termitoxenien entdeckt! Ent. Tijskr., pp. 58-61. **1923.** Hyphænosymphilie, eine neue merkwürdige Art von Myrmekophilie bei einem neuen myrmekophilen Schmetterling (*Wurthia aurivillii* n. sp.) aus Java-beobachtet. Ark. f. Zool. 15, pp. 1-28, 4 pls. **1925a.** Larva termitovorax, eine merkwürdige parasitische Fliegenlarve, die im Kopfe von Termitensoldaten lebt, etc. *Ibid.* 17a, pp. 1-15, 1 pl., 2 figs. **1925b.** Weitere Beobachtungen über das Genus Gnathotermes Holmgr., das sich als auf parasitierte Termes-Individuen begründet erwiesen hat. Ent. Tidskr. 3, pp. 157-163, 3 figs. **1926a.** Zur Kenntnis der Staphyliniden Larven II. Die Lebensweise und die parasitische Entwicklung der echten Aleochariden. *Ibid.*, pp. 133-170, 5 pls., 4 figs. **1926b.** Ueber die Zucht der Larve einer echten Termitoxenia. 3te Internat. Ent. Kongr. Zürich 2, pp. 389-404, 1 pl., 2 figs.

KIEFFER (J. J.): **1907.** Hymenoptera. Fam. Dryinidæ, in Wytsman's "Genera Insectorum" 33 pp., 2 pls. **1908.** Fam. Bethylidæ. *Ibid.*, 50 pp., 3 pls. **1909.** Description de Nouveaux Microhyménoptères du Brésil. Ann. Soc. Ent. France 78, pp. 287-348.

KING (J. L.): **1916.** Observations on the Life-History of *Pterodontia flavipes* Gray. Ann. Ent. Soc. Amer. 9, pp. 309-321, 2 pls.

KINSEY (A. C.): **1919.** Fossil Cynipids. Psyche 26, pp. 44-49. 2 figs.

KIRKALDY (G. W.): **1911.** Some Remarks on the Reduviid Subfamily Holoptilinæ and on the Species *Ptilocerus ochraceus* Montandon. Tijdschr. Ent. 54, pp. 170-174, 1 pl.

KLEBS (R.): **1910.** Ueber Bernsteineinschlüsse im allgemeinen und die Coleopteren meiner Bernsteinsammlung. Schrift. physik. ökonom. Ges. Königsberg 51, pp. 217-242.

KLEIN: **1904.** Futterbrei und weibliche Bienenlarve. Die Bienenpflege 26.

KNOWER (H. McE.): **1894.** Origin of the "Nasutus" (soldier) of Eutermes. Johns Hopkins Univ. Circ. 13, pp. 58-59.

KOBER (L.): **1921.** Der Bau der Erde. Berlin, Gebr. Bornträger.

KOHL (F. F.): **1896.** Die Gattungen der Sphegiden. Ann. k. k. naturh. Hofmus. Wien 11, pp. 233-516, 7 pls., 90 figs.

KOLBE (H. J.): **1913.** Die Differenzierung der zoogeographischen Elemente der Kontinente. Trans. 2nd Internat. Congr. Ent. (1912) 2, pp. 433-476.

KORNHAUSER (S. I.): **1919.** The Sexual Characteristics of the Membracid *Thelia bimaculata* (Fabr.) 1. External Changes Induced by *Aphelopus theliæ* (Gahan). Journ. Morph. 32, pp. 531-635, 54 figs.

KOSMINSKY (P.): **1924a.** Ueber Erzeugung von Intersexen bei *Stilpnotia salicis* L. im Temperaturexperiment. Biol. Zentralbl. 44, pp. 15-16. **1924b.** Der Gynandromorphismus bei Lymantria dispar L. unter der Einwirkung äusserer Einflüsse. *Ibid.*, 44, pp. 66-68, 3 figs.

KRONER (R.): **1913.** Zweck und Gesetz in der Biologie. Tübingen, J. C. B. Mohr. **1919.** Das Problem der historischen Biologie, Heft 2 in Schaxel's " Abhandl. theoret. Biologie ", Berlin, Gebr. Bornträger.

KÜHNLE (K. F.): **1913.** Untersuchungen über das Gehirn, die Kopfnerven und die Kopfdrüsen des gemeinen Ohrwurms. Jen. Zeitschr. Naturg. 50, pp. 147-276, 5 pls.

KÜNCKEL-d'HERCULAIS (J.): **1869.** Observations relatives aux *Volucella zonaria, inanis, pellucens* et *bombylans*. Ann. Soc. Ent. France (4) 9, Bull., pp. 17, 20 and 23. **1870.** Observations sur les Métamorphoses des Conops. *Ibid.* (4) 10, p. 63. **1875-81.** Recherches sur l'Organisation et le Développement des Volucelles, Insectes Diptères de la Famille des Syrphides. 1me Partie, 1875, pp. 1-208, 12 pls.; 2e Partie 1881, 13 pls. **1895.** Notices sur les Titres et Travaux de M. J. Künckel d'Herculais. Paris, Ed. Crété.

KUO (Z. Y.): **1924.** A Psychology without Heredity. Psychol. Rev. 31, pp. 427-448.

KUTTER (H.): **1913a.** Zur Biologie von *Formica rufa* und *Formica fusca* i. sp. Biol. Centralbl. 33, pp. 703-707. **1913b.** Ein weiterer Beitrag zur Frage der sozialparasitischen Koloniegründung von *F. rufa* L. Zeitschr. wiss. Insektenbiol. 9, pp. 193-196. **1917.** Myrmekologische Beobachtungen. Biol. Zentralbl. 37, 249-437. **1920.** *Strongylognathus huberi* r. *alpinus* Wh. eine sklavenraubende Ameise. *Ibid.* 40, pp. 528-538, 1 fig. **1923.** Der Sklavenräuber *Strongylognathus huberi* Forel s. sp. *alpinus* Wheeler. Rev. Suisse Zool. 30, pp. 387-424. **1927.** Ein myrmekologischer Streifzug durch Sizilien. Folia Myrmecolog. Termitolog. 1, pp. 135-136.

LAMBORN (W. A.): **1913.** The Fly *Bengalia depressa* Walk. (Calliphorinæ) robbing *Dorylus nigricans* on the March. Proc. Ent. Soc. London, pp. cxxv.-cxxxii. **1914.** On the Relationship between certain West African Insects, especially Ants, Lycænidæ and Homoptera. Trans. Ent. Soc. London, pp. 436-524, 4 pls. **1919.** Further Notes on the Habits of the Fly Bengalia. Proc. Ent. Soc. London, pp. lii.-lviii.

LAMEERE (A.): **1909.** L'Origine des Sociétés d'Insectes. Ann. Soc. Ent. Belg. 53, pp. 507-515.

LANGHOFER (A.) : **1897.** Beiträge zur Kenntnis der Mundteile der Hymenoptera I. Apidæ, Arb. Akad. Wiss. Kunst Agram. 80, pp. 25-93, 3 pls. (Autoreferat in Biol. Centralbl., 1898, pp. 607-613.)

DE LANNOY (F.) : **1908.** Notes sur le *Lasius niger* et le *Lasius fuliginosus*. Ann. Soc. Ent. Belg. 52, pp. 47-53.

LE BON (G.) : **1925.** Psychologie des Foules. 31me Edit. Paris, F. Alcan.

LE CONTE (J.) : **1884.** Elements of Geology. (First edn., 1878. N.Y., Appleton & Co.

LE DANTEC (F.) : **1918.** L'Egoisme seule Base de toute Société. Paris, E. Flammarion.

LEGEWIE (H.) : **1922.** Beiträge zur Biologie der Bienengattung Halictus. Mitteil. Bad. Landw. Naturk. Naturgesch. Freiburg i. Br. N. F. 1, pp. 235-237.

LEIBY (R. W.) : **1922.** The Polyembryonic Development of *Copidosoma gelechiæ*, with Notes on its Biology. Journ. Morph. 37 pp. 195-284, 18 pls.

LEPELETIER DE ST. FARGEAU (A. L. M.) : **1827.** Encyclopédie Méthodique 10. Paris, p. 183.

LESPÈS (C.) : **1856.** Recherches sur l'Organisation et les Mœurs du Termite lucifuge. Ann. Sc. Nat. Zool. (4) 5, pp. 227-282, 3 pls.

LEUCKART (R.) : **1887.** Neue Beiträge zur Kenntniss des Baues und der Lebensgeschichte der Nematoden. Abh. math. phys. Cl. kgl. Sächs. Ges. Wiss. 13, pp. 567-704, 3 pls.

LICHTENSTEIN (J. L.) : **1921.** Sur la Biologie d'un Chalcidien. C. R. Acad. Sc. Paris, 173, pp. 733-735. 1 fig.

LICHTENSTEIN (J. L.) and PICARD (F.) : **1917.** Etude Morphologique et Biologique du *Sycogaster Lavagnei* Pic. et Licht. Bull. Sc. France Belg. 51, pp. 440-474, 33 figs.

LICHTENSTEIN (J. L.) and RABAUD (E.) : **1922.** Le Comportement des " Polysphincta ", Ichneumonides parasites des Araignées. Bull. Sc. France Belg. 55, pp. 267-287, 11 figs.

LINEBURG (B.) : **1924.** The Feeding of Honey-bee Larvæ. Bull. No. 1222 U.S. Dept. Agric., pp. 25-37, 2 figs.

LOMNICKI (J.) : **1921.** Sur la Micrandrie chez la Fourmi sanguine. (Polish) Kosmos, Bull. Soc. Polon. Natural. Leopol., pp. 98-99. **1922.** Ueber den Anfang der Kolonien der glänzendschwarzen Holzameise. Entom. Anzeig. 2, pp. 79-80.

LONGMAN (H. A.) : **1923.** The Zoogeography of Marsupials, with Notes on the Origin of the Australian Fauna. Mem. Queensland Mus. 8, pp. 1-15.

LOVEJOY (A. O.) : **1922.** The Length of Human Infancy in Eighteenth Century Thought. Journ. Philos., 19, pp. 381-385.

LUBBOCK (Sir J.) : **1861.** On *Sphærularia bombi* Nat. Hist. Review 1, pp. 44-57, 1 pl. **1863.** On two Aquatic Hymenoptera, one of which uses its wings in swimming. Trans. Linn. Soc., London, 24, pp. 135-141, 6 figs. **1888.** Ants, Bees and Wasps. London, Kegan Paul & Co., Ltd.

LUCAS (H.): **1878.** (Une note sur la nidification d'un Hyménoptère du Genre Euglossa). Bull. Soc. Ent. France 8, pp. cxlii.-cxliii.

LUDWIG (F.): **1904.** Nest und Vorratskammern der Loñalap von Ponape. Allgem. Zeitschr. Ent. 9, pp. 225-227, 1 fig.

LÜDERWALDT (H.): **1911.** Nestbau von Neocorynura erinnys Schrottky. Zeitschr. wiss. Insekten-biol. 7, pp. 94-96, 1 fig.

MACGILLIVRAY (A. D.): **1914.** The Immature Stages of the Tenthredinoidea. 44th Ann. Rep. Ent. Soc. Ontario (1913), pp. 54-75, 1 pl..

MANN (W. M.): **1923.** New Genera and Species of Termitophilous Coleoptera from Northern South America. Zoologica 3, pp. 323-366, 13 figs.

MARCHAL (P.): **1890.** Formation d'une Espèce par le Parasitisme. Etude sur le *Sphecodes gibbus*. Rev. Scient. 45, pp. 199-204. **1894.** Le Parasitisme des Sphecodes. Bull. Soc. Ent. France 63, p. cxv. **1896.** La Reproduction et l'Evolution des Guêpes Sociales. Arch. Zool. Expér. Gén. (3) 4, pp. 1-100, 8 figs. **1897.** La Castration Nutricial chez les Hyménoptères Sociaux. C. R. Soc. Biol., 2 pp. **1900.** Sur un nouvel Hyménoptère aquatique, le *Limnodytes gerriphagus*. Ann. Soc. Ent. France 69, pp. 171-176. **1905.** Observations biologiques sur un parasite de la Galéruque de l'Orme, le *Tetrastichus xanthomelænæ* (Rond.) Bull. Soc. Ent. France, pp. 64-68. **1906.** Recherches sur la biologie et le développement des Hyménoptères parasites. II. Les Platygasters. Arch. Zool. Expér. Gén. (4) 4, pp. 485-640, 8 pls. **1909.** La ponte des Aphelinus et l'intérêt individuel dans les actes liés a la conservation de l'Espèce. C. R. Acad. Sc. Paris 148, pp. 1223-1225.

MARSHALL (G. A. K.): **1898.** Notes on the South African Social Spiders (Stegodyphus). Zoologist (4) 2, pp. 417-422.

MARTIN (E. D.): **1920.** The Behavior of Crowds. N.Y. and London, Harper Bros.

MARTYNOV (A.): **1925.** To the Knowledge of Fossil Insects from Jurassic Beds in Turkestan. Bull. Acad. Sc. de l'Urss. 19, pp. 753-762, 20 figs.

MASSART (J.) and VANDERVELDE (E.): **1893.** Parasitisme Organique et Parasitisme Social. Bull. Sc. France Belg. 25, pp. 1-68.

MATHESON (R.) and CROSBY (C. R.): **1912.** Aquatic Hymenoptera in America. Ann. Ent. Soc. Amer. 5, pp. 65-71, 3 figs.

MATHEWS (A. P.): **1921.** Physiological Chemistry. N.Y., Wm. Wood & Co.

MATTAUSCH (I.): **1909.** Gynandromorphic Membracidæ. Journ. N.Y. Ent. Soc. 17, p. 165, 1 pl. **1911.** The Effects of Parasitic Castration in Membracidæ. *Ibid.* 19, pp. 194-196, 1 pl.

MATTHEW (W. D.): **1915.** Climate and Evolution. Ann. N.Y. Acad. Sci. 24, pp. 171-318.

MAYR (G.): **1867.** Vorläufige Studien über die Radoboj-Formiciden. Jahrb. k. k. geolog. Reichsanst. Wien 17, pp. 47-62, 1 pl. **1868.** Die Ameisen des baltischen Bernsteins. Schrift. physik. ökonom. Ges. Königsberg 1, 102 pp., 5 pls.

McDougall (W.): **1920.** The Group Mind. Cambridge Univ. Press, New York, the Macmillan Co., Ltd.

McIndoo (N. E.): **1914a.** The Olfactory Sense of Insects. Smithson. Misc. Coll. 63, No. 9, pp. 1-63, 6 figs. **1914b.** The Olfactory Sense of Hymenoptera. Proc. Acad. Nat. Sci. Phila., pp. 294-341, 2 pls., 3 figs. **1914b.** The Olfactory Sense of the Honey Bee, Journ. Exper. Zool. 16, pp. 265-346, 24 figs. **1915.** The Olfactory Sense of Coleoptera. Biol. Bull. 28, pp. 407-460, 2 pls. 3 figs. **1916.** The Sense Organs of the Mouthparts of the Honey Bee. Smithson. Misc. Coll. 65, No. 14, pp. 1-55, 10 figs. **1917.** The Olfactory Organs of Lepidoptera. Journ. Morph. 29, pp. 33-54. **1918.** The Olfactory Organs of Diptera. Journ. Comp. Neurol. 29, pp. 457-484. **1920.** The Olfactory Organs of Orthoptera. *Ibid.* 31, pp. 457-484. **1923.** Glandular Structure of the, Abdominal Appendages of a Termite Guest (Spirachtha). Zoologica 3, pp. 367-380, 3 figs.

Mehling (E.): **1915.** Ueber die gynandromorphen Bienen des Eugsterschen Stockes. Verh. phys. med. Ges. Würzburg 43, pp. 173-236, 8 pls.

Mehlis (G.): **1915.** Lehrbuch der Geschichtsphilosophie. Berlin, J. Springer.

de Meijere (J. C. H.): **1909.** Drei myrmekophile Dipteren aus Java. Tijdschr. Ent. 52, pp. 165-174, 1 pl. **1910.** Ueber drei von Jacobson auf Java bei *Pheidologeton diversus* Jerdon beobachtete Fliegen. *Ibid.*, 53, pp. 336-340. **1911.** Zur Metamorphose der myrmekophilen Culicide *Harpagomyia splendens* de Meij. *Ibid.* 54, pp. 162-167, 1 pl.

Melander (A. L.) and Brues (C. T.): **1903.** Guests and Parasites of the Burrowing Bee Halictus. Biol. Bull. 5, pp. 1-27, 7 figs.

Melin (E.): **1925.** Untersuchungen über die Bedeutung der Baummykorrhiza. Jena, G. Fischer.

Metzer (C.): **1910.** Die Verbindung zwischen Vorder- und Mitteldarm bei der Biene. In E. Zander's "Studien über die Honigbiene (Apis mellifica)" III. Zeitschr. wiss. Zool. 96, pp. 539-571, 2 pls., 2 figs.

Meyer (A.): **1923.** Naturalismus und Historismus als Leitideen der modernen Naturwissenschaft. Verh. naturw. Ver. Hamburg (4) 1, pp. 69-93. **1926.** Logik der Morphologie im Rahmen einer Logik der gesamten Biologie. Berlin, J. Springer.

Meyer (E.): **1927.** Die Ernährung der Mutterameise und ihrer Brut während der solitären Koloniegründung. Biol. Zentralbl. 47, pp. 264-307, 5 figs.

Middleton (W.): **1917.** Notes on the Larvæ of Some Cephidæ. Proc. Ent. Soc. Washington, 19, pp. 174-179.

Mik (J.): **1882.** Zur Biologie von *Gonatopus pilosus* Thoms. Wiener Ent. Zeitg. 1, pp. 215-221, 1 pl.

Milliken (F. B.): **1921.** Results of Work on Blister Beetles in Kansas. Bull. No. 967, U.S. Dept. Agric. 26 pp., 22 figs.

Minnich (D. E.): **1921.** An Experimental Study of the Tarsal Chemoreceptors of two Nymphalid Butterflies. Journ. Exper. Zool., 33, pp. 173-203, 6 figs. **1922a.** The Chemical Sensitivity of the Tarsi of the Red Admiral Butterfly, *Pyrameis atalanta* Linn. *Ibid.* 35, pp. 57-81. **1922b.** A Quantitative Study of Tarsal Sensitivity to Solutions of Saccharose, in the Red Admiral Butterfly, *Pyrameis atalanta* Linn. *Ibid.* 36, pp. 445-457, 1 fig. **1924.** The Olfactory Sense of the Cabbage Butterfly *Pieris rapæ* Linn., an Experimental Study. *Ibid.* 39, pp. 339-356, 1 fig. **1926.** The Chemical Sensitivity of the Tarsi of Certain Muscid Flies. Biol. Bull. 51, pp. 166-178, 2 figs., 4 tables.

MÖBIUS (K.) : **1856.** Die Nester der geselligen Wespen. Abh. Ver. Hamburg 3, pp. 121-171, 19 pls.

MORAWITZ (F.) : **1864.** Ueber *Vespa austriaca* Panz. und drei neue Bienen. Bull. Soc. Imp. Nat. Moscou 37, pp. 439-449.

MORGAN (C. L.) : **1923.** Emergent evolution. London, Williams and Norgate.

MORICE (R. F. D.) : **1901.** Observations on Sphecodes. Ent. Month. Mag. 12, pp. 53-58.

MORLEY (C.) : **1910.** On the Position of the Rhopalosomidæ, with the Description of a Second Species. Trans. Ent. Soc. London, pp. 386-387, pl. 50, fig. 15.

MORSTATT (H.) : **1922.** Ueber Pilzgärten bei Termiten. Ent. Mitteil. 11, pp. 94-99.

MRÁZEK (A.) : **1908.** Myrmekologické poznámky. III. Brachypterni mermithogyny u *Lasius alienus.* Acta Soc. Ent. Bohem. 5, pp. 1-8, 4 figs. **1909.** Fleischfressende Blattwespen. Zeitschr. wiss. Insektenbiol. 5, p. 245. **1916.** Der Polymorphismus der sozialen Hymenopteren. Sbornik Zool. 1, pp. 27-57.

MÜLLER (H.) : **1872.** Anwendung der Darwinschen Lehre auf Blumen. Verh. naturh. Ver. preuss. Rheinl. u. Westf. 29, pp. 1-96, 2 pls. **1881.** Die Entwickelung der Blumenthätigkeit bei den Insekten. Kosmos 5.

MÜLLER (W.) : **1889.** Ueber *Agriotypus armatus.* Zool. Jahrb. Abt. Syst. 4, pp. 1132-1134.

MUESEBECK (C. F. W.) : and DOHANIAN (S. M.) : **1927.** A Study of Hyperparasitism, with particular reference to the Parasites of *Apanteles melanoscelus* (Ratzeburg). Bull. No. 1487, U.S. Dept. Agr., 35 pp., 10 figs.

MUIR (F.) : **1918.** (Harpagomyia recorded from Formosa.) Proc. Ent. Soc. London, p. xxxviii.

MURRAY (A.) : **1870.** Note on the egg of *Rhipiphorus paradoxus.* Ann. Mag. Nat. Hist. 6, pp. 326-328.

MYERS (J. G.) : **1925.** Biological Notes on *Arachnocoris albomaculatus* Scott. Journ. N.Y. Ent. Soc. 33, pp. 136-146, 1 pl. **1927.** The Habits of *Alysia manducator.* Bull. Ent. Research 17, pp. 219-229.

NANGLE : **1905.** (On Bengalia.) Journ. Bombay Nat. Hist. Soc. 16, p. 747.

NELSON (J. A.) and STURTEVANT (A. P.) : **1924.** The Rate of Growth of the Honeybee Larva. Bull. No. 1222, U.S. Dept. Agric., 24 pp., 11 figs., 4 tables.

NEWELL (W.) : **1909.** The Life History of the Argentine Ant. Journ. Econ. Ent. 2, pp. 174-192, 3 pls., 4 figs.

NEWELL (W.) and BARBER (T. C.) : **1913.** The Argentine Ant. Bull. No. 122. Bur. Ent. U.S. Dept. Agric., 98 pp., 12 pls., 13 figs.

NIELSEN (J. C.) : **1902.** Biologiske Studier over danske enlige Bier og deres Snyltere. Videsk. Meddel. Natur. För. Kjöbenhavn, pp. 75-106, 16 figs. **1903.** Om Bislaegten Sphecodes Latr. Ent. Meddel. (2) 2, pp. 21-30. **1903-'05.** Om *Perisemus fulvicornis* Curt. En. Overgangsform naellem Snylte-og Gravehvepsene. *Ibid.* (2) 2, pp. 105-109.

ONIONS (G. W.) : **1912.** South African "Fertile" Worker Bees. Agric. Journ. Union S. Afric. 3, pp. 720-728. **1914.** South African "Fertile" Worker Bees. *Ibid.* 7, pp. 44-46.

OTTRAMARE (J. H.) : **1919.** Quelques Réflexions à propos de l'action de l'obscurité sur les êtres vivantes. C. R. Soc. Biol. Paris 82, pp. 190-191.

PACKARD (A. S.) : **1865.** The Humble Bees of New England and their Parasites. Proc. Essex Inst. 4, pp. 107-140, 1 pl.

PANTEL (J.) : **1909.** Recherches sur les Diptères à Larves Entomobies I. Caractères parasitiques aux points de vue biologiques, éthologiques et histologiques. La Cellulle 26, pp. 27-216, 5 pls., 26 figs.

PARETO (V.) : **1917.** Traité de Sociologie Générale. Trans. by P. Boven. 2 vols., Paris, Payot et Cie.

PARKER (G. H.) : **1912.** The Relation of Smell, Taste, and the Common Chemical Sense in Vertebrates. Journ. Acad. Nat. Sci. Phila. 15, pp. 221-234. **1922.** Smell, Taste and Allied Senses in the Vertebrates. Philadelphia and London, Lippincott & Co. **1924.** Some Implications of the Evolutionary Hypothesis. Philosoph. Review 33, pp. 593-603.

PARKER (H. L.) : **1924.** Recherches sur les Formes postembryonnaires des Chalcidiens. Ann. Soc. Ent. France, 93, pp. 261-379, 38 pls.

PARKER (J. B.) and BÖVING (A. G.) : **1924.** The Blister Beetle *Tricrania sanguinipennis.* Biology, Descriptions of Different Stages and Systematic Relationship. Proc. U.S. Natl. Mus. 6, pp. 1-40, 5 pls.

PATTEN (W.) : **1894.** On the Morphology and Physiology of the Brain and Sense Organs of Limulus. Quart. Journ. Micr. Sci. 35, pp. 1-96, 5 pls.

PATTERSON (J. T.) : **1918.** Studies on the Biology of Paracopidosomopsis. IV. The Asexual Larvæ. Biol. Bull. 35, pp. 362-376, 3 pls. **1921.** The Development of Paracopidosomopsis. Journ. Morph. 36, pp. 1-68, 12 pls.

PEARL (R.) : **1924.** Studies on Human Biology. Baltimore, Williams and Wilkins Co.

PÉREZ (J.) : **1884.** Les Apiaires Parasites au point de vue de la Théorie de l'Evolution. Bordeaux. **1895.** Sur la prétendue Parthénogenèse des Halictus. Act. Soc. Linn. Bordeaux 48, pp. 145-157.

Pérez (C.) : **1912.** Rapport sur les Travaux Exécutés en 1911. Caisse des Recherches Scient. Rapports 1912, pp. 693-704. **1920.** Titres et Travaux Scientifiques. Laval, Barnéoud et Cie.

Perez (T. De-Stephani) : **1902.** Osservazioni biologiche sopra un Braconide acquatico, *Giardinaia urinator*, e descrizione di due altri Imenotteri nuovi. Zool. Jahrb. Abt. Syst. 15, pp. 625-634, 1 pl.

Perkins (R. C. L.) : **1887.** Notes on Some Habits of Sphecodes Latr. and Nomada Fabr. Ent. Month. Mag. 23, pp. 271-274. **1889.** Is Sphecodes Parasitic? *Ibid.*, 25, pp. 206-208. **1905.** Leaf-Hoppers and their Natural Enemies I. Dryinidæ. Bull. No. 1, Exper. Stat. Hawaii. Sugar Planters' Ass., 69 pp.

Perris (E.) : **1857.** Nouvelles Excursions dans les grandes Landes. Ann. Soc. Linn. Lyon (2) 4, pp. 172-173.

Petrucci (R.) : **1906.** Origine polyphylétique, Homotypie et non Comparabilité directe des Sociétés Animales. Notes et Mém. Inst. Solvay, Fasc. 7, 126 pp.

de Peyerimhoff (P.) : **1897.** La Variation Sexuelle chez les Arthropodes. Ann. Soc. Ent. France 66, pp. 245-260. **1910.** Sur un Cas de Pœcilandrie discontinue observé chez un Bythinus. Bull. Soc. Ent. France, pp. 287-290, 3 figs.

Phillips (E. F.) : **1907.** Wax Moths and American Foul Brood. Bull. No. 75, Bur. Ent. U.S. Dept. Agric., pp. 19-22, 3 pls.

Picard (F.) : **1919.** Contribution à l'Etude du Peuplement d'un Végétal. La Faune Entomologique du Figuier. Ann. Serv. Epiphyt. 6, pp. 35-174, 36 figs. **1922.** Contribution a l'Etude des Parasites de *Pieris brassicæ* L. Bull. Biol. France Belg. 56, pp. 54-130. **1923.** Recherches Biologiques et Anatomiques sur *Melitobia acasta* Walk. *Ibid.* 57, pp. 269-508.

Piéron (H.) : **1910.** La Genèse des Instincts esclavagistes et parasitaires chez les Fourmis. Rev. Gén. Sciences 21, pp. 726-736, 769-779.

Pietschker (H.) : **1910.** Das Gehirn der Ameise. Jen. Zeitschr. Naturw. 47, pp. 1-72, 3 pls., 16 figs.

von Planta (A.) : **1888.** Ueber den Futtersaft der Bienen. Zeitschr. physiol. Chem. 12, pp. 327-354.

Plath (O. E.) : **1922.** Notes on Psithyrus, with Records on Two American Hosts. Biol. Bull. 43, pp. 23-44, 1 pl.

Pocock (R. J.) : **1903.** Notes on the Commensalism subsisting between a Gregarious Spider, Stegodyphus sp. and the Moth *Batrachedra stegodyphobius* Wlsn. Ent. Month. Mag. (2) 14, pp. 167-170.

Poisson (R.) : **1924.** Contribution a l'Etude des Hémiptères Aquatiques. Bull. Biol. France Belg., pp. 49-305, 13 pls., 35 figs.

Popovici-Baznosanu (A.) : **1910a.** Experimentelle Untersuchungen über *Osmia rufa*. Zeitschr. wiss. Insektenbiol. 6, pp. 224-228, 3 figs. **1910b.** Relation entre la taille de l'adulte et la quantité de nourriture absorbée par les larves chez *l'Osmia rufa* et *Osmia cornuta*. C. R. Soc. Biol., Paris, 69, pp. 480-481.

POULTON (E. B.): **1927**. The Effect of Mechanical Shock upon the Development of the Secondary Sexual Characters of *Papilio dardanus*. Proc. Ent. Soc. London, 1, pp. 81-93, 2 pls.

PRICER (J. L.): **1908**. The Life History of the Carpenter Ant. Biol. Bull. 14, pp. 177-218, 7 figs.

RABAUD (E.): **1911**. Le Déterminisme des Changements de Milieu. Bull. Scient. France Belg. 45, pp. 169-185. **1917**. Essai sur la Vie et la Mort des Espèces. *Ibid.* 50, pp. 287-380. **1922a**. L'Adaptation et l'Evolution. Paris, E. Chiron. **1922b**. Le Contraste entre le Régime Alimentaire des Larves et celui des Adultes chez divers Insectes. Bull. Biol. France Belg. 56, pp. 230-243. **1927**. Les Rassemblements de Mâles d'Halictes et le Phénomène Social chez les Insectes. *Ibid.* 61, pp. 163-185. **1927**. L'Orientation Lointaine et la Reconnaissance des Lieux. Paris, Alcan. English translation, London, Kegan Paul & Co., Ltd.

RAINES (M. A.): **1922**. Vegetative Vigor of the Host as a factor influencing susceptibility and resistance to certain rust diseases of the higher plants. Amer. Journ. Botany 9, pp. 183-238, 2 pls.

VOM RATH (O.): **1894**. Ueber die Nervenendigungen der Hautsinnesorgane der Arthropoden nach Behandlung mit der Methylenblau- und Chromsilbermethode. Ber. Naturf. Ges. Freiburg i. Br. 9, pp. 1-28, 1 pl.

RAU (P.): **1926**. The Ecology of a Sheltered Clay Bank; a Study in Insect Sociology. Trans. Acad. Sci. St. Louis 25, pp. 157-260, 8 pls.

RÉAUMUR (R. A. F.): **1742**. Mémoires pour Servir à l'Histoire des Insectes. T. VI., Paris, Impr. Roy.

REICHENBACH (H.): **1902**. Ueber Parthenogenese bei Ameisen und andere Beobachtungen an Ameisenkolonien in künstlichen Nestern. Biol. Centralbl. 22, pp. 461-465.

REICHENSPERGER (A.): **1921**. Symphilie, Amikal-Selektion, Trophallaxis und fremddienliche Zweckmässigkeit. Biol. Centralbl. 41, pp. 279-288.

RENNIE (J.), HARVEY (E.) and WHITE (P. B.). **1921**. Isle of Wight Disease in Hive Bees. Trans. Roy. Soc., Edinburgh 52, pp. 737-779, 3 pls.

REUTER (O. M.): **1918**. Lebensgewohnheiten und Instinkte der Insekten bis zum Erwachen der sozialen Instinkte. Berlin, F. Friedländer u. Sohn.

REYNOLDS (E.) and MACOMBER (D.): **1921a**. Defective Diet as a Cause of Sterility. Journ. Amer. Med. Assoc. 77, pp. 169-175. **1921b**. Certain Dietary Factors in the Causation of Sterility in Rats. Amer. Journ. Obstet. Gynæcol., 2. **1923**. Defective Diet as a Cause of Sterility. Final Report. Journ. Amer. Med. Assoc. 80, pp. 978-980.

RICHARDSON (C. H.): **1913**. Studies on the Habits and Development of a Hymenopterous Parasite, *Spalangia muscidarum* Richardson. Journ. Morph. 24, pp. 513-557, 4 pls.

RICKERT (H.): **1921.** Die Grenzen der naturwissenschaftlichen Begriffsbildung, 3te u. 4te Aufl. Tübingen, J. C. B. Mohr.

RILEY (C. V.). **1893.** Parasitism in Insects. Proc. Ent. Soc. Washington 2, pp. 1-35.

ROBERTSON (C.): **1926.** Phenology of Inquiline and Nest-making Bees. Psyche 33, pp. 116-120.

ROBSON (C.): **1898.** *Vespa austriaca*, a Cuckoo Wasp. Science Gossip (n. s.) 5, pp. 69-73.

ROHER (S. A.): **1912.** Studies in the Woodwasp Superfamily Oryssoidea, with Descriptions of New Species. Proc. U.S. Natl. Mus. 43, pp. 141-158, 2 pls., 6 figs. **1913a.** Notes on the Feeding Habits of Two Adult Sawflies. Proc. Ent. Soc. Washington 15, pp. 148-149. **1913c.** (Remarks on Mr. Wood's Paper on *Rhopalosoma poeyi*). *Ibid.* 15, pp. 147-148. **1915.** A Remarkable New Genus of Cephidæ. *Ibid.* 17, pp. 114-117, 1 pl. **1916.** Vespoidea in "The Hymenoptera of Connecticut". Conn. State Geol. and Nat. Hist. Survey, Bull. No. 22, pp. 606-644.

ROHWER (S. A.) and CUSHMAN (R. A.): **1917.** Idiogastra, a New Suborder of Hymenoptera with Notes on the Immature Stages of Oryssus. Proc. Ent. Soc., Washington 19, pp. 89-98, 2 pls.

ROOT (F. M.): **1924.** Parasitism among Insects. Scient. Month. 19, pp. 479-495.

VON ROSEN (K.): **1913a.** Die Fossilen Termiten: Eine kurze Zusammenfassung der bis jetzt bekannten Funde. Trans. 2d Internat. Congr. Ent. (1912), pp. 318-335, 6 pls. **1913b.** Studien am Sehorgan der Termiten nebst Beiträgen zur Kenntnis des Gehirns derselben. Zool. Jahrb. Abt. Anat. Ont. 35, pp. 625-664, 3 pls., 11 figs.

ROUBAUD (E.): **1908.** Gradation et perfectionnement de l'instinct chez les Guêpes Solitaires d'Afrique, du genre Synagris. C. R. Acad. Sc. Paris 147, pp. 695-697. **1910a.** Evolution de l'instinct chez les Guêpes Sociales d'Afrique du genre Belonogaster Sauss. *Ibid.* 151., pp. 553-556. **1910b.** Recherches sur la Biologie des Synagris, Évolution de l'instinct chez les Guêpes Solitaires. Ann. Soc. Ent., France 79, pp. 1-21. Transl. Ann. Rep. Smithson. Inst. Washington, 1910, pp. 509-525, 4 pls. **1916.** Recherches Biologiques sur les Guêpes Solitaires et Sociales d'Afrique. Ann. Sc. Nat. Zool. (10) 1, pp. 1-160, 34 figs. **1917.** Observations Biologiques sur *Nasonia brevicornis* Ashm., Chalcide parasite des Pupes des Muscides. Bull. Scient. France Belg. 50, pp. 425-439, 1 fig. **1918.** Le Venin et l'Evolution paralysante chez les Hyménoptères Prédateurs. *Ibid.* 51, pp. 391-419. **1924.** Histoire des Anacamptomyies, Mouches Parasites des Guêpes Sociales d'Afrique. Ann. Sc. Nat. Bot. Zool. 7, pp. 197-248, 2 pls., 15 figs.

ROUGIER (L.): **1925.** La Scolastique et le Thomisme. Paris, Gauthier-Villars et Cie.

ROUSSEAU (E.): **1907.** Les Hyménoptères Aquatiques avec description de deux espèces nouvelles par W. A. Schulz. Ann. Biol. Lacust. 2, pp. 388-401.

RÜSCHKAMP (F.): **1924.** Instinktmodifikation in einer Ameisen-Adoptionskolonie. Zeitschr. wiss. Insektenbiol. 19, pp. 176-178.

SANTSCHI (F.): **1906.** A propos de Mœurs parasitiques des Fourmis du Genre Bothriomyrmex. Ann. Soc. Ent. France 72, pp. 363-392. **1920.** Fourmis du Genre "Bothriomyrmex" Emery. (Systématique et Mœurs). Rev. Zool. Afric. 7, pp. 201-224, 2 pls., 2 figs. **1922.** Description de Nouvelles Fourmis de l'Argentine et Pays Limitrophes. An. Soc. Cient. Argentina 94, pp. 241-262, 2 figs. **1927.** Deux Notices sur les Fourmis. Bull. Ann. Soc. Ent. Belg. 66, pp. 327-330.

SAULCY (F.): **1891.** Ueber die Antennen. Ann. Soc. Ent. France (6) 10, Bull., pp. 154-155.

DE SAUSSURE (H.): **1852.** Monographie des Guêpes Solitaires ou de la tribu des Euméniens. Paris, Masson. 6 parts, 286 pp., 22 pls. **1853.** Note sur la tribu des Masariens et principalement sur le *Masaris vespiformis*. Ann. Soc. Ent. France (3) 1, pp. 17-21. **1875.** Synopsis of American Wasps. Smithson. Misc. Coll. No. 254, Washington, xxxv. + 385 pp., 4 pls.

SCHAXEL (J.): **1922.** Grundzüge der Theorienbildung der Biologie. 2te Aufl. Jena, G. Fischer.

SCHIEMENZ (P.): **1883.** Ueber das Herkommen des Futtersaftes und die Speicheldrüsen der Biene, nebst einem Anhange über das Riechorgan. Zeitschr. wiss. Zool. 38, pp. 71-135, 3 pls.

SCHIMMER (F.): **1909.** Beitrag zu einer Monographie der Gryllodeengattung Myrmecophila Latr. Zeitschr. wiss. Zool. 93, pp. 409-534, 3 pls., 29 figs. **1910.** Ueber die Wasmannsche Hypothese des "Duldungsinstinktes" der Ameisen gegenüber synöker Myrmekophilen. Zool. Anzeig. 36, pp. 81-95.

SCHIÖDTE (J. G.): **1853.** On Some Staphylinidæ, found in the Nests of Termites. Proc. Zool. Soc., London 21, pp. 101-103. **1856.** Observations sur des Staphylins vivipares qui inhabitent chez les Termites. Ann. Sc. Nat. Zool. (4) 5, pp. 169-183, 1 pl.

SCHMIDT (H.): **1910.** Beitrag zur Biologie der Steinobst-Blattwespe (*Lyda nemoralis* L.). Zeitschr. wiss. Insektenbiol. 6, pp. 17-23, 86-92, 4 figs.

SCHMIEDEKNECHT (O.): **1907.** Die Hymenopteren Mitteleuropas. Jena, G, Fischer.

SCHNEIDER (A.): **1883-'85.** Ueber die Entwickelung von *Sphærularia bombi*. Zool. Beiträge 1, 1883, pp. 1-9, 1 pl., 1885, pp. 247-251.

SCHÖNFELD (P.): **1886.** Die physiologische Bedeutung des Magenmundes der Honigbiene. Arch. Anat. Physiol. Phys. Abt., pp. 451-458, 1 fig.

SCHROTTKY (C.): **1922.** Soziale Gewohnheiten bei solitären Insekten. Zeitschr. wiss. Insektenbiol. 17, pp. 49-57.

SCHUCHERT (C.): **1914.** Climates of Geologic Times. Chapter 21 in E. Huntington, "The Climatic Factor as Illustrated in Arid America." Carnegie Inst. Washington, 341 pp., 90 figs.

SCHULZ (W. A.): **1902.** Zur Kenntnis der Nestweise von *Euglossa cordata* L. Allgem. Zeitschr. Ent. 7, pp. 153-154. **1905.** Hymenopteren-Studien. Leipzig, W. Engelmann. **1907a.** Family Trigonaloidæ, in Wytsman's "Genera Insectorum," 24 pp., 3 pls. **1907b.** Schwimmende Braconiden. Ann. Soc. Ent.

Belg. 51, pp. 164-173. **1910a.** Neuer Beitrag zur Kenntnis der Wasserimmen. Ann. Biol. Lacust. 4, pp. 187-193, 4 figs. **1910b.** Süsswasser-Hymenopteren aus dem See von Overmiere. *Ibid.* 4, pp. 194-210, 2 pls., 2 figs.

SCHWARZ (E. A.) : **1904.** (On the Social Spider *Uloborus republicanus* Simon in Cuba.) Proc. Ent. Soc. Washington 6, pp. 147-148.

SCOTT (H.) : **1920.** Notes on (I.) the Parasitic Staphylinid *Aleochara algarum* Fauvel, and its Hosts, the Phycodromid Flies; (II.) A Case of Supposed Parasitism in the Genus Homalota. Ent. Month. Mag. (3) 6, pp. 148-157, 2 figs.

SCUDDER (S. H.) : **1883.** The Fossil White Ants of Colorado. Proc. Amer. Acad. Arts Sci. 19, pp. 133-145. **1890.** The Tertiary Insects. In Hayden's Rep. U.S. Geol. Surv. Terr. 13, 734 pp., 28 pls.

SELLARS (R. W.) : **1922.** Evolutionary Naturalism. Chicago, Open Court Publ. Co.

SEMICHON (L.) : **1912.** Parasitisme provoqué entre deux larves d'*Halictus quadricinctus* Fabricius. Bull. Soc. Ent. France, pp. 90-92.

SEVERIN (H. C.) : **1920.** The Plum Web-spinning Sawfly. Techn. Bull. No. 1. Agric. Exper. Stat. South Dakota, pp. 1-53, 11 figs.

SHARP (D.) : **1899.** Insects., Vol. VI., Cambridge Natural History. London, Macmillan & Co.

SHELFORD (R.) : **1906.** Studies of the Blattidæ V-VII. Trans. Ent. Soc. London, pp. 487-519, 1 pl. **1907.** Blattoidea in Sjöstedt's "Kilimandjaro-Meru Expedition." No. 17, 2, pp. 13-48, 2 pls.

SHELFORD (V. E.) : **1913.** The Life-History of a Bee-Fly (*Spogostylum anale* Say) parasite of the Larva of a Tiger Beetle (*Cicindela scutellaris* Say var. *lecontei* Hald.). Ann. Ent. Soc. Amer. 6, pp. 213-225.

SICHEL (J.) : **1865.** Revision Monographique, Critique et Synonymique du Genre Mellifère Sphecodes. Ann. Soc. Ent. France (4) 5, pp. 397-466.

SIEBER (N.) and METALNIKOV (S.) : **1904.** Ueber Ernährung und Verdauung der Bienenmotte (*Galleria melonella*). Arch. Ges. Physiol., pp. 269-286.

VON SIEBOLD (C. T. E.) : **1836.** Bericht über die Leistungen im Gebiete der Helminthologie während des Jahres 1837. Wiegmann's Arch. 4, p. 305. **1858.** Ueber *Agriotypus armatus* in *Trichostoma picicorne*. Ber. Versam. Naturf. Carlsruhe, 1858, p. 211. **1864.** Ueber Zwitterbienen. Zeitschr. wiss. Zool. 14, pp. 73-80.

SILVESTRI (F.) : **1901.** Operai ginecoidi di Termes, con osservazioni intorno l'origine delle varie caste nei Termitidi. Real Accad. Lincei (5) 10, pp. 479-484. **1902.** Contribuzione alla Conoscenza dei Meliponidi del Bacino del Rio de la Plata. Riv. Patol. Veget. 10, pp. 121-170, 3 pls., 19 figs. **1903.** Contribuzioni alla Conoscenza dei Termitidi e Termitofili dell'America Meridionale. Redia 1, pp. 1-235, 6 pls. **1904.** Contribuzioni alla Conoscenza della Metamorfosi e dei Costumi della *Lebia scapularis* Fourc. *Ibid.* 2, pp. 68-84, 5 pls. **1906.** Contribuzioni alla

Conoscenza Biologica degli Imenotteri Parasiti. I. Biologia del *Litomastix truncatellus* (Dalm.). Ann. R. Scuol. Sup. Agric. Portici 6, pp. 1-51, 5 pls., 13 figs. **1907.** Contribuzioni alla Conoscienza degli Insetti Dannosi all' Olivo. La Tignola dell' Olivo. Boll. Lab. Zool. Gen. Agrar. Portici 2, pp. 83-184, 68 figs. **1909.** Isoptera, in W. Michaelsen and R. Hartmeyer, " Die Fauna Südwest-Australiens," 2, pp. 279-314, 6. pls. **1910.** Contribuzioni alla Conoscenza degli Insetti Dannosi e dei loro Simbionti. I. Galerucella dell Olmo (*Galerucella luteola* F. Müll.). Boll. Lab. Zool. Gen. Agrar. Portici 4, pp. 246-288, 25 figs. **1911.** Contribuzioni alla Conoscenza degli Insetti Dannosi e dei loro Simbionti. II. *Plusia gamma* (L.). *Ibid.* 5, pp. 287-319, 26 figs. **1913.** Descrizione di un Nuovo Ordine di Insetti. *Ibid.* 7, pp. 193-209, 3 figs. **1920.** Contribuzioni alla Conoscenza dei Termitidi e Termitofile dell, Africa Occidentale. II. Termitofile. *Ibid.* 14, pp. 265-319, 32 figs. **1927.** Contribuzioni alla Conoscenza dei Grylloblattidæ. *Ibid.* 20, pp. 107-121, 8 figs.

SIMON (E.) : **1891.** Observations Biologiques sur les Arachnides. I. Araignées Sociables. Ann. Soc. Ent. France 60, pp. 5-14, 4 pls. **1892-'95.** Histoire Naturelle des Araignées (2me ed.), 2 vols. Paris, Roret. **1909.** Sur l'Araignée Mosquero. C. R. Acad. Sc. Paris 148, pp. 736-737.

SJÖSTEDT (Y.) : **1925.** Revision der Termiten Afrikas. 3. Monographie. K. Svensk. Vetens. Akad. Handl. (3) 3, pp. 3-419, 16 pls., 81 figs.

SLADEN (F. W. L.) : **1895.** Observations on *Sphecodes rubicundus* v. Hag. Ent. Month. Mag. 31, p. 256. **1899.** Bombi in Captivity and Habits of Psithyrus. *Ibid.* 35, pp. 230-234. **1902.** A Scent-producing Organ in the Abdomen of Worker of *Apis mellifica*. *Ibid.* 38, pp. 208-211, **1912.** The Humble-bee, its Life History and how to Domesticate it. London, Macmillan Co. **1915.** Inquiline Bumble-Bees in British Columbia. Canad. Ent. 47, p. 84. **1918.** The Genus Vespa in Canada. Ottawa Naturalist, 32, pp. 71-72.

SMITH (F.) : **1855.** Catalogue of British Hymenoptera in the Collection of the British Museum, Part I., Apidæ. London. **1866.** Notes on Some Hymenopterous Insects collected by Mr. Peckold at Catagallo, South Brazil. Trans. Ent. Soc., London (3) 5, pp. 323-327.

SMITH (H. S.) : **1912.** The Chalcidoid Genus Perilampus and its Relations to the Problem of Parasite Introduction. Techn. Ser. No. 19, Pt. IV., U.S. Dept. Agric. Bur. Ent., pp. 33-69, 8 figs. **1916.** The Habit of Leaf-Oviposition among the Parasitic Hymenoptera. Science, n.s. 44, pp. 925-926. **1917.** The Habit of Leaf-Oviposition among the Parasitic Hymenoptera. Psyche, 24, pp. 63-68, 4 figs.

SMITH (J. B.) : **1901.** Notes on Some Digger Bees. Journ. N.Y. Ent. Soc. 9, pp. 29-40, 52-72, 3 pls., 3 figs.

SMULYAN (M. T.) : **1916.** The Hyperparasitic Chalcidoid Planidium on Aphids. Journ. Econom. Ent. 9, p. 510.

SMUTS (J. C.) : **1926.** Holism and Evolution. New York, Macmillan Co.

SNODGRASS (R. E.): **1926.** The Morphology of the Insect Sense Organs and the Sensory Nervous System. Smithson. Misc. Coll. 77, 80 pp., 32 figs.

SNYDER (T. E.): **1924.** " Adaptations " to Social Life : The Termites (Isoptera). Smithson. Misc. Coll. 76, No. 12, 14 pp., 3 pls. **1925.** Notes on Fossil Termites with peculiar Reference to Florissant, Colorado. Proc. Biol. Soc. Washington 38, pp. 149-166, 1 chart. **1926.** The Biology of the Termite Castes. Quart. Review Biol. 1., pp. 522-552, 15 figs.

SPRINGER (F.): **1915.** Ueber den Polymorphismus bei den Larven von *Miastor metrolaos.* Zool. Jahrb. Abt. Syst. 40, pp. 57-118, 2 pls.

STÄGER (R.): **1918.** Myrmekologische Beobachtungen auf Belalp. Ber. Naturf. Ges. Bern. **1923.** Ergebnisse meiner neuen Beobachtungen an *Formicoxenus nitidulus* Nyl. Zeitschr. wiss wiss. Insektenbiol. 18, pp. 290-292. **1925.** Das Leben der Gastameise (*Formicoxenus nitidulus* Nyl.) in neuer Beleuchtung. Zeitschr. Morph. Ökol. Tiere 3, pp. 452-476.

STAERCKE (A.): **1921.** Psycho-analysis and Psychiatry. Internat. Journ. Psycho-anal. 2, pp. 361-415.

STÖCKHERT (E.): **1923.** Ueber Entwicklung und Lebensweise der Bienengattung Halictus Latr. und ihrer Schmarotzer. 1, Teil. Die Biologie der Gattung Halictus Latr. Konowia, 2, pp. 48-64, 146-165, 216-247.

STOLL (O.): **1898.** Zur Kenntnis der geographischen Verbreitung der Ameisen. Mitteil. schweiz. Ent. Ges. 10, pp. 127-151.

STRICKLAND (E. H.): **1912.** The Pezomachini of North America. Ann. Ent. Soc. Amer. 5, pp. 113-140, 5 figs.

STUMPER (R.): **1918a.** Zur Kenntnis des Polymorphismus der Formiciden. Ges. Luxemburger Naturfreunde 7 pp., 8 figs. **1918b.** *Formicoxenus nitidulus* Nyl. 1, Beitr. Biol. Zentralbl. 38, pp. 160-179, 14 figs. **1921.** Etudes sur les Fourmis III. Recherches sur l'Ethologie du *Formicoxenus nitidulus* Nyl. Bull. Soc. Ent. Belg. (7) 3, pp. 90-97, 1 fig.

STURTEVANT (A. H.): **1925.** Notes on the Ant Fauna of Oak Galls in the Woods Hole Region. Psyche 32, pp. 313-314. **1927.** The Social Parasitism of the Ant *Harpagoxenus americanus. Ibid.* 34, pp. 1-9.

SURCOUF (F.) and GUYON (L.): **1912.** Nouvelles Espèces de Calliphorines de l'Afrique Occidentale. Bull. Mus d'Hist. Nat. Paris, pp. 418-428.

SWAMMERDAM (J.): **1737-38.** Biblia Naturæ, sive Historia Insectorum, etc. 2 vols. Leydæ, Severin, etc.

SWENK (M. H.): **1911.** Notes on Some Insects Injurious in Nebraska in 1910. Journ. Econom. Ent. 4, p. 286.

SWEZEY (O. H.): **1903.** Observations on Hymenopterous Parasites of Certain Fulgoridæ. Ohio Naturalist 3, pp. 444-451, 2 pls. **1915.** A Preliminary List of the Hymenopterous Parasites of Lepidoptera in Hawaii. Proc. Hawaii. Ent. Soc. 3, pp. 99-109.

TANNER (J. E.): **1892.** *Oecodoma cephalotes.* Second Paper. Trinidad Field. Nat. Club 1, pp. 123-127.

TANQUARY (M.): **1911.** Experiments on the Adoption of Lasius, Formica and Polyergus Queens by Colonies of Alien Species. Biol. Bull. 20, pp. 281-308. **1913.** Biological and Embryological Studies on Formicidæ. Bull. No. 9, Illinois State Lab. Nat. Hist., pp. 417-477, 8 pls.

TARDE (G.): **1921.** Les Lois de l'Imitation. 7me Ed. Paris, Alcan.

TAYLOR (F. B.): **1910.** Bearing of the Tertiary Mountain Belt on the Origin of the Earth's Plan. Bull. Geol. Soc. Amer. 21, pp. 179.

TAYLOR (G. W): **1898.** Note on *Trigonalys canadensis* Hargtn. Canad. Ent. 30, pp. 14-16.

THOMPSON (C. B.): **1913.** A comparative Study of the Brains of Three Genera of Ants, with Special Reference to the Mushroom Bodies. Journ. Comp. Neurol. 23, pp. 515-570, 6 pls. **1916.** The Brain and the Frontal Gland of the Castes of the "White Ant" *Leucotermes flavipes*, Kollar. *Ibid.* 26, pp. 553-602, 5 pls. **1917.** Origin of the Castes of the Common Termite, *Leucotermes flavipes* Koll. Journ. Morph. 30, pp. 83-152, 42 figs **1919.** The Development of the Castes of Nine Genera and Thirteen Species of Termites. Biol. Bull. 36, pp. 379-398, 10 figs. **1922.** The Castes of Termopsis. Journ. Morph. 36, pp. 495-535 3 pls., 9 figs.

THOMPSON (C. B.) and SNYDER (T. E.): **1919.** The Question of the Phylogenetic Origin of Termite Castes. Biol. Bull. 36, pp. 115-132, 2 pls., 5 figs. **1920.** The "Third Form," the Wingless Reproductive Type of Termites: Reticulitermes and Prorhinotermes. Journ. Morph. 32, pp. 591-532, 3 pls., 20 figs.

THOMPSON (W. R.): **1915.** Contribution à la Connaissance de la Larve Planidium. Bull. Scient. France Belg. (7) 48, pp. 1-31, 5 figs.

THURSTONE (L. L.): **1924.** The Nature of Intelligence. New York. Harcourt, Brace & Co.; London, Kegan Paul & Co., Ltd.

TILGHER (A.): **1921.** La Crisi Mondiale e Saggi Critici di Marxiano e Socialismo. Bologna, N. Zanichelli.

TILLYARD (R. J.): **1918.** The Panorpoid Complex. Part I. Proc. Linn. Soc. N.S. Wales 43, pp. 265-319, 2 pls., 16 figs. **1923.** The Lower Permain Insects of Kansas. Prelim. Announcement Ent. News 34, pp. 292-295. **1924.** Kansas Permian Insects. Part 3. The New Order Protohymenoptera. Amer. Journ. Sci. (5) 8, pp. 111-122, 1 pl., 5 figs. **1926a.** Kansas Permian Insects. Part 6. Additions to the Orders Protohymenoptera and Odonata. *Ibid.* 11, pp. 58-73, 9 figs. **1926b.** Fossil Insects in Relation to Living Forms. Nature, 117, p. 828, 5 figs.

TIMBERLAKE (P. H.): **1912.** Experimental Parasitism, a Study of the Biology of *Limnerium validum* (Cresson). Techn. Ser. No. 19, U.S. Dept. Agric. Bur. Ent., pp. 71-92, 10 figs.

TOSI (A.): **1895.** Osservazioni sulla valvola del cardias in varii generi della familia delle Apidi. Ric. Lab. Anat. Univ. Roma, 5, pp. 5-26, 3 pls. **1896.** Di un nuovo genere di Apiaria fossile nell' Ambra di Sicilia (*Meloponoryctes succini, M. sicula*) Revist. Ital. Paleont. 5 pp., 1 pl., 9 figs.

TOWNER (R. H.): **1923.** The Philosophy of Civilization. 2 vols. New York and London, Putnam's Sons.

TRÄGÅRDH (I.): **1907a.** Description of Termitomimus, a New Genus. of Termitophilous, Physogastric Aleocharini, with Notes on its Anatomy. Zool. Studier, Upsala, pp. 172-190, 1 pl., 10 figs. **1907b.** Notes on a Termitophilous Tineid Larva. Ark. Zool. 3, pp. 1-17, 1 pl.

TROELTSCH (E.): **1922.** Der Historismus und seine Probleme. I. Bd. Tübingen (Bd. 3 of his Gesammelte Schriften).

TROUVELOT (B.): **1921.** Observations biologiques sur l'*Habrobracon johansenni*. Soc. Biol. 85, pp. 1022-1024.

TURNER (C. H.): **1899.** Notes on the Mushroom Bodies of the Invertebrates. Zool. Bull. 2, pp. 155-160, 6 figs.

TURNER (R. E.) and WATERSTON (J.): **1917.** Notes on the Hymenopterous Families Bethylidæ and Rhopalosomidæ. Ann. Mag. Nat. Hist. (8) 20, pp. 101-108, 1 fig.

ULMER (G.): **1912.** Die Trichopteren des baltischen Bernsteins. Beitr. Naturk. Preuss. herausgeg. physik. ökonom. Ges. Königsberg, 10, 380 pp., 480 figs.

VANDEL (A.): **1927.** Modifications déterminées par un Nématode du Genre " Mermis " chez les Ouvrières et les Soldats de la Fourmi " *Pheidole pallidula* " Nyl. Bull. Biol. France Belg. 61, pp. 38-48, 3 figs.

VERHOEFF (C.): **1891.** Biologische Aphorismen über einige Hymenopteren, Dipteren, und Coleopteren. Verh. naturh. Ver. preuss. Reinl. 48, pp. 1-80, 3 pls. **1892.** Beiträge zur Biologie der Hymenopteren. Zool. Jahrb. Abt. Syst. 6, pp. 680-751, 2 pls., 7 figs. **1897.** Zur Lebensgeschichte der Gattung Halictus (Anthophila), insbesondere einer Uebergangsform zu sozialen Bienen. Zool. Anzeig. 20, pp. 369-393, 21 figs.

VERLAINE (L.): **1926.** Les Reines fécondées des Hyménoptères Sociaux peuvent-elles normalement engendrer des Mâles ? Ann. Bull. Soc. Ent. Belg. 66, pp. 287-318.

VIALLANES (H.): **1893.** Etude Histologique et Organologique sur les Centres Nerveux et les Organes des Sens des Animaux Articulés. Ann. Soc. Nat. Zool. Pal. 14, pp. 405-455, 2 pls.

VIEHMEYER (H.): **1904.** Experimente zu Wasmann's Lomechusa-Pseudogynen-Theorie. Allgem. Zeitschr. Ent., pp. 334-344. **1906.** Beiträge zur Ameisenfauna des Königreiches Sachsen. Abhandl. naturw. Ges. Isis, Dresden, pp. 55-69, 1 pl. **1908.** Zur Koloniegründung der parasitischen Ameisen. Biol. Centralbl., pp. 18-32. **1909.** Beobachtungen und Experimente zur Koloniegründung von *Formica sanguinea*. Zeitschr. wiss. Insektenbiol. 5, pp. 353-356, 390-394. **1910a.** Bemerkungen zu Wasmann's neuester Arbeit: Ueber den Ursprung des sozialen Parasitismus, der Sklaverei und der Myrmecophilie bei den Ameisen. Zool. Anzeig. 35, pp. 450-457. **1910b.** Ontogenetische und phylogenetische Betrachtungen über die parasitische Koloniegründung von *Formica sanguinea*. Biol. Centralbl. 30, pp. 569-580. **1912.** Ueber die Verbreitung und die geflügelten Weibchen von *Harpagoxenus sublævis* Nyl. Ent. Mitteil. 1, pp. 193-197. **1921.** Die mitteleuropäischen Beobachtungen von Harpagoxenus. Biol. Zentralbl. 41, pp. 269-278. **1923.** Polymorphismus und Ernährung bei den Ameisen. Jen. Arch. Naturg. 89, pp. 1-12.

VOUKASSOVITCH (P.) : **1924.** Sur la Biologie d'un Ichneumonide (*Dicælotus erythrostoma*) parasite de l'*Eudemis* (*Polychrosis*) *botrana* Schiff. Bull. Biol. France Belg. 58, pp. 495-499, 2 figs.

WADSWORTH (J. T.) : **1915.** On the Life-history of *Aleochara bilineata* Gyll., a Staphylinid parasite of *Chortophila brassicæ* Bouché. Journ. Econom. Ent. 10, pp. 1-27, 2 pls., 1 fig.

WAGNER (W.) : **1907.** Psychobiologische Untersuchungen an Hummeln mit Bezugnahme auf die Frage der Geselligkeit im Tierreiche. Zoologica 19, pp. 1-239, 1 pl., 136 figs.

DE WALCKENAER (C. A.) : **1817.** Mémoires pour servir a l'Histoire Naturelle des Abeilles Solitaires qui composent le genre Halicte. Paris, Didot. 90 pp., 1 pl.

WALKER (E. M.) : **1914.** A New Species of Orthoptera, forming a New Genus and Family. Canad. Ent., 46, pp. 93-99, 1 pl.

WALKER (F.) : **1836.** *Agriotypus armatus.* Ent. Mag., p. 412.

WALSINGHAM (LORD) : **1903.** Description of a New Species of Batrachedra associated with Spiders in South Africa. Ent. Month. Mag. (2) 4, pp. 166-167.

WARREN (E.) : **1919.** Termites and Termitophiles. Presidential Address. South Afric. Journ. Sci., pp. 1-20, 3 pls. **1920.** Observations on the Comparative Anatomy of the Termitophilous Aleocharine *Paracorotoca akermani* (Warren). Ann. Natal Mus. 4, pp. 297-366, 6 pls., 5 figs.

WASMANN (E.) : **1891.** Die zusammengesetzten Nester und gemischten Kolonien der Ameisen. Münster, Aschendorff, 262 pp., 2 pls., 16 figs. **1895.** Die ergatogynen Formen bei den Ameisen und ihre Erklärung. Biol. Centralbl. 15, pp. 606-646. **1900-01a.** Termitoxenia, ein neues flügelloses physogastres Dipterengenus aus Termitennestern. 1. Theil. Zeitschr. wiss. Zool. 57, 1900, pp. 599-617, 1 pl., 2 Theil. *Ibid.* 70, 1901, pp. 289-298. **1910b.** Zur Lebensweise der Ameisengrillen (Myrmecophila) Natur. u. Offenbar. 47, pp. 129-152. **1901c.** Neues über die zusammengesetzten Nester und gemischten Kolonien der Ameisen. Allgem. Zeitschr. Ent. 6, pp. 23-24 ; 7, pp. 1-21. **1903.** Zur näheren Kenntnis des echten Gastverhältnisses (Symphilie) bei den Ameisen- und Termitengästen. Biol. Centralbl. 23, pp. 63-72, 195-207, 232-248, 261-276, 298-310. **1904.** Contribução para o estudio dos hospedes de abelhas brazileiras. Revist. Mus. Paulista 6. p. 482-487, 6 figs. **1905.** Ursprung und Entwickelung der Sklaverei bei den Ameisen. Biol. Centralbl. 25, pp. 117-127, 129-141, 161-169, 256-270, 273-392, 2 figs. **1908a.** Zur Kastenbildung und Systematik der Termiten. *Ibid.* 28, pp. 68-73. **1908b.** Weitere Beiträge zum sozialen Parasitismus und der Sklaverei bei den Ameisen. *Ibid.* 28, pp. 257-271, 289-306, 321-333, 353-382, 417-441. **1909a.** Die psychischen Fähigkeiten der Ameisen. 2. Aufl. Stuttgart. **1909b.** Ueber gemischte Kolonien von Lasius Arten. Zool. Anzeig. 35, pp. 129-141. **1909c.** Zur Geschichte der Sklaverei und des sozialen Parasitismus bei den Ameisen. Naturw. Wochenschr. n. F. 8, pp. 401-407, 5 figs. **1909d.** Ueber den Ursprung des sozialen Parasitismus, der Sklaverei und der Myrmekophilie bei den Ameisen. Biol. Centralbl. 29, pp. 692, figs. 1 and 2. **1910a.** Ueber das Wesen und den Ursprung der Symphilie. *Ibid.* 30, pp. 97-102 ; 129-138, 161-181. **1910b.** Nachträge zum sozialen Parasitismus und der Sklaverei bei den Ameisen. *Ibid.* 30, pp. 453-464, 475-496,

515-524. **1911.** Die Ameisen und ihre Gäste. 1er Congr. Internat. d'Ent. (1910), pp. 209-234, 6 pls. **1915a.** Neue Beiträge zur Biologie von Lomechusa und Atemeles, mit kritischen Bemerkungen über das echte Gastverhältnis. Zeitschr. wiss. Zool. 114, pp. 233-402, 2 pls., 1 chart. 2 figs. **1915b.** *Anergatides Kohli*, eine neue arbeiterlose Schmarotzerameise vom oberen Congo. Ent. Mitteil. 4, pp. 279-288, 2 pls. **1915c.** Das Gesellschaftsleben der Ameisen. 1 Bd. 2te Aufl. Münster i. W., Aschendorff. **1917.** Neue Anpassungstypen bei Dorylinengästen Afrikas. Zeitschr. wiss. Zool. 117, pp. 257-360, 4 pls. **1919.** Ueber unsere Kenntnis der fossilen Paussiden. Tijdschr. Ent. 62, pp. xl.-xliv. **1920.** Die Gastpflege der Ameisen, ihre biologischen und philosophischen Probleme. Heft 4. J. Schaxel's Abhandl. Theoret. Biol. xvii. + 176 pp., 2 pls. **1923a.** Eine interessante Instinktregulation bei Ameisen (*Lasius mixtus* Nyl.). Atti Pontif. Accad. Romana, pp. 255-259. **1923b.** Die Larvenernährung bei den Ameisen und die Theorie der Trophallaxis, *Ibid.* (2) 6, pp. 3-23, 1 fig.

WEBSTER (F. M.) and REEVES (G. J.) : **1909.** The Wheat Strawworm (*Isosoma grande* Riley). Circ. No. 106, U.S. Dept. Agric. 15 pp., 13 figs.

WEGENER (A.) : **1924.** The Origin of Continents and Oceans. Transl. by J. G. A. Skerl. New York, Dutton & Co.

WEISMANN (A.) : **1892.** Das Keimplasma. Eine Theorie der Vererbung. Jena, G. Fischer.

WESMAEL (C.): **1835.** Observations sur les Espèces du Genre Sphécode. Bull. Acad. Roy. Sc. Bruxelles, 2, pp. 279-287.

WESTWOOD (J. O.) : **1874.** Thesaurus Entomologicus Oxoniensis, or Illustrations of New, Rare and Interesting Insects. Oxford, Clarendon Press. **1881.** Observations on the Hymenopterous Genus Scleroderma Klug, and Some Allied Groups. Trans. Ent. Soc. London, pp. 117-140, 4 pls.

WHEELER (E. H.) : **1923.** Some Braconids Parasitic on Aphids and their Life History. Ann. Ent. Soc. Amer. 16, pp. 1-29, 9 figs.

WHEELER (G. C.) : **1928.** The Larva of Leptanilla. Psyche 35 (in press).

WHEELER (G. C. and E. H.) : **1924.** A New Species of Schizaspidia (Eucharidæ) with Notes on a Eulophid Ant Parasite. Psyche 31, pp. 49-56, 2 figs.

WHEELER (W. M.) : **1900a.** A New Myremecophile from the Mushroom Gardens of the Texan Leafcutting Ant. Amer. Natural. 34. pp. 851-862, 6 figs. **1900b.** The Habits of *Myrmecophila nebrascensis* Bruner. Psyche 9, pp. 111-115. **1901a.** An Extraordinary Ant Guest. Amer. Natural. 35, pp. 1007-1016, 2 figs. **1901b.** The Parasitic Origin of Macroergates among Ants. *Ibid.* 35, pp. 877-886, 1 fig. **1901c.** The Compound and Mixed Nests of American Ants. *Ibid.* 35, pp. 431-448, 513-539. 701-724, 791-818, 20 figs. **1903a.** The Origin of Female and Worker Ants from the Eggs of Parthenogenetic Workers. Science n.s. 18, pp. 830-833. **1903b.** Some New Gynandromorphous Ants, with a Review of the Previously Recorded Cases. Bull. Amer. Mus. Nat. Hist. 19, pp. 853-683, 11 figs. **1903c.** Ethological Observations on an American Ant (*Leptothorax Emersoni* Wheeler). Arch. Psychol. Neurol. 2, pp. 1-31, 1 pl. **1904a.**

The Phylogeny of Termites. Biol. Bull. 8, pp. 29-37. **1904b.** A New Type of Social Parasitism among Ants. Bull. Amer. Mus. Nat. Hist. 20, pp. 347-375. **1904c.** Three New Genera of Inquiline Ants from Utah and Colorado. *Ibid.* 20, pp. 1-17, 2 pls. **1905.** An Interpretation of the Slave-Making Instincts of Ants. *Ibid.* 21, pp. 1-16. **1906.** On the Founding of Colonies by Queen Ants, with Special Reference to the Parasitic and Slave-making Species. *Ibid.* 22, pp. 33-105, 7 pls. **1907a.** The Polymorphism of Ants, with an Account of Some Singular Abnormalities Due to Parasitism. *Ibid.* 23, pp. 1-93, 6 pls. **1907b.** The Origin of Slavery among Ants. Pop. Sci. Month., pp. 350-359. **1907c.** Notes on a New Guest Ant, *Leptothorax glacialis*, and the Varieties of *Myrmica brevinodis* Emery. Bull. Wisconsin Nat. Hist. Soc. 5, pp. 70-83. **1908a.** Comparative Ethology of the European and American Ants. Journ. Psychol. Neurol. 13, pp. 404-435, 4 pls., 6 figs. **1908b.** Honey Ants, with a Revision of the North American Myrmecocysti. Bull. Amer. Mus. Nat. Hist. 24 pp. 345-397, 28 figs. **1908c.** The Ants of Casco Bay, Maine, with Observations on two Races of *Formica sanguinea* Latr. *Ibid.* 24, pp. 619-645. **1909.** Observations on Some European Ants. Journ. N. Y. Ent. Soc. 17, pp. 173-187, 2 figs. **1910a.** Ants, their Structure, Development and Behaviour. New York, Columbia Univ. Press. **1910b.** The Effects of Parasitic and Other Kinds of Castration in Insects. Journ. Exper. Zool. 8, pp. 377-438, 6 figs. **1911a.** The Ant Colony as an Organism. Journ. Morph. 22, pp. 307-325. **1911b.** An Ant-Nest Coccinellid (*Brachyacantha quadripunctata* Mels.). Journ. N. Y. Ent. Soc. 19, pp. 169-174, 1 fig. **1914.** The Ants of the Baltic Amber. Schrift. physik. ökonom. Ges. Königsberg 55, 142 pp., 66 figs. **1916.** Four New and Interesting Ants from the Mountains of Borneo and Luzon. Proc. New Engl. Zool. Club 6, pp. 9-18, 4 figs. **1917a.** The Temporary Social Parasitism of *Lasius subumbratus* Viereck, Psyche 24, pp. 167-176. **1917b.** The Pleometrosis of Myrmecocystus. *Ibid.* 24, pp. 180-182. **1917c.** The Mountain Ants of Western North America. Proc. Amer. Acad. Arts Sci. 52, pp. 457-569. **1918a.** A Study of Some Ant Larvæ, with a Consideration of the Origin and Meaning of the Social Habit among Insects. Proc. Amer. Philos. Soc. 57, pp. 293-343, 12 figs. **1918b.** The Ants of the Genus Opisthopsis. Bull. Mus. Comp. Zool. 62, pp. 343-362, 3 pls. **1919a.** The Parasitic Aculeata, a Study in Evolution. Proc. Amer. Philos. Soc. 58, pp. 1-40. 6 tables. **1919b.** Two Gynandromorphous Ants. Psyche 26, p. p1-8, 2 figs. **1919c.** The Phoresy of Antherophagus. *Ibid.* 26, pp. 145-152, 1 fig. **1920.** The Sub-families of Formicidæ and Other Taxonomic Notes. *Ibid.* 27, pp. 46-55, 3 figs. **1921a.** A New Case of Parabiosis and the "Ant Gardens" of British Guiana. Ecology 2, pp. 89-103, 3 figs. **1921b.** Chinese Ants collected by Prof. C. W. Howard. Psyche 28 pp. 110-115, 2 figs. **1921c.** A study of Some Social Beetles in British Guiana and of their Relations to the Ant-Plant Tachigalia, Zoologica 3, pp. 35-126, 5 pls., 12 figs. **1922a.** Ants of the American Museum Congo Expedition. A Contribution to the Myrmecology of Africa, with the Collaboration of J. Bequaert, I. W. Bailey, F. Santschi and W. M. Mann. Bull. Amer. Mus. Nat. Hist. 45, 1139 pp. 45 pls., 47 maps, 103 figs. **1922b.** Observations on *Gigantiops destructor* Fabricius and Other Leaping Ants. Biol. Bull. 42, pp. 185-201, 3 figs. **1923.** Social Life among the Insects. New York, Harcourt, Brace

& Co. 1925. A New Guest Ant and Other New Formicidæ from Barro Colorado Island, Panama. Biol. Bull. 49, pp. 150-181, 8 figs. 1926a. Social Habits of Some Canary Island Spiders Psyche 33, pp. 29-31. 1926b. Emergent Evolution and the Social. Science n.s. 44, pp. 433-440; Psyche (London) No. 27, pp. 28-40; Proc. 6th Internat. Congr. Philosophy, pp. 33-46. 1927. The Ants of the Canary Islands. Proc. Amer. Acad. Arts Sci. 62, pp. 93-120, 3 pls.

WHEELER (W. M.) and BAILEY (I. W.): 1920. The Feeding Habits of Pseudomyrmine and Other Ants. Trans. Amer. Philos. Soc. n. s. 22, pp. 235-279, 5 pls., 6 figs.

WHEELER (W. M.) and MANN (W. M.): 1923. A Singular Habit of Sawfly Larvæ. Psyche 30, pp. 9-13, 1 fig.

WHEELER (W. M.) and MCCLENDON (S. F.): 1903. Dimorphic Queens in an American Ant (*Lasius latipes* Walsh). Biol. Bull. 4, pp. 149-163, 3 figs.

WHEELER (W. M.) and TAYLOR (L. H.): 1921. *Vespa arctica* Rohwer, a parasite of *Vespa diabolica* De Saussure. Psyche 28, pp. 135-144, 3 figs.

WHITE (G. F.): 1906. The Bacteria of the Apiary, with Special Reference to Bee Diseases. Techn. Ser. No. 14, Bur. Ent. U. S. Dept. Agric. 50 pp.,

WHITING (P. W.): 1914. Observations on Blow Flies; Duration of the Prepual Stage and Colour Determination. Biol. Bull. 26, 184-194 pp. 1921a. Rearing Meal Moths and Parasitic Wasps for Experimental Purposes. Journ. Hered. 12, pp. 255-261, 9 figs. 1921b. Studies on the Parasitic Wasp, *Habrobracon brevicornis* (Wesmael), Genetics of an Orange-eyed Mutation and the Production of Mosaic Males from Fertilized Eggs. Biol. Bull. 41, pp. 42-54.

WIELAND (G. R.): 1924. Fossil Plants as Evidence for Resistance to Environment. Chapter V. in "Organic Adaptation to Environment," New Haven, Yale Univ. Press.

WILLE (J.): 1920. Biologie und Bekämpfung der deutschen Schabe (*Phyllodromia germanica* L.) Monogr. angewandt. Ent. No. 5 (Zeitschr. angewandt. Ent.) iv. + 140 pp., 2 pls., 53 figs.

WILLIAMS (F. X.): 1919a. *Epyris extraneus* Bridwell (Bethylidæ), a Fossorial Wasp that Preys on the Larva of the Tenebrionid Beetle, *Gonocephalum seriatum* (Boiduval). Proc. Hawaii. Ent. Soc. 4, pp. 55-63, 2 pls., 8 figs. 1919b. Philippine Wasp Studies. Bull. No. 14, Exper. Stat. Hawaiian Sugar Planters' Assoc., 186 pp., 106 figs.

WILLIAMS (F. X.) and HUNGERPORD (H. B.): 1914. Notes on Coleoptera from Western Kansas. Ent. News 25, pp. 1-9, 2 pls.

WINKLER (H.): 1920. Verbreitung und Ursache der Parthenogenesis im Pflanzen -und Tierreiche. Jena, G. Fischer.

WODSEDALEK (J. E.): 1917. Five Years of Starvation of Larvæ. Science n.s. 46, pp. 366-367.

YUASA (H.): 1922. A Classification of the Larvæ of the Tenthredinoidea. Illinois Biol. Monogr. Vol. 1, 172 pp., 14 pls.

ZANDER (E.): 1911. Der Bau der Biene. Stuttgart. iv. + 182 pp., 20 pls., 149 figs.

ZIEGLER (H. E.): 1912. Die Gehirne der Insekten. Naturw. Wochenschr. n. F. 21, pp. 433-442, 18 figs. 1920. Der Begriff des Instinktes einst und jetzt. 3te Aufl. Jena, G. Fischer.

INDEX